*The Emerging Consensus
in Social Systems Theory*

The Emerging Consensus in Social Systems Theory

Kenneth C. Bausch

Research Director/CEO
Ashley Montagu Institute
Los Angeles, California

Research Fellow and Advisor
CWA Ltd.
Paoli, Pennsylvania

Kluwer Academic / Plenum Publishers
New York   Boston   Dordrecht   London   Moscow

Library of Congress Cataloging-in-Publication Data

Bausch, Kenneth C., 1936–
    The emerging consensus in social systems theory/Kenneth C. Bausch.
      p.   cm
    Includes bibliographical references and index.
    ISBN 0-306-46539-6
     1. Social systems—Philosophy.   2. Systems theory.   I. Title.

HM701.B38 2001
306′.01—dc21

                                      00-067109

If you find errors, especially misstatements of fact or mistakes of interpretation, please inform the author:

Ken Bausch, Ph.D.
Ongoing Emergence
5430 Bannergate Drive
Atlanta, GA 30022
770-849-0891
kenbausch@mindspring.com
fax 707-929-1114

ISBN: 0-306-46539-6

©2001 Kluwer Academic / Plenum Publishers
233 Spring Street, New York, N.Y. 10013

http://www.wkap.nl/

10 9 8 7 6 5 4 3 2 1

A C.I.P. record for this book is available from the Library of Congress

Printed in the United States of America

Dedicated to the memory of
Carl and Pearl Bausch

FOREWORD

The acceleration of contemporary culture wreaks havoc on the traditional standards, structures, and stories that gave meaning to our lives in earlier times. It demands decisions and actions of us that do not fit old paradigms and do not wait for philosophical reflection. In our "postmodern" culture, we lack generally accepted "metanarratives" that would place our lives and decisions in meaningful contexts.

In earlier periods of history, we were at home in our traditions. We dwelled in religious, ethnocentric and scientific worldviews that defined our place in the world. Stories of enlightenment and progress fueled our imaginations well into the 20th century. We gloried in the accomplishments of Western civilization. We proclaimed the superiority of our democracy, science, and capitalism.

The barbarity of the Second World War radically challenged this faith in our enlightenment, especially in Europe. Thinkers, such as Jean Paul Sartre, Albert Camus, and Theodor Adorno, brought to print the dark underbelly of our civilization. Adorno expressed this disillusion in a remarkable essay, "After Auschwitz," that declared the Enlightenment to be dead.

Today, the old stories are passé. Old verities manifest themselves as quaint, even sinister, remnants of past hubris. Ethnocentricity grudgingly, often violently, yields to multiculturism. We lack accepted narratives (other than unbridled self-interest) that would give context to our lives. We do not want to go back, but we miss the support that cultural meaning used to give us. We need a meaningful new cultural narrative.

In the mid 20th century General Systems Theory offered a new narrative that promised to unite the sciences. Ludwig von Bertalanffy and others identified similar systemic functions in the different branches of science and described a comprehensive worldview based on those elements. Erich Jantsch, for example, in *The Self-Organizing Universe*, spelled out how self-transcending processes generate physical, biological, social, and psychological systems.

GST received a cool reception from the established sciences. It was criticized for dealing in metaphors, for being philosophical speculation, and for being incapable of falsification. As a result, the claims of GST were not taken seriously in the courts of academia and public opinion.

In spite of the eclipse of GST, the insights of systems theory survived and thrived. They were integrated into the fields of evolutionary theory, cybernetics, information theory, organization theory, social theory, and family therapy. With the advent of computers, new sciences of self-organization have arisen in the fields of evolution, artificial intelligence, synergetics, catastrophe theory, chaos, complexity, dissipative structures, self-organized criticality, and so on. The richness of these sciences urges us to find some unity among them.

Ken Bausch takes a large step toward such unification in this book. He presents the results of an extensive investigation into the works of major systemic authors. He condenses their pivotal works. He generates the five syntheses presented here by working with those digests, abstracting key concepts and comparing them.

The results are remarkable. Bausch applies management theory to the theory of management. With the methodology of Interactive Management, he generates an enhancement pattern that illustrates the relational influences among the multiple standards that are proposed for the designing process.

He offers a major advance in social evolutionary theory. He distills three interrelated narratives from the thoughts of these authors that describe the self-organization of our social world. The discourse/lifeworld story draws together the thoughts of Habermas, Mead, Durkheim, Maturana, Varela, Bickerton, and Goertzel that describe how the social world is generated through evolving processes of discourse. The autopoiesis story (principally Maturana, Varela, and Luhmann) describes how expectations drive social evolution and generate the structures of the social world. The component-system story (Csanyi, Kampis, Goertzel) demonstrates the process nature of our social world, how it exists as patterns of information. These stories all portray the social world as a system of processes that maintain themselves as structures through their constant reproduction.

Bausch uncovers a deep foundation for information science in the thoughts of these authors. He traces communication from the rudimentary ideas of interconnection and structural coupling, through stages of non-symbolic signs and semiotic interaction. He traces the origin of representations and language. He clarifies the origins of meaning and the evolution of cognitive maps. In addition, he probes the reality of cognitions, the circularity of our knowledge, and the bases of epistemological rigor.

This book does not produce a metanarrative for our postmodern age. It does, however, reveal the terrain upon which such a narrative will run. The narrative will not be ethnocentric like the past metanarratives of feudal Christianity, Western superiority, and scientific enlightenment. This change. It will provide a general understanding of how we move into an uncertain future. To a surprising extent, this narrative will capture outstanding elements of the archetypal myths while providing a dynamic context for rapid decision-making.

This book is well researched, well thought out, and well written. It advances conceptual integration in a bifurcating world. From beginning to end it is a treat for the thoughtful reader.

Bela H. Banathy
Professor Emeritus, Saybrook Graduate School
President International Systems Institute
And
Alexander Christakis
President, CWA, Ltd.

ACKNOWLEDGMENTS

I am especially indebted to Bela H. Banathy and Alexander Christakis. Bela was the first to encourage my research into Systems Theory and he has stuck with me ever since. Aleco came into my research at a critical time and provided the methodology that enabled its success. They graciously consented to write the Foreword to this book.

Alexander Laszlo contributed considerable editorial, strategic, and emotional support; he stood by me at crucial times. Charles Webel enforced a rigorous standard of academic quality. Allan Combs has supported me at every opportunity.

Diane Conaway has supplied impeccable technical services with a smile. Candace Kaspars contributed invaluable comments on an earlier version of this work. Ken Derham provided enduring editorial support. Numerous other people generously read parts of this material and offered helpful comments.

The amazing authors whose works are encapsulated in this book deserve credit for much of the illumination that it might provide. The brilliance is theirs. The obfuscations are my own.

Finally, Marie Kane has anchored me during the years that I worked on this book. She has been my sounding board, my goad to ongoing effort, and my emotional support.

CONTENTS

PART TWO:
INCORPORATING HUMAN PARTICIPATION
INTO SYSTEMS THEORY AND DESIGN

PART THREE:
ADVANCES IN THE AREAS
OF SOCIAL, COGNITIVE,
AND EVOLUTIONARY THEORY

PART FOUR:
FIVE EMERGING SYNTHESES

INTRODUCTION

THE EMERGING SYSTEMIC PARADIGM

Systemic thinking is integrative. Unlike many sciences that arrange themselves in horizontal strata of disciplines (such as physics, chemistry, biology, psychology, and sociology), systems theory has a vertical orientation. It finds and correlates parallel expressions, approaches, and theories in the various levels of science and tests them for coherence and compatibility.

      This kind of thinking has an ancient pedigree. Under the name of "philosophy," it was the dominant paradigm of Western thought even in the time of Galileo and Newton. In modern times, the old integrative style of thinking was supplanted by the paradigm of analytic science. In the words of Descartes, the analytic idea is: "To divide up each of the difficulties [of a problem] . . . into as many parts as possible . . . in order that it might be resolved in the best manner possible" (1952; 1619, p. 47). This analytical emphasis of Descartes combined with the empirical emphasis of thinkers such as Francis Bacon to generate modern science and its ongoing technological revolutions.

      In our age the analytic scientific paradigm is clearly victorious. The citadel of its old academic adversary, philosophy, is reduced to the rubble of postmodernism. The scientific and technological juggernaut rushes on to deeper insights and more societal transformation.

      And yet, we begin the millennium with some apprehension. We seem to move at ever increasing speed in all directions at once. We lack a common metanarrative, a cultural compass. Perhaps we have thrown out the baby with the bath water.

      The feeling of millennial malaise is not just personal; it is palpable in scientific academia. Early in the 20th century, the physics of relativity and the quantum superseded Newtonian ideas on the nature of the universe. Heisenberg's uncertainty principle sent tremors throughout all modern thinking and culminated in postmodernism. Godel's incompleteness theorem demolished the effort to construct

1

a logically coherent and comprehensive theory of mathematics and logic. While the locomotive of science plunges along, it has lost its theoretical tracks.

What is next? Is there some innovation on the horizon that can bring coherence into this burgeoning theoretical chaos? There is. It is the growing realization that nature is creative. Matter/energy spontaneously generates forms of order and organization. Living things maintain themselves as ever-evolving processes of self-reproduction. Societies organize themselves. In social situations of rapidly accelerating change, the self-organizing processes of participatory democracy provide the best survival strategy.

This new realization has a strong base in mathematics just as the new thing of the 17th century had its base in the calculus of Newton and Leibniz. This new mathematics has various names: complexity theory, nonlinear dynamics, fractal geometry, or chaos theory. This new mathematics, abetted by our modern computing capabilities, demonstrates the inevitability of self-organization among units engaged in random motion.

This new realization also manifests in the fundamental physics of far-from-equilibrium thermodynamics in which linear processes turn turbulent and generate more complex processes. It manifests in the biologically based theories of component-systems and autopoiesis in which living systems constantly reproduce themselves in ever-adapting incarnations. It manifests in theories and experiments with cellular automata, neural nets, and artificial intelligence. It manifests in practices of stakeholder design and interactive management that supplant old hierarchical and linear modes of governance.

This emerging paradigm brings innovation and coherence to evolutionary science. It renews the project of integration begun by General Systems Theory fifty years ago. Goertzel expresses this latter thought as follows:

> What does modern complex systems science have that General Systems Theory did not? The answer, I suspect, is remarkably simple: computing power (1994, p. 6).

The emerging systemic paradigm is integrating information systems, cybernetics, communication theory, second-order cybernetics, organizational design and management, and evolutionary theories (general, life, cognitive, social, linguistic, and psychological) into a coherent second-order vision of our world.

This second-order vision does not focus on the way *things* exist in our world. Instead, it focuses on how dynamic and evolutionary *processes* work. The emerging paradigm is also a paradigm of emergence. It explains how processes evolve in complex environments. It gives us new tools for anticipating our future.

FIVE CONTEXTS

This book grew out of my fascination with applying systems theory to social processes. As I was examining the works of systemic theorists to construct this book, I took special note of the social aspects of their thinking. In the process of synthesizing, I found that my notes fell into five thematic areas: designing social systems, the structure of the social world, communication, cognition, and epistemology.

These five areas are foundational for a theoretic and practical systemic synthesis. They were topics of contention in a historic debate between Habermas and Luhmann in the early 1970's, *Theory of Society or Social Technology: What Does Systems Research Accomplish?* (1971). They continue to be contentious topics for social philosophy and philosophical sociology.

Since the 1970's, systemic thinking has taken great strides in the areas of mathematics, physics, biology, psychology, and sociology. These advances radically alter the theoretic perspective for considering design, social structure, communication, cognition, and epistemology.

This book presents a spectrum of those theoretical advances. It synthesizes what various strains of contemporary systems science have to say about social processes and assesses the quality of the resulting integrated explanations. In this book, I ascertain what a widely divergent sampling of present-day systemic theorists has to say (1) in general and (2) with regard to social processes, and (3) create integrated representations of their messages in the above-mentioned five areas. Finally, I (4) evaluate the validity of these integrated representations within the self-referential epistemology that emerges in this book.

The thoughts of the authors considered in this book range far beyond the ideas synthesized in the final chapters. In the antecedent chapters, the authors' thoughts are more robust and understandable. The directions taken by these authors indicate the far-flung ramifications of systemic thinking.

In the following chapters, I first present a background that includes the historical context of previous systems theory, contemporary theories of evolution, and the debate between Habermas and Luhmann. Second, I relate developments in sociological theory and in the practices of systemic design that accommodate Habermasian objections in ways that approximate an "ideal speech situation." Third, I describe advances in social, cognitive, and evolutionary theory in the past thirty years, which includes the mature theory of Luhmann, information theory, the cognitive equation, and the omnipresence of metaphor.

Finally, the five chapters of synthesis show the converging areas of consensus among these outstanding thinkers. They sketch the emerging systemic paradigm in the areas of design, social theory, communication, knowledge, and epistemology. Advances in these fields lay foundations for broad consensual progress in our theoretical and social endeavors.

PART ONE

BACKGROUND

BACKGROUND

The early history of systems theory and the criticism it provoked provide the historical background behind the question: whether systems theory can be validly applied to social processes. Chapter 1 provides a brief sketch of the early history of systems theory.

Systems theory and evolution are closely entwined because systems theory focuses on the commonalities of physical, biological, social, and cognitive processes. The three following chapters provide the evolutionary background for discussion of whether and how systems theory can be applied to social processes. Chapter 2 presents current thinking on physical and life evolution. Chapter 3 discusses theories of social and cognitive evolution.

The Habermas/Luhmann debate (Chapter 4) crystallizes many of the disputes concerning the use of systems theory in dealing with social processes. Habermas's more recent thinking (Chapter 5) provides a guide for some recent sociological and design theory and a foil for other forms of systemic thought.

CHAPTER 1

THE HISTORICAL CONTEXT

ABSTRACT

Systems theory grew out of the confluence of certain theoretical positions drawn from biology, cybernetics, sociology, and systems engineering after World War II, but it has long cultural antecedents. As General System Theory (Von Bertalanffy), it sought to provide an overarching terminology and a generic description of processes that were common to differing scientific disciplines.

The Habermas/Luhmann debate of the late 1960's and early 1970's focused attention on several problems involved with applying systems theory to social processes. The subsequent theories of Habermas, Luhmann, and the burgeoning of diverse systems approaches prompt us to reconsider the question: To what degree can systems theory be applied to social processes in theoretically and ethically sound ways? Of particular importance are complexity/ bifurcation/ component systems thinking and autopoiesis. In large part, this book provides an explication of these two theories and their relevance to social processes.

GENERAL USAGE

We talk about systems in numerous contexts: the solar system, the endocrine system, the system of complex numbers, computer systems, capitalist systems, bureaucratic systems, school systems, personality systems, and so on. The overriding concept of "system" is "a regularly interacting or interdependent group

9

of items forming a unified whole" (*Webster's Ninth New Collegiate Dictionary*, p. 1199).

The question arises whether human activity systems are "systems" in the same sense as a computer system or a digestive system. In what ways are these "systems" alike? In what ways do they differ?

EARLY SYSTEMS THINKING

The idea of systems was common to the thinking of ancient Greeks, many of whom held that a whole (for example, a body) was greater than the sum of its parts. The body was a common metaphor for a social system (for example, St. Paul called the church "the body of Christ"). The idea of the body as a system was reinforced when Harvey discovered the circulatory system (1628). Hobbes employed the metaphor of society as an organism as the title of his *Leviathan*, 1651. In 1933, Durkheim said that the division of labor is "a phenomenon of generalized biology" (quoted by Habermas, 1987a, p. 114). Most, if not all, of these early conceptions considered systems to be self-sufficient and therefore closed.

Parsons

Talcott Parsons spearheaded the application of systems theory to society in the United States. Parsons, over his long and distinguished career, considered most of the promising theoretical ideas about social issues that were available to him and tried to integrate them into a consistent representation. He persisted in writing essays "*in* systematic theory," while realizing that he was unable, given the then current state of social and systemic thought, "to present a system *of* theory" (Parsons, 1951, pp. 536-537). He explains his effort in *The Social System* as follows:

> It attempts to represent the best attainable in the present state of knowledge with respect to the theoretical analysis of a carefully defined class of empirical [social] systems. It is fully recognized that this theory is fragmentary and incomplete. But at the same time, the concept of system as a guiding conceptual scheme is of the first importance as an organizing principle and a guide to research (1951, p. 537).

The concept of systems that Parsons was working with was that of an open system; that is, an input-output machine.

Parsons carefully differentiated social systems theory from personality theory and a theory of culture. In his conception, these three theories combine in an overall theory of action, in which we can understand how we actually do human (social) activity. For him these three systems are intricately related: both independent of each other and interdependent on each other. He saw the inadequacy of treating social systems as either extensions of psychology or manifestations of culture. The psychological point of view, he says, "is clearly inadequate most fundamentally because it ignores the organization of action about the exigencies of social systems as systems." He further states that it "is equally unacceptable . . . to treat social systems as only 'embodiments' of patterns of culture, as a certain trend of thought common among anthropologists has tended to do" (1951, p. 539). Parsons evidently believed that his conception of systems theory provided a valuable, if partial, model for understanding social structures and processes.

In wrestling with ideas such as these, Parsons found himself caught between horns of various dilemmas from which he was never able to completely extricate himself, as numerous commentators have pointed out. Habermas (1989, pp. 199-300), for example, examines Parson's long struggle to integrate his action theory of society and a theory of society as an autonomous system. The action theory emphasizes the importance of communication, norms, and other communal developments, that is, the interdependence of society with individuals and their culture. The systems theory emphasizes the "functional" autonomy of societal processes, which are independent of lifeworld concerns such as these. In the end, Parsons did not resolve the tensions in his thinking; in other words, he did not produce a system of theory (as he himself recognized).

According to Buckley, any critique "of the Parsonian framework must contend with its loose conceptual structure" (1967, p. 23). Alexander finds strength as well as confusion in the tension of Parsons' thought, which is "internally contradictory," and which "provides support for a variety of partial, often mutually antagonistic interpretations" (Alexander, 1983, p. 309). Parsons attempted to combine the best of the past with the newly forming corpus of systems thought. He describes the historical roots of his theorizing as:

> a convergence from at least four sources: Freud's psychology, starting from a medical-biological base; Weber's sociology, which worked to transcend the problems of the German intellectual tradition concerning idealism-materialism; Durkheim's analysis of the individual actor's relations to the "social facts" of his situation; and the social psychology of the American "symbolic interactionists" Cooley and Mead, who built upon the philosophy of pragmatism (1977, p. 179).

In his later work, Parsons describes his action theory of cultural-level systems as having "four generic types of subsystems," which are the organism, the social system, the cultural system, and the personality (1977, p. 178; originally published 1968). Any human cultural activity then results from a combination of the effects of organic, social, cultural and personality effects. In this four-function schema, activities in these different areas mutually influence, or "interpenetrate" each other.

Parsons uses this idea of interpenetration to explain how two free, and therefore unpredictable, human beings can engage each other in mutually advantageous behaviors. He calls this situation the problem of "double contingency." In this situation, ego makes her action contingent upon the expected action of alter, and alter makes his action contingent upon the expected action of ego. This situation of double contingency creates the "problem": ego and alter are dependent upon each other to satisfy their self-determined needs, but are unable to depend on (cannot determine) each other's actions. Unless this problem is solved, according to Parsons, there can be no concerted social action (cf. Parson and Shils, 1951, p. 16; and Parsons, 1968, p. 436). Parsons solves this problem by postulating a cultural consensus concerning action that is passed down to ego and alter.

The idea of consensus plays a major role in Parsons' conception of society, which he conceives as coalescing around accepted values and norms of behavior. Parsons deals with how a society maintains its values and norms in the face of deviant behavior, and thereby maintains its equilibrium. "The solidarity of a community is essentially the degree to which (and the ways in which) its collective interest can be expected to prevail over the unit interests of its members wherever the two conflict" (1977, p. 182). Then he tries to explain how change is possible within a consensual society that restrains discordant voices. In his reliance upon consensus to solve the problem of double contingency and to provide the foundation of society, Parsons leans to an action and lifeworld explanation of society.

In his explanation of how societies evolve through differentiating themselves, however, he leans toward an impersonal systemic explanation. In differentiating themselves, social systems create subsystems to deal with functions that in their earlier stages of evolution were handled in a global, undifferentiated way. In this way, as societies become more complex, they differentiate out religious functionaries, military leaders, administrative bureaucracy, economic activities, and judicial functions. In modern societies, some of these differentiated functions have become semi-autonomous ways of coordinating activity which Parsons calls "symbolic media of exchange" (1977, pp. 188-189).

The pre-eminent medium of exchange is *money* in the economic sphere, but there also exist: *power*, which is invested in authority for the maintenance of political order; *commitments* to valued institutions, which articulate cultural systems; and *influence*, which justifies group consensus on the basis of specialized expertise (Parsons, 1977, p. 189-190). In dealing with differentiation and media of exchange, Parsons explains society in a systemic, impersonal way that excludes the need for consensus and other lifeworld concerns.

The Systems Movement

The systems movement, as such, took hold in postwar America with the convergence of ideas drawn from biology, systems engineering, cybernetics, and sociology. Biology, with its evolutionary emphasis, contributed the ideas of emergence and hierarchy among and within living systems. Systems engineering contributed the "hard systems" approach to problem solving that grew out of the experience of engineers trying to develop weapons and logistical systems during World War II. Cybernetics introduced the idea of control through information (feedback) in operational systems. The names most notable in the creation of systems theory are: from biology, Von Bertalanffy, "who in the mid 1940's generalized organismic thinking . . . into thinking concerned with systems in general" (Checkland, 1981, p. 77); from hard systems, the RAND corporation, which developed a prominent problem solving methodology; from cybernetics, Norbert Weiner who defined cybernetics as "the entire field of control and communication theory, whether in the machine or in the animal" (quoted from Checkland, 1981, p. 84); and from sociology, Parsons.

The Mechanical Model

Walter Buckley (1967) provides a convenient summary of the antecedents of systems theory. He provides three social systems models: mechanical, organic, and process. Following Sorokin (1928), he describes the origin in the 17th century of "Social Physics," that considered a person to be an elaborate *machine*.

> The physical concepts of space, time, attraction, inertia, force, power-which must be recognized as anthropomorphisms originally borrowed from every day human experience-were borrowed back in their new connotative attire and applied to man and society (p. 8).

Vilfredo Pareto (1916/1936) supplemented the mechanical model with the idea of equilibrium. In his concept of equilibrium, "any moderate changes in the elements or their interrelationships [of a system] away from the equilibrium position are counterbalanced by changes tending to restore it" (Buckley, p. 9). According to Buckley, this concept of systemic equilibrium was "taken over almost unchanged" by many sociologists following Pareto, especially by Talcott Parsons.

The Organic Model

Herbert Spencer (1897) dealt with social systems in terms of organic evolution. He recognized that societies and living bodies bore similarities and dissimilarities. He stated, "there exist no analogies between the body politic and a living body, save those necessitated by that mutual dependence of parts which they display in common" (quoted by Buckley, p. 15). He contrasted the social organism with biological organisms in the following terms: discrete vs. concrete, asymmetrical vs. symmetrical, and sensitive in all its units vs. having a single sensitive center. Still, he used the functions of the human body as illustrations of structures and function that also apply in society. He is credited as the originator of the serious scientific usage of the *organic model*.

Some of the followers of Spencer were not so modest. They alleged social analogues of the heart, brain, circulatory system, and so forth. This more elaborate form of the organic model is often called the organismic model. James Miller has produced the most sophisticated organismic model in *Living Systems* (1995; 1978). Miller identifies 20 critical subsystems of a living system within 8 levels of organization: cell, organ, organism, group, organization, community, society, and supranational system (1995, pp. xvii and xix).

Spencer also coined the phrase "survival of the fittest," and spearheaded the development of social-Darwinism, which applied evolutionary and quasi-scientific ideas indiscriminately to social theory and social practice. On Spencer's model, evolution could proceed on an antagonistic "militaristic" manner or in a cooperative "industrial" manner. Spencer himself held that the degree of evolution among organic and social organisms revealed itself in their degree of cooperation. Other social Darwinists, such as William Graham Sumner and Andrew Carnegie (cf. Hofstadter, 1945, pp. 18-51), held an opposite emphasis based upon "the 'natural' and inevitable rise of the 'fittest' individuals in the competitive social struggle" (Buckley, p. 13). Sumner, for example, said

> Let it be understood that we cannot go outside of this alternative: liberty, inequality, survival of the fittest; not-liberty, equality, survival of the unfittest. The former carries society forward and favors all its best members; the latter carries society downwards and favors all its worst members (quoted in Hofstadter, 1945, p. 37).

Parsons in his adaptation of the biological model emphasized stability, order, cooperation, and consensus. He said, for example, "Order-peaceful coexistence under conditions of scarcity-is one of the very first of the functional imperatives of social systems . . . this order must have a tendency to self-maintenance, which is very generally expressed in the concept of equilibrium" (Parson and Shils, 1951, p. 180).

Walter Cannon (1939) coined the word "homeostasis" to clearly distinguish between physical and biological equilibrium. Physical equilibrium operates in a closed system and results in a static equivalence of forces at the lowest common denominator. Homeostasis operates in open systems that interact with their environments in a kind of stable instability and maintain a degree of integrated complexity. In Buckley's terms, homeostasis describes "the dynamic, processual, potential-maintaining properties of *basically unstable* physiological systems" (Buckley, p. 14).

The organic model and homeostasis are basic for many cybernetic explanations of societal functioning. They are the background for numerous discussions of organization and information. Organic analogies are inadequate as models of social systems. This was recognized by Karl Deutsch (1956) ten years before Buckley's (1967) publication. He criticized the narrow conception of homeostasis and the organic model.

> Homeostasis is not a broad enough concept to describe either the *internal restructuring* of learning systems or the combinatorial findings of the solutions. It

is too narrow a concept because it is change rather than stability which we must account for (quoted by Buckley, p. 15).

Since 1967, systems thinking has differentiated into several specialties and combined with novel thinking in the areas of physics, mathematics, and biology. Among these developments can be listed: non-linear thermodynamics, complexity theory, information theory, component-systems, and autopoiesis. Buckley was acutely aware of the antecedents of these advances, which blur the distinctions between mechanical, organic, and societal processes. He lists some of them under what he calls the *process model* of society.

The Process Model

In Buckley's version of the process model,

> [Society is] a complex, multifaceted, fluid interplay of widely varying degrees and intensities of association and dissociation. In this model, "structure" is not "something distinct from the ongoing interactive process but rather a temporary, accommodative representation of it at any one time (Buckley, p. 18).

Georg Simmel (c. 1908) is credited by Buckley with anticipating many of the principles of the process model. Simmel described his work as a "struggle for life," and treated "topics such as power, conflict, group structure, individuality, and social differentiation" (Schnabel, 1985, p. 751). In the United States, George Herbert Mead and John Dewey, in their model of "social interactionism," proposed a continuous process of socialization in the complexity and fluidity of interaction. Also Gordon Allport (1961) defined personality as "a complex of elements in mutual interaction" (quoted by Buckley, p. 22).

The general systems perspective grafts the idea that matter is neither living nor non-living, but both (la Mettrie) onto the process model. With this idea, it places the idea of matter's *organization* at the center of systems research. In this way, it fuses "both organicism and mechanism in cybernetics and general systems theory" (Buckley, p. 37). This perspective attempts to explain how societies create more complicated relations on the basis of open systems that interchange with their environments across their boundaries.

Buckley concludes his summary with the observation that systems theory "can be seen as a culmination of a broad shift in scientific perspective striving for dominance over the last few centuries" (1967, p.37). Many of the basic ideas of systems theory have a long pedigree. This summary also indicates the turbulent state of systems and social systems theorizing in the 1960's before the appearance and coalescence of the two grand unifying theories of present-day systems thinking: complexity/ bifurcation/component systems and autopoiesis.

Self-Organization and Autopoiesis

These two strands of thinking advance systems theory beyond the bounds of mechanical (closed) models and organic (open) models and move it into the arena of emergent models. Component-system thinking, which is propounded by Csanyi, Kampis, and (to some extent) Goertzel, is an outgrowth of Bertalanffy's General Systems Theory (GST). GST "enabled one to interrelate the theory of the organism, thermodynamics, and evolutionary theory" (Luhmann, 1995, pp. 6-7). Component-system theory loosely includes the bifurcation thinking of Prigogine, the molecular biology of Eigen, the complexity thinking of Kauffman and Gell-Mann, the physics of information theory, and the sociology of cognitive maps. It describes the processes that generate increasing unity and complexity in specific details that are alleged to have universal application.

Autopoiesis in its biological form, proposed by Maturana and Varela, considers organisms as systems that are closed in their internal organization, but open on the level of their structural composition and metabolism. Autopoiesis in its sociological form, proposed by Luhmann, focuses on the difference between system and environment and identifies autopoietic systems with the unity of contradiction that derives from their being simultaneously autonomous from their environment and totally dependent upon it. In our thinking about autopoietic and component-systems, we discover vistas of new and possibly fruitful explanations of physical, organic, social, and cultural processes. It turns out that these ideas comprise the bulk of the ideas that are considered and evaluated in this research.

Prior to the development of systems theory, Simmel's idea of differentiation had led European sociology to develop ideas about how societies evolved by differentiating generic social functions, like clan leader, into diversified functions, like chief, priest, healer, hunter, and farmer, among others. Max Weber and George Herbert Mead considered this evolution from the standpoint of actors involved in the "lifeworld" (*Lebenswelt*). Emile Durkheim considered it from the viewpoint of an observer who describes "objectively" what occurs within society-that is, within a proto-systems context, which shows how activities within a society are interrelated.

HABERMAS/LUHMANN

The attempted sociological synthesis of Talcott Parsons was particularly influential because of the ambitious scope of his theorizing. As Habermas has shown (1989, pp.199-300), Parsons alternated between lifeworld and systems explanations for societal processes; he probably gave dominance to the systems approach. Among the dominant ideas of Parsons, beyond action systems, equilibrium, and the consensus constitution of society, was a particular brand of differentiation theory and its steering media.

Luhmann fastens upon the differentiation project of Parsons and enlivens it with his phenomenological account of meaning. He stresses the idea that differentiation and the steering media replace lifeworld activity with an automatic coordination of activity. He also gives a dialectical cast to many sociological conceptions by stressing the importance of contradiction, conflict, and noise for generating self-reference and a living society.

The ideas of Parsons and Luhmann have advanced our understanding of social realities, but they have not, by any measure, made them transparent. On the contrary, they have generated the controversy of lifeworld versus systems.

Systems theories have found a formidable but receptive adversary in Jurgen Habermas. Habermas confronted Luhmann in debate in the early 1970's. As a result of their debates, Habermas and Luhmann came to appreciate each other's viewpoints. When Habermas analyzed Parsons's work in volume two of *The Theory of Communicative Action* (1989), he relied upon Luhmann's insights. Habermas, however, continues to voice serious concerns about the systems approach to social processes.

In turn, Luhmann has expanded his concept of self-reflective systems that create meaning in the face of complexity. Among other things, he now proposes that systems "use the difference between system and environment within themselves for orientation and as a principle for creating information" (1995, p. 9). He defines social systems as autopoietic systems that reproduce themselves through communication (1990, p. 3). He solves the problem of double contingency (the situation facing two autopoietic systems in which neither party can predict the behavior of the other) in terms of the trust that autopoietic systems extend to each other (1995, pp. 125-134). He also develops a self-referential epistemology that he calls "functional analysis" (1995, pp. 52-58).

SUBSEQUENT SYSTEMS HISTORY

The Club of Rome

In the 1970's, the Club of Rome (Meadows et al, 1972) released its first report. The original report met rebuttals (Cole et al., 1971; Richter, 1974), and generated sequels (Mesarovic and Pestel, 1964; Meadows et al., 1974a and 1974b). The controversy is well summarized by Neurath (1994/1976).

The scientists and philosophers of the Club took a systemic look at the development of present-day civilization by considering the interactions of global subsystems in the areas of population growth, agricultural production, dwindling resources, and pollution. On the basis of their computer simulation of the future course of world ecology, they predicted worldwide calamity by the year 2025.

After concluding that our present course of accelerating extensive growth is not sustainable, they advocated experiments in alternative growth possibilities including birth control, recycling, pollution control, and contingency planning. Their point of view has strong advocates within the general systems community. In recent years, it has increased awareness by calling the world's attention to ecological effects such as global warming and environmental pollution. It has also gained coherence by employing the principles of non-linear thermodynamics.

Recent Advances

Systems theory has progressed in the past twenty years. In the area of systems practice, cybernetics, synergetics, and "soft systems" have expanded the hard systems methodology to deal with greater complexity. "Hard systems" developed as a refinement of the time-honored method of engineers. This method precisely defines a problem and systematically proposes and evaluates possible solutions with an eye to their ramifications. Cybernetics, especially as the Viable Systems Model (Stafford Beer), equips a system with strategies that enable it to cope with "environmental changes even if those changes could not have been foreseen at the time the system was designed" (Jackson, 1991, p. 294). Synergetics, as proposed by Herman Haken, shows the dynamic power of intercommunication. Operational research (hard systems) was developed and expanded by C. West Churchman, Russell Ackoff, and Peter Checkland, among others. This same triumvirate created the "soft systems" methodology to deal with social situations in which there are no clear-cut objective(s) and there is just an ill-defined sense of something being wrong. Soft systems theorists consider hard systems methodology to be a special application of systems theory in situations where objectives are not in question.

Churchman laid the foundations for soft systems in the 1970's when he defined it as an inquiring system (that is, a research methodology) similar to the inquiring systems of Kant and Hegel. He also faced head-on the ethical considerations that bear on decision-making. He deals, for instance, with the following apparently inconsistent consequences of democratic decision-making:

1. Ultimately all interested persons should be partners in social decisions.
2. Good decision-making depends on the selection of relevant information and its proper utilization.
3. Only a few people have the time and means to obtain relevant information and to ascertain its proper use (1961, p. 7).

Ackoff, among his other achievements, is remembered for coining the technical word "messes" to describe the domain of soft systems.

Finally, Checkland formalized the soft systems methodology into clear steps. The process starts with a real world description of the problematic situation. It moves to an interpretive analysis that creates abstract systemic models of the situation. Then it moves back into the real world to select a model that most adequately matches the problem situation. The first cycle of the process ends with action on the basis of the model. The process continues on to further iterations of this methodology until a problem is clearly defined and addressed. At that point, hard methodologies might be employed to further address the situation. Soft systems methodology is flexible. Work can and often does start at intermediate steps, move both backward and forward, and jump to intuitive conclusions out of sequence. Checkland felt, in 1981, that this methodology was a value-neutral means for reaching decisions, and that it approximated the "ideal speech situation" of Habermas (cf. 1981, p. 283). Checkland's assumptions on this score were soon to be challenged.

Critical systems theorists began challenging Checkland's claim of value neutrality in the 1980's. They showed that the practice of soft methodology has a regulative and managerial bias because it neglects power inequalities in its decision-making processes. They develop methodologies and social theories to correct those inequalities. In so doing, they explicitly take the observations of Habermas into account.

Other important contributions to systems theory come from the field of non-linear thermodynamics. Chaos theory charts the changes that systems experience as they accelerate from linear to periodic to turbulent motion, and the peculiar laws that apply to such systems. Complexity theory explores self-ordering physics and mathematics (Kauffman), the "order from noise" principle (Von Foerster), and the ramifications of information theory.

In addition, the theory of dissipative structures (Prigogine) explains the leaps to greater complexity in non-living and living structures in terms of entropy. The theory of hypercycles (Eigen) carefully explicates the steps toward organic life while the theory of autopoiesis (Maturana and Varela) describes the evolution of living organisms in terms of their self-generating autonomous organization. Csanyi and Kampis develop earlier insights, from Jantsch and others, into a general theory of evolution on both descriptive and mathematical levels. Goertzel develops a full-blown cybernetic construction of human psychology. Laszlo, Banathy, Warfield, and others propose models and designs for moving into a new era of social organization.

CONCLUSION

The serious questions raised in the Habermas/Luhmann debate twenty-five years ago, deserve reconsideration in light of (1) the subsequent development of their thought and (2) and the growth of systems theory. The new systems thinking that has just been mentioned provides theories and methods that are far more sophisticated than those available to Buckley, Habermas, and Luhmann in the 1960's and early 1970's. These new developments empower us to more adequately describe social processes, to design them, and to assess the truth, validity and ethics of such descriptions and designs. These systems developments and many of their applications are detailed in the following chapters.

CHAPTER 2

PHYSICAL AND LIFE EVOLUTION

ABSTRACT

Darwinism and neo-Darwinism were formulated, by necessity, on purely biological grounds because physics was, until recently, hostile to evolution. Today, strong new theories of evolution are possible because (a) the physics of far-from-equilibrium thermodynamics has explained how evolution is inherent in material processes, (b) our computational power has brought to light the wonders of complexity theory, and (c) molecular biology has established the reasonability of life's arising from pre-biotic processes.

   Darwinian evolution does not make use of these advances in physics, computation, and molecular biology. Two newer theories of evolution explain evolution in more powerful ways. The component system model of Vilmos Csanyi and George Kampis expands the principles of non-equilibrium thermodynamics to explain pre-biotic and biotic evolution. The theory of autopoiesis proposed by Humberto Maturana and Francisco Varela explains how living things maintain their lives and grow by structurally coupling with their environments. These two theories are complementary as intersystemic and intrasystemic descriptions of the same phenomena

KAUFFMAN

In *The Origins of Order* (1993), Stuart Kauffman describes the self-ordering processes in complex dynamic systems and relates that self-ordering to the process of Darwinian selection. He develops a second-order theory of evolution in which selection (1) operates on the ordered complexity which nature presents to it, and (2) picks out the optimal conditions that make self-ordering possible.

With regard to autocatalysis, he discusses polymer chemistry. Polymers are strings of bonded chemicals. Such polymers can bond with each other and form parallel Boolean systems which exhibit "and," "or," "only if," and other relationships. They exist as massively parallel Boolean systems. Therefore, we might expect that they generate order. Using these properties of polymers, Kauffman shows "that self-reproduction and homeostasis, basic features of organisms, are *natural collective expressions of polymer chemistry*" (p. 298).

In addition, autocatalytic polymeric systems "afford a crystalline founding example of functional wholeness, hence functional integration" (p. 388). They are prime examples of how existing systems organize themselves.

> Once a catalytic set of polymers emerges, it is a coherent whole by virtue of achieving catalytic closure. Given the underlying model chemistry and catalytic closure, the functional role of each polymer or monomer in the continued existence and proliferation of the autocatalytic set is clear (p. 388).

These polymers and monomers set up relations with each other that differentiate a subset of essential variables. These essential variables must be kept within bounds. If the system-cum-environment (called System) does not push these variables out of their limits, the system changes nothing. If the System pushes the system out of bounds, the system makes a jump to a different attractor in hopes of keeping its essential variables within bounds. In this situation, unsuccessful jumps and staying in place lead to death; successful jumps lead to survival and likely evolutionary progress.

In polymeric self-reproduction, the primitive notion of "agency" arises because polymers work together for a common goal of survival. As Kauffman says, "In this non-conscious sense, an autocatalytic set becomes a locus of agency" (p. 388).

LASZLO

In *The Interconnected Universe* (1995), Ervin Laszlo argues that the process of structuration that has generated the universe cannot be the result of *pure* chance. In order to bring evolution by chance into the realm of possibility, there must be some physical principal that interconnects events in time. The four recognized physical fields (gravity, electromagnetism, the strong and weak nuclear forces) are sufficient to explain happenings in the classical universe that lacks an arrow of time. They are incapable of explaining a universe that exhibits bifurcations in the direction of greater complexity.

He postulates a field of universal interconnectivity that provides the needed connectivity. This fifth field exists in the zero-point energies of the quantum vacuum as a universal three-dimensional holographic memory composed of longitudinally propagated (Tesla) waves.

Without some sort of universal memory, such as that described by Laszlo, self-organization would be impossible. With such a memory, the universality of self-organization is plausible and obvious. If the profound connectivity of the universe has evolved into human societies, then those societies themselves must be profoundly interconnected. As a corollary, human societies must function best when they honor each other and the unity that they constitute.

PRIGOGINE

Theology and philosophy, prior to the scientific revolution of Copernicus, Galileo, Descartes, and Newton, conceived of space in Aristotelian terms. For Aristotle, space changed constantly as countless entities sought their own ends: rocks sought to be at rest; plants wanted to grow; animals wanted to move; and humans wanted to know. Aristotelian space was not stable; it changed as the entities that composed it changed.

Classical physics broke free of Aristotelian thought when it began to conceive of the world in Euclidean terms; that is, as a timeless abstract space that did not change over time. This abstract space is the basis of Euclid's *Elements* and the Cartesian coordinates. Within this space, every action is mathematically reversible. Classical dynamic theory holds that *all* processes in the natural world are also reversible. It holds, for example, that chemical processes work just like the mathematical formulations that express them:

$$2\{H\} + O \rightarrow H2O, \text{ and } H2O \rightarrow 2\{H\} + O.$$

In other words, classical physics denies an arrow of time.

Charles Darwin drafted his theory of natural selection in an intellectual world dominated by classical dynamics, a timeless Euclidean space in which every process is reversible. It was inhospitable for a biology that proposed non-reversible progressive change. Darwin proposed a progressive arrow of time in a scientific climate that either denied there was an arrow (classical dynamics) or held that the arrow pointed toward the dissipation of all energy (the second law of thermodynamics). Accordingly, Darwin proposed his theory on a narrow biological basis that defied the physics of his day.

The new physics of dissipative structuring restores time to physical space. It describes a more biology-friendly space in which many actions, like the regularity of chemical clocks, are irreversible in the direction of greater complexity and communication. This new space is Aristotelian rather than Euclidean.

The Aristotelian space of the new physics is a space that is, at any moment, different from the same space at another moment. The new physics embraces the aphorism of Heraclitus: "You cannot step into the same river twice." Prigogine says,

> Modern science was born when Aristotelian space, for which one source of inspiration was the organization and solidarity of biological functions, was replaced by the homogeneous space of Euclid. However, the theory of dissipative structures moves us closer to Aristotle's conception (Prigogine and Stengers, 1984, p. 171).

Darwin accepted evolution as a fact. He developed a theory that explained how evolution could come about in spite of classical dynamics. He thereby institutionalized a radical divergence between biology and physics. While physics continued to dwell in Euclidean space, biology returned to its more hospitable Aristotelian confines.

Darwin's mechanism of natural selection explained the evolution of adaptive, purposive change by joining three undeniable facts of life with two basic deductions. Jonathan Schull phrases his argument succinctly, as follows:

> For a given species in a given environment, the number of individuals tends to remain constant (Fact 1). And yet, in each generation there tend to be more offspring than parents (Fact 2). Therefore in each generation some offspring die without reproducing, while others live to become parents. There is, in short, competition among individuals in populations, and losers do not leave offspring (Deduction 1). Finally, individuals vary, and offspring tend to resemble their parents (Fact 3), and so in each generation the offspring will tend to resemble the more successful parents of the previous generation (where "success" or "fitness" means reproductive success; Deduction 2; Schull, 1991, p. 61).

The argument of natural selection shows that subtle but purely physical environmental factors lead to inheritable characteristics in populations of organisms. Darwinian natural selection explains, in part, how evolution is possible. It does this by presupposing an accumulation of minor useful changes.

Neo-Darwinists incorporate ideas of genetics, like dominant and allele genes, genotypes and phenotypes, and, eventually, DNA and RNA, into natural selection. They hold that "evolution is simply the slow but continuous change of the gene frequencies of populations" (Csanyi, 1989, p. 96). They emphasize the role of chance in genetics and the necessity of reproductive survival in the ontogeny of organisms.

Objections have been raised against neo-Darwinian orthodoxy, and alternate theories of selection have been proposed. Molecular biologists uncovered redundancies in DNA sequences and variant alleles that neo-Darwinism cannot explain. Paleontologists (Gould and Eldridge, 1977) came up with theories of "punctuated equilibrium," which describe sudden great changes that are followed by an equilibrium state lasting for millions of years. The whole concept of adaptation and fitness was profoundly criticized by Ho and Saunders (1979) who found that "*relaxation* of selection favors an increase of variability and thus the origin of new species" (quoted in Csanyi, 1989, p. 103). They proposed that "the most important source of evolutionary change is interaction between environmental conditions and the epigenetic processes of ontogenesis" (Csanyi, p. 103) If a relaxation of selection fosters a florescence of species, then a design needs to be formulated that enables leaps of evolution in relaxed environmental constraints. Biological theories employing the idea of bifurcations offer promise of providing such a theory.

Evolution has never had an agreed upon physical theory. From Darwin's time, biologists have been of split minds. As evolutionists and optimistic Victorians they thought, as Browning put it, "Every day in every way, we're getting better and better." As scientists, however, they identified with the accomplishments of classical physics and professed its doctrine that there is no arrow of time.

Biologists no longer have to defend their theories against the Euclidean space assumptions of classical dynamics because physicists have rejoined biologists within Aristotelian space. How are they to conceive evolution when physics has moved "from Euclidean to Aristotelian space," where "life, far from being outside the natural order, appears as the supreme expression of the self-organizing processes that occur" (Prigogine and Stengers, p. 171)? They no longer have to defend evolution. They have only to describe how it works.

In non-linear thermodynamics, "matter is no longer the passive substance described in the mechanistic world view but is associated with spontaneous activity" (Prigogine and Stengers, p. 9). Physicists now explain the self-organizing processes that are constantly occurring in nature as the result of autocatalytic, transcatalytic, and autoinhibitive feedback loops. The Belousov-Zhabotinsky chemical reaction, for example, has been well publicized (cf. Prigogine and Stengers, p. 152; and Prigogine, p. 200).

The theory of dissipative structuring is briefly this: When an open dynamic system (like a chemical compound) is pushed far-from-equilibrium by free energy like sunlight) it loses its normal periodicity and starts to act erratically. In the graph of this situation, the system ceases behaving in a linear fashion, and its path bifurcates; that is, at some point, two radically different future courses become equally probable for it. If the system jumps to the course that is radically different from itself, it becomes a new dissipative structure. A very simple example of this process is observed when a flat shallow pan of water is evenly heated from underneath. Spontaneously, a honeycomb of bubbles lines the bottom of the pan. This phenomenon is called the "Bènard instability."

Structures like the Bènard instability are called "dissipative structures" because they are based upon the dissipation (or entropy) of energy. Free energy from the burner turns the bottom layer of water into steam, and the chemical and physical properties of the water molecules capture the steam in a honeycomb. The instability is dependent upon a continuous even flow of energy under the pan.

Many dissipative structures are not instabilities; in fact, every living thing is a relatively stable dissipative structure. A fish, for example, remains a fish through its lifetime and the lifetimes of its progeny. A fish is really metastable; it is a dissipative structure because it creates a niche in its environment through its metabolism with it. It absorbs free energy; converts that energy into its own dynamic structure (replaces molecules in its cells, produces proteins, and so on). It also casts off old and depleted molecules, proteins, and energy. back into the environment. It constantly assembles and disassembles itself by metabolism with its environment.

Prigogine talks of a new coherence and a mechanism of "communication" among molecules that arise in far-from-equilibrium conditions. He notes that "such communication seems to be the rule in the world of biology" (1984, p. 13). Biologists find mutations toward greater complexity and order spontaneously occurring at the levels of the genome, ontogeny, and social relations. The termite mounds, discussed by Prigogine (1984, pp. 86-87), are a prime example. The same self-organization is shown in economic geography by the Christaller model described by Prigogine (1984, pp. 197-203).

Findings such as these have conjoined physical and biological speculation and produced strong new theories of evolution that explain the leaps of evolution and its playfulness. New theories chronicle the increase of embodied information and communication in pre-biotic and living entities. They give evidence of universal systemic evolution.

EIGEN

Manfred Eigen in *Steps Toward Life* (1992) recaps a lifetime of work on the processes that inevitably generated life. He fashions a natural selection theory on the basis of the amount of information that self-organizing chemical systems employ. He analyzes how replication (itself a cyclic process) generates additional reinforcing cycles (thus forming "hypercycles") and intensifies by "compartmenting" itself within cell walls. The resulting "compartmented hypercycle" is a living cell, the building block of all life. He details how replicative errors and bifurcations gradually push this process along.

WHAT IS LIFE?

What is a living thing? How do living things differ from non-living ones? These are not easy questions. Eigen defines a living cell, the basis of all living things, as a compartmented hypercycle. Csanyi defines life as an autogenetic development of replicating chemical networks (RCNs). Maturana and Varela define living things as autopoietic unities, that is, as things that constantly re-create themselves. What are the connections among these three answers to the question, "What is life?"

The connections can be shown by some related examples as they are explained by Prigogine, Eigen, Csanyi, Maturana and Varela.

Prigogine/Eigen

Consider two complex polymer molecules: one is a nucleic acid; the other is a protein. Neither of them is a living thing, but they are both dissipative structures (and RCNs). As such, they are not stable. They are maintained only by chaotic cycles of chemical and thermal energy. If the energy exchange fails, they start to disintegrate.

The nucleic acid has two special chemical functions: it is autocatalytic (it reproduces itself), and it acts as a catalyst for the synthesis of the protein. The protein, on the other hand, catalyzes the self-reproduction of the nucleic acid. The conjunction of the nucleic acid and the protein results in a hypercyclic (or supercyclic) organization.

In this hypercycle, the autocatalytic cycle of the nucleic acid is encircled by a cycle composed of the nucleic acid and the protein that are transcatalyzing each other. Such a hypercycle experiences accelerating reproduction and hyperbolic growth. It is not yet, however, a living thing.

For the protein/nucleic acid hypercycle to become alive, it must have an accurate way of reproducing itself and it must have boundaries. These two conditions are related. If this hypercycle gets boundaries it will be able to control the quality of its reproductions. Within boundaries the hypercycle can more easily match up parts of itself before it splits them into two (cf. Prigogine, pp. 108-109; Prigogine and Stengers, pp. 190-191; Eigen, pp. 64-79).

In actual biology, the cell wall provides the boundaries for the hypercycle; within the cell wall molecules pair themselves up for replication. The paired molecules are the basis of the genetic code. The information of these paired molecules is eventually encoded in DNA and RNA molecules.

Csanyi

A cell is, in Csanyi's terminology, a component-system. It replicates itself and its component nucleic acids and proteins constantly as it metabolizes molecules and calories from its environment. The cell organizes this replication. Through all of its structural changes (in protein-nucleic acid, moment-to-moment replication), the cell maintains its supercyclic and compartmented organization. This metabolism of a cell is its replication in time.

The cell also replicates in space through the process of mitosis. In mitosis, a cell pairs up its proteins, nucleic acids, and other molecules; then it splits the pairs and the whole cell into two parts. The two parts grow to become whole cells that can start the process all over again. Mitosis is made possible by the development of "identical replication."

Maturana and Varela

Autopoietic organizations exist on a different ontic plane from non-living reproductive chemical networks (RCNs) because of a difference in their continuities. Non-living RCNs are distinct from their environment only momentarily; in the second moment, some other RCN replaces them. Non-living RCNs are totally dependent on the thermal and chemical conditions that effect them; they have no autonomy.

The existence of autopoietic organizations (living unities) on the other hand, separates them continuously from their environments. Autopoietic organizations have a past; they endure; they do not fluctuate even though their component parts do. They continuously maintain separation from their environments. In this way, they maintain their autonomy amid the chemical and energy fluctuations of their structures.

COMPLEMENTARY THEORIES

Maturana and Varela explain evolution as the byproduct of the contortions and leaps that organisms make in their quest to stay alive. Csanyi and Eigen explain evolution in terms of thermodynamics, fluctuations, information, complexity, communication, replication, and selection. The autopoietic description takes an embodied point of view, that of the dancing organism. The replicative model takes an objective view that tries to choreograph the whole production of evolution.

Autopoietic theory has several major advantages. It expresses epistemological sophistication and does not assume a discredited Cartesian dualism or a naive positivism. It clearly recognizes the embodied character of its theorists, its observers, and its actors. Within the methodology of biological phenomenology it explains the origin of evolutionary changes including language, the ego, and the observer. It provides a context that makes the dualist/positivist "objective observer" appear in evolution.

Among the advantages of the replicative model of evolution is its continuity with existing theory in the natural and social sciences. By providing teleonomic descriptions similar to those of Prigogine and Eigen to biological, social, and cultural situations, this model achieves a long-standing goal of General Systems Theory: to provide a common vocabulary and framework for interpreting change on multiple levels and in various aspects of "reality." In addition, the replicative model describes how an autopoietic thing can come about in the first place.

From a methatheoretic viewpoint, the replicative and autopoietic descriptions of living things and their evolution are complementary. Both descriptions account for the basic phenomena of biological evolution. Both also describe evolution as a learning process: supercycles in terms of information; autopoiesis as co-adaptation. By being aware of the difference of standpoint between the two theories, we can explore complementary explanations for any phenomenon that might otherwise be described in only one way.

DANCING AND CHOREOGRAPHING

If evolution were a jazz dance, autopoietic theory would describe how the dancers relate to the music and to each other. The replicative theorists would try to chart their choreography in relation to the story, the music, the stage, and the audience. Autopoiesis attends to what the dancers do. Replicative theory attends to the overall production.

The term "saltatory" is descriptive of an action that contains dancing or leaping. Saltatory evolution describes a progression that both dances and leaps. Structural coupling, in autopoietic theory, is a dance between two or more organisms that results in metacellular unity. The same activity, in replicative theory, is described as a new dissipative structure that is forming as the involved organisms go far-from-equilibrium, react cyclically, and compartmentalize because of their interactions or environmental pressures. Both conceptions, structural coupling and dissipative structuring, involve mutual involvement (the dance) in interaction and lowering of boundaries. They also account for leaps into a higher level of complexity and communication.

The kind of learning that is involved in evolution is also saltatory. It is like the activity of a jazz musician who plays with a riff, jams with fellow musicians, and generates something extraordinary. Such learning involves dissipative structuring: the shattering of boundaries, movement into unknown territory, massive dissipation of energy, and a leap into higher composition and communication with an accompanying relaxation and enjoyment of the final product.

How the dancing and the choreographing of evolution complement each other can be described accordingly.

> *Dance*: Two or more cells exist in the same neighborhood. One cell acts, the other reacts (they are environments for each other). Each cell is actively adjusting its structure to maintain its autopoiesis. If these reactions become stabilized in recurrent interactions, the cells achieve *structural coupling* and create a new biological unity: the *metacellular* structure. As cells continue their mutual interactions they work out different patterns of co-existing through their structural adaptations. Some of their co-adaptations leap to become new dance forms: multicellular organisms, reproduction in space, and more structural couplings.

> *Choreography*: The cells are pushed far from equilibrium by environmental influences (which includes the other cells as part of any one cell's environment). They reestablish their equilibrium by establishing a new, more complex metabolism with their environment (which includes those other cells). Successful metacellular structures establish supercyclic

relationships that embody greater information. Metacellular structures that happen to nucleate (or compartmentalize) establish quality control over their replicative processes by increasing the reliability of their replicative information. They are likely candidates to become the new hit productions on Broadway: multicellular organisms that reproduce in space and continue to evolve.

COMPONENT-SYSTEMS

Vilmos Csanyi (1989; Csanyi and Kampis, 1991) develops a robust theory of the origins of life. This theory of selection assumes the profligate self-organization of matter (and organisms) into states of higher complexity and order. It defines the processes that enable and steer prebiotic and biotic replicative activity.

The Conditions of Evolution

Csanyi prefaces his account of evolution by saying that the law of energy conservation and irreversible thermodynamics "express the necessary but *not sufficient* conditions for the origin of the living systems" (Csanyi, 1989, p. 31); in other words, living systems cannot be explained as mere dissipative structures.

Csanyi says that dissipative structures become biological systems through the presence of certain constraints that are similar to those proposed by Eigen; they are called supercycles (hypercycles) and boundary conditions (compartments). He also proposes several other conditions that will soon be discussed. These biological systems stabilize when they attain minimum entropy and maximum free-energy content.

Orderliness and Stored Energy

A dissipative structure needs energy to maintain its existence. As a system becomes more complicated, it slows the flow of free energy that passes through it; the energy spends more time in the system; the system absorbs more of the free energy's entropy; it *stores* energy; and it uses that energy to become more ordered. "By producing more complex molecules, the system itself becomes more complex, and its *orderliness* increases at the expense of the energy flowing through it" (Csanyi, p. 34). And, "All changes tending to increase free energy's storage are incorporated into the system and become a stable part of it" (p. 41). In this situation, "*the degree of order is proportional to the mean residence time of the energy* in the system while flowing from the source to the sink" (Csanyi, p. 34).

The Components and Functions of Organization

The organization of a component-system transcends mere orderliness. It requires a special mode of connection between complexity and order that Csanyi expresses on the basis of several key definitions.

> A *component system* has (1) physical components that are constantly assembling and disassembling (for example, proteins); and (2) an energy flux running through it that can excite some of the components. In dynamic systems, some components have abilities (*functions*) which influence the *probability of the genesis or survival* of other components of the system, due to their relationships to the component-producing or component-decaying processes (Csanyi and Kampis, p. 78).

Functions, then, are activities of components that affect the reproduction of other components. They are the *organizing agents* at work that turn processes of ordering into processes of organization.

The Start of Life Evolution

The situation of primeval earth was the "ground state" of the molecular system immediately preceding the start of evolution. It contained the basic bioelements: Carbon, Hydrogen, Nitrogen, Oxygen, Phosphorus, and Sulfur (CHNOPS) which were irradiated by visible and ultraviolet light. This situation was a "zero system." A *zero-system* is a system that has not yet developed functions; it has only non-dynamic structure. Csanyi describes a zero-system that might have begun evolution as follows:

> The molecular zero-system should (a) contain a sufficient variety of reactive atoms that form chemical bonds (CHNOPS supplemented with metals, halogens, etc.), (b) have a temperature that never exceeds a limit (certainly below 100 C), and (c) is charged continuously with energy by the sun's radiation . . . In such a zero-system a series of chemical reactions immediately begins: more complex molecules are synthesized, material cycles develop, autocatalytic reaction chains appear, and molecules of a higher level of complexity (primarily various polymers) are formed (Csanyi, p. 44).

In environments that have the three requirements listed above, reproductive chemical networks (RCNs) appear spontaneously in the form of cyclic biochemical reactions. With the appearance of RCNs, *selective processes* emerge in the system. "This selection is not Darwinian or competitive . . . The reproductive chemical network as a *system*, as a proto-organization, will select those processes that will not destroy the system" (Csanyi, p. 48).

At the beginning of evolution, constructive and destructive forces were at equilibrium; RCNs appeared and disappeared without an arrow of time. Their reproduction was not at all reliable. Only when some components started influencing the reproduction of other components did protofunctions arise. The constraints provided by protofunctions increased the reliability of the information that was passed on to succeeding generations. Accordingly, replicands began to be more similar to their replicators; RCNs moved from nonidentical replication toward identical replication. When RCNs increased their replicative fidelity to "somewhat higher than 0.5" (Csanyi, p. 54), then they began the stage of identical replication.

The Evolution of Life

At about this stage, control mechanisms organized themselves into genetic codes that are the mechanisms of identical replication. As an RCN uses a code, the fidelity of its replication slowly approaches unity. "As a result of identical replication, changes in structures that provide various selective advantages became irreversibly fixed in the system's components" (Csanyi, p. 60).

There is an expansion and contraction at work in this evolution. In non-identical replication, divergence rules; in identical replication, variability decreases and there is "a convergence of the control systems" (Csanyi, p.60). As a result of convergence, the components of a system (that have already been fixed in the previous stage of evolution) come to cooperate and form the organization of a component system.

The sequence of evolution that Csanyi traces goes as follows. A zero system spontaneously generates RCNs. Protofunctions (influences of one RCN affecting the replication of other RCNs) produce a progressive ordering of replication. Protofunctions become more effective control mechanisms and develop a genetic code when their replication fidelity quotient surpasses fifty percent.

Thereafter replicant RCNs progressively more closely resemble their replicating RCNS. First, they form stable components. Then these stable components coordinate their activities to such an extent that they form an organization. An organization, by definition, is "an interrelated network of components and component-producing processes" (Csanyi and Kampis, p. 79). Over time, these organizations themselves progress from nonidentical replication to identical replication. These organizations then can become components of a larger organization. And so on. Csanyi calls this modular form of evolution a "spontaneous *autogenetic* process" (Csanyi, p. 65).

Further Evolution

Csanyi continues his systems-general description of evolution by observing that further development is possible for compartments that are already replicating with high fidelity. These compartments can develop functional relationships to each other. Then a new component system comes into being *on a new level*, of which the components are the compartments of the former organizational level. The fidelity of replication within the former compartments can be high, but on the next level, in the new component-system, it can still be low. On the new level a new autogenetic process can commence. As a result, compartments of compartments come into being in the course of replicative coordination. Eventually the whole system begins to replicate as a unity, with more and more perfect fidelity on all its levels (Csanyi, 1991, p. 81).

MATURANA AND VARELA

Humberto Maturana and Francisco Varela offer a different perspective on evolution, one that departs from the mainstream "objective" scientific viewpoint. They give a phenomenological description of life from the viewpoint of a living organism. They start their considerations with a definition of living things as *autopoietic* unities; that is, as machines that generate their own generative processes. They are saying, in other words, that living things are constantly reconstituting their constitutions.

Autopoiesis

The definition of life for Maturana and Varela is *autopoiesis* (Greek: self + producing). This conception is similar to "organization" as described by Csanyi. A cell, for example, has autopoietic organization because it is a closed system whose sole activity is its self-generation. A cell also has structure(s) (nucleic acids, for example) which are the components of its organization. These structures are open systems that are constantly modified by their interactions with their environments; they assemble, disassemble, and reorder themselves constantly. The cell, then, has a closed organization and an open structure. Its structure varies; its organization stays constant and constrains the activities of its structures.

The unique quality of a living thing is its self-generativity. It lives only to live and reproduce itself by generating its components and component-producing processes. Its doing is its being. It is a homeostatic machine that keeps constant a network of self-regenerative processes. It continually re-creates itself. In its interactions with its environment, it must maintain its autopoietic organization. If it does not, it dies. For this reason, autopoietic systems have a *closed organization* within an *open* metabolic *structure*.

Three elements, then, can be distinguished in the existence of an autopoietic unity: its fluctuating environment, its invariant organization, and its variable structure. Logically, because of metabolism with its environment, an autopoietic unity has to change; and, if its organization cannot change, its structure must. Living unities keep themselves alive by continually altering their structures to accommodate perturbations.

Maturana and Varela describe an organism's evolution and learning from the organism's point of view. In that perspective, the organism just couples with elements of its environment. By such "structural coupling" the organism engages in a natural drift. Natural drift eventually results in the historic appearance of "an observer," a human, who is able to describe behaviors from a detached viewpoint. In this way, Maturana and Varela account for their own origins and for the very possibility of scientific observation.

Csanyi's autogenesis is very similar to autopoiesis, and the development of component systems on new levels is analogous to the process of structural coupling. Csanyi, however, has taken biological theory to a systems-general level of abstraction. In so doing, he frames autopoietic thinking within an overall evolutionary process; and he propounds a systems theory that promises to explain the evolution of social systems and societies.

What Csanyi does not attempt to do, of course, is propound his theory from the autopoietic standpoint of an embodied organism (like a sponge) that would describe *its* phenomenological domain. He cannot do this because the phenomenological domain of his discourse is determined by his observer standpoint that views organisms simultaneously with their temporal and spatial environments.

The domain of autopoietic observation, on the other hand, is limited to the viewpoint of the dancing organism itself. From the viewpoint of the organism, the activities of cognition, autonomous autopoiesis, reproduction, and coupling have a simple meaning in terms of nomic consequences (staying alive). The organism reacts point-to-point and moment-to-moment to maintain autopoiesis. An autopoietic unity does not require the teleonomic concepts of information and function; such concepts are applied to these unities when we observe what they do in their temporal and spatial environments.

Three Phenomenologies

Phenomena are our immediate experiences of reality. They are not the physical realities themselves. The blue color in a black and white shadow of my finger, for example, is not physical color; it is a biological and conscious phenomenon (cf. Maturana and Varela, 1987, pp. 19-23). Such a phenomenon is not imaginary either; it is biological and conscious reality; it is what we see.

Phenomenological discipline sharpens our immediate experience of phenomena. It trains us to attend directly to our experience *before* we consciously place it within categories. In our attending to unreflective experience, we stop overly quick categorization and maintain fluidity of description. With sustained discipline, we produce more accurate descriptions of our experiences.

What we perceive is co-determined by external reality and our perceptual processes. It is, therefore, a mistake to say that perceptions are caused unilaterally by external objects. They are also caused by the perceptive ability of their subjects.

In other words, the perceptive ability of a perceiver determines phenomena equally with the external reality that is perceived.

Suppose I am walking down a road and a baby blue Mercedes convertible races by with a stunning blond at the wheel. In this situation, I might attend to the car, to the blond, to the beautiful package, or to my terror of almost being run over. The physical circumstances provide likely distinctions; I provide the faculties (eyes, ears, admiration, terror) that focus my attention (make the distinctions) that co-create the phenomenon.

Kinds of Distinctions

Many things distinguish themselves in the evolution of life. Gases and liquids distinguish themselves from inert solids. Systems far-from-equilibrium (dissipative structures) distinguish themselves from equilibrium structures. RCNs distinguish themselves from dissipative instabilities. Compartments distinguish themselves from their environment. Hypercycles distinguish themselves from non-accelerating RCNs. Living RCNs distinguish themselves from non-living ones by their autopoiesis, continuity, and autonomy.

All these "real world" distinctions are not created solely by their objects. They are co-created by the scientists and philosophers who possess the faculties (physical organs, physical mechanisms, mentality, cultural history, theoretical basis, appropriate language, and attention) to make such distinctions. Phenomena are created by the operations of distinction that occur between entities and observers.

It is in this facet of co-determination that Maturana and Varela extend phenomenology beyond human consciousness. They speak of biological and even physical phenomenology. The formation of a unity always determines a number of phenomena associated with the features that define it; thus we may say that each class of unities specifies a particular phenomenology.

> This is so, not because autopoietic unities go against any aspect of physical phenomenology-since their molecular components must fulfill all physical laws-but because the phenomena they generate in functioning as autopoietic unities depend on their organization and the way this organization comes about, and not on the physical nature of their components (which only determine their space of existence; 1987, p. 51).

Such phenomenologies have the logical possibilities of (1) being known from the inside, (2) being described from the inside, or (3) being described from the outside. We know our own phenomena (1) from the inside as experience; (2) from the outside as descriptions of our experience; or, more remotely, (3) as descriptions of our behavior (for example, dream behavior recorded in REM measurements). [In the following paragraphs, the abbreviations (1), (2), and (3) refer to these different phenomenologies.]

Organisms have an inside to the extent that they have structural memories and co-adapt (structurally couple) with their environments. We do not (1) know the inside phenomenal experiences of biological unities-of either a frog or our own internal organs. We can only (2) reconstruct biological experience or (3) describe biological behavior from the outside.

Non-living things have no inside awareness. We can (2) hardly reconstruct (or even imagine) the (1) internal experience of non-living things. We can, however, (3) describe their behavior from the outside.

Even highly complex non-living reproductive chemical networks (RCNs) (for example, nucleic acids, proteins, and non-compartmented hypercycles) have no stable interior viewpoint. They are moment-to-moment vulnerable to whatever chemicals they contact. They have little or no control over what they might become in the next moment. Their temporal and spatial reproductions are haphazard. RCNs have no autonomy. Their identities change with each of their momentary reproductions. The phenomenology of RCNs, therefore, is approachable only (3) from the outside as descriptions of behavior.

The topological plane of biological perception is intermediate to and parallel with the planes of physical and conscious awareness. An organism has no awareness of the immense mathematical complexity and contingency of its physical structure. In that, it resembles ourselves who also have no direct awareness of our physical momentariness. They do not experience reality as an observer as we do, either. Their experience is a moment-to-moment co-adaptation with their environments.

Prigogine, Eigen, and Csanyi describe the behavior of living things from the viewpoint of the intersystemic observer. They talk of a cell's generation in far-from-equilibrium chemical systems, its component RCNs, its information content, its genetic code, its metabolism with its environment, and so on. They offer (3) external descriptions of the exterior relations of living systems.

Maturana and Varela, in contrast, (2) reconstruct the interior phenomenology of biological unities. They claim no inside awareness of that inner phenomenology. They deny the possibility of such awareness because the domains of biological and conscious phenomenology are experientially exclusive and exist on parallel topological planes. What they do offer is descriptions (from the topological plane of the observer) of the autopoietic activities of living things. They refuse to mix intersystemic perspectives into their descriptions because intersystemic perspectives are unavailable to the biological organism. They call their method of explaining life, cognition, and evolution "biological phenomenology."

Biological Phenomenology

Because of their self-generativity, autopoietic systems distinguish themselves from their environments. A cell, for example, in its very formation, separates itself from its background. It creates both itself as a unity and its environment as what is not-its-unity. In its every interaction, a cell adjusts its structure and thereby marks a perturbation in a certain way that facilitates future reactions to similar perturbations.

> Consider the remarkable behavior of Acrasiales amoebas in their formation of slime molds. When the environment in which these amoebas live and multiply becomes poor in nutrients, they undergo a spectacular transformation . . . Starting as a population of isolated cells, they join to form a mass composed of several tens of thousands of cells. This "pseudoplasmodium" then undergoes differentiation, all the while changing shape (Prigogine and Stengers, p. 159).

This process is quite complex; it provides for the survival of the mold by the diffusion of spores and several subsequent cyclical processes. In this process, an ameba distinguishes between rich and poor nutrients and signals the lack of nutrients by emitting cyclic AMP. Other amoebas respond to the cAMP and form a slime mold.

The activities of these amebas are dissipative structures of communication. At some point in evolution, a distressed (far-from-equilibrium) ameba emitted cyclic AMP (a novel dissipative structure). That cAMP acted as an attractor to other amoebas (creating a hypercyclic reinforcement), and so on. The action of the distressed ameba was its learning. The emission of cAMP became part of the structural memory of its species.

The creation of self and environment, or unity and background, is the first step of *biological phenomenology*. The history of an organism's structural interaction with its environment is simultaneously its history of cognitions and its incorporation of those cognitions into its structure.

Knowing Is Doing

For the Acrasiales amoebas, knowledge of the lack of water/nutrient is defined by distress and the emission of cAMP. In general, for any organism, its knowing is its doing. Also, for a living being, its doing is its being because autopoietic activity is its essence. Therefore, knowledge was present at the origin of life.

Cognition, in this very general sense is defined as: "an *effective action,* an action that will enable a living being to continue its existence in a definite environment as it brings forth its world. Nothing more, nothing less" (Maturana and Varela, 1987, pp. 29-30).

Cells gradually increase their knowledge by flexibly orienting to their environments. Acrasiales amoebas, for example, know their environments more thoroughly than other species of amoebas that do not make slime molds. Every increase in an organism's "perceptual" ability enables it to enact couplings with additional perturbations.

As cells increase the complexity of their interactions and their knowledge, they sometimes couple up with other living things. In coupling, cells become metacellular organisms. Some metacellular organisms become multicellular organisms as they learn to reproduce sexually.

Multicellular organisms also develop ways to communicate internally. Plants rely on neuro-peptides, like cAMP, for their internal communication and their external coupling. Animals supplement this form of communication with nervous systems.

With their nervous systems, animals accelerate their internal communications and interactions by many orders of magnitude. When a lizard is stalking a fly, both lizard and fly engage in a complicated dance involving stealth, sensitivity, anticipation, and speed. The lizard inches its way towards the fly, and stops the moment that the fly shows any agitation. It inches forward again; eventually the lizard snaps the fly with its tongue or the fly darts away unharmed.

Both lizard and fly act with rapidity toward external and internal perturbations. The multitude of their doing results in abundant learning through a process of operant conditioning. Each successful pattern of behavior adds to their structural memories. The lizard, with its superior nervous system, exhibits learned behaviors that human hunters utilize even today.

The nervous system (including the neuro-peptide system) embodies an animal's survival instruments and its structural memory. It enables an animal to react with great flexibility and sensitivity to external and internal perturbations. When an animal adapts its structure to retain its autopoiesis, its nervous system supplies the brunt of its adaptation by adjusting its structure point-by-point to the perturbation imposed upon it. By this interaction of perturbation and adjustment, a nervous system adds to its memory and increases its ability to respond to future perturbations. A nervous system is itself an autopoietic system, whose cognitive domain "is the domain of all the deformations it may undergo without losing its autopoiesis" (Maturana and Varela, 1980, pp. 118- 119).

Structural Coupling

Maturana and Varela hold that selection for evolutionary adaptation is not made solely by the environment. Organisms, mutant or not, survive by structural coupling with their environments. In their structural coupling, both organisms and their environments undergo transformations. Co-adaptation is a dynamic process; it "is a necessary consequence of that unity's structural coupling with that environment" (1987, p. 102).

To an observer, this co-adaptation seems to be a "selection" by the environment on the basis of compatibility. In the actual dynamics, a living system works out its relationship with its environment through mutual interaction. Maturana and Varela say it this way:

> Conservation of autopoiesis and conservation of adaptation are necessary conditions for the existence of living beings; the ontogenetic structural change of a living being in an environment always occurs as a structural drift congruent with the structural drift of the environment (1987, p. 103).

In brief, an observer sees selection by the environment, but the organism sees ways to stay alive.

Summary

Some of the main points of Maturana and Varela concerning evolution are expressed as follows:

> Autopoietic unities increase their survival chances by increasing their flexibility to environmental perturbations. To this end they create a novel dissipative structure: *reproduction*. Through self-reproduction an organism produces another with a similar organization to its own, through a process that is coupled to the process of its own specifications (Maturana and Varela, 1980, pp. 137-138).

And,

> [The phenomenon of reproduction] results in the separation of two unities with the same organization but with different structures of the original unity; [therefore] it conserves organization and gives rise to structural variation . . . Those aspects of the initial structure of the new unity which we evaluate as identical to the original unity are called *heredity*; those aspects of the initial structure of the new unity which we evaluate as different from the original unity are called reproductive *variation* 1987, p. 68).

And,

> [There are] many different lineages . . . [which are] distinct ways of preserving ontogenetic coupling in the environment (1987, p. 81).

And,

> Evolution occurs as a phenomenon of structural drift under ongoing phylogenetic selection (1987, p. 115).

EMERGENCE OF THE OBSERVER

"With no law other than the conservation of an identity and the capacity to reproduce, we have all emerged" (Maturana and Varela, 1987, p. 117).

New theories of evolution portray a chain of evolution connecting chemical order with the emergence of the human observer. The long chain of evolution leads to more extensive knowledge and finer discrimination. At the beginning of that chain is an organism with an absolute need to maintain its self-regenerative organization. It varies any or all of its non-essential components and structures to meet environmentally caused fluctuations. In this process, organisms (like Acrasiales amebas) produce new organic structures (like slime molds). Evolution is the joint result of such ontogenetic changes and of the selective processes of heredity that favor successful survival strategies.

Living things introduce temporal stability to evolution because living things endure by maintaining their organizations. An organism has a degree of autonomy and an ability to grow. It learns because it changes even while it stays the same. It thereby creates an enduring and cumulative frame of reference through its recursive responses to perturbations.

In time, recursive communications have produced incredibly intricate communication networks involving neuropeptides and nerves. In one particular branch of evolution, millions of years of inventive innovation culminate in the emergence of the human observer.

CONCLUSION

Both autopoiesis and the replicative model are theories that bring evolution up to date. Autopoiesis reflects the philosophical sophistication of hermeneutic and critical philosophy. The replicative model builds upon the time-revolution in physics and the findings of molecular biology. In the dialogue between proponents of these theories, a rounded and convincing portrayal of evolution, as both dance and choreography, emerges.

CHAPTER 3

SOCIAL AND
COGNITIVE EVOLUTION

ABSTRACT

Sociality, cognition, communication, and language are united in human living; they escape explanation by any one theory. This chapter considers two evolutionary accounts of these realities. The first is Maturana and Varela's idea of "knowing is doing." On this foundation, Varela and his associates develop a theory of enactive cognition and contrast it with representational theories of cognition. The second theory, of Derek Bickerton, proposes that language has developed not from communications but from representations, and traces the evolution of representations.

FIRST ORDER AUTOPOIESIS

Single cells are autopoietic in the original first-order sense. We know enough about their molecular processes to see how they form themselves as autopoietic unities. They maintain their unities even as they metabolize with their environments. They exist as metastabilities through energy exchange with their environments.

In addition to accepting calories from their environments, cells selectively accept ions, such as sodium and calcium, which they incorporate as elements of regenerated proteins and nucleic acids. In this way, cells incorporate parts of their environments into their autopoiesis. At the same time, they refuse to accept other ions, such as cesium or lithium, which would disrupt their autopoiesis. Cells exist because previous cells in their phylogeny accepted sodium and rejected cesium. They continue to exist by following the knowledge that was learned by their progenitors.

SECOND ORDER AUTOPOIESIS

Cells selectively couple with their environments to maintain their autopoiesis. Because that environment also includes other cells, a cell sometimes couples its structure with other cells in its proximity. The result is a metacellular unity, like the *Physarum* (cf. Maturana and Varela, 1987, pp. 76-78), whose spores can become paramecia or amoebas depending on humidity, but also come together to form a slime mold which produces a macroscopic fructiferous body that produces spores.

> A simple structural coupling among paramecia and amoebas, such as this, forms a metacellular unit capable of giving origin to a lineage by reproducing through single cells. It originates a phenomenology different from the phenomenology of the cells [that make it up] . . . This metacellular or *second-order unity* will have a structural coupling adequate to its structure as a composite unity (Maturana and Varela, 1987, p. 78).

The *physarum* has a macroscopic ontogeny unlike the microscopic ontogeny of its cells. To accomplish this macroscopic ontogeny, the cells of the new metacellular body make structural changes that complement each other.

Metacellular unities like the *physarum* or the whale are second-order autopoietic unities. Like cells, they also maintain their organization at all costs while they metabolize and creatively couple with their environments. Are second-order unities also first-order unities? Maturana and Varela call this question a "stickler." They ask, "How can we possibly describe in an organism the components and relations that make it a molecular autopoietic system?" (1987, p. 88).

THIRD ORDER AUTOPOIESIS

As metacellular bodies are formed by coordinating cellular behaviors, societies are formed by the coordination of social behaviors. Multicellular organisms, when their interactions take on a recurrent nature, also enter into structural couplings with each other.

> When this happens, the co-drifting organisms give rise to a *new phenomenological domain*, which may become particularly complex when there is a nervous system. The phenomena arising . . .[are] *third-order structural couplings* (Maturana and Varela, 1987, pp. 180-181).

Animals come together to mate, to rear their young, and for other functions. Social insects, for instance, form colonies in which there are a variety of worker females, male drones, and a single queen. A principal method of communication that holds the colony together is trophallaxis (insects exchanging chemical information by sharing their stomach contents).

Vertebrates, with their larger nervous systems, form more diverse social groupings utilizing many different forms of communication. Herds of antelope, for example, have family groupings, but also role assignments for the whole group, especially when a predator is stalking the herd. A dominant male takes the lead; females and children follow, males bring up the rear; and a single male trails the group to keep an eye on the stalker.

Individuals and societies are co-constituted. Wherever third-order unities (societies) are constituted, be it among insects, ungulates, or primates, individual lives are brought about by being linked with other lives in the society that they jointly are. The individual ontogenies of all the participating organisms occur fundamentally as part of the network of co-ontogenies that they bring about in constituting third-order unities (1987, p. 193).

There is no contradiction between selfishness and altruism for animals, like the trailing antelope, that put themselves at risk for their society. These animals act to maintain the individuality that they attain in the group.

Cells are the pivotal first-order autopoietic unities; they "produce components which make up the network of transformations that produced them" (1987, p. 44). They are the basis of multicellular organisms (second-order couplings) and societies (third-order couplings) which are autopoietic because of their cellular make-up. Second- and third-order autopoietic unities: (1) are composed of cells which are first-order unities; (2) exist in metabolic interaction with their environments; (3) maintain an invariant organization; and (4) alter their structures as needed to maintain their organizational unity (cf. 1987, p. 89). Second-and third-order autopoietic unities are not necessarily also first-order unities like cells because "we are still ignorant of the molecular processes that would constitute those metacellulars as autopoietic unities comparable to cells" (1987, p. 88).

LINGUISTIC DOMAINS

Vertebrates have a special way of communicating socially; they imitate each other. Because of their mimicking behavior, they share the ontogenetic growth of one individual with the whole society. The imitative behavior of the macaque monkeys of Japan has become famous through the parable of the "hundredth monkey." Through such mimicry, one generation passes down its ontogenetically acquired behaviors to subsequent generations.

Through communication, cognitive domains become mutual; individual interactions with environments become standardized; cognitive domains become linguistic domains. Even before the advent of language, these mutually shared cognitive (linguistic) domains constitute a primitive socio-cultural realm. These domains are dubbed "linguistic" because social communication among animals can be described (by an observer) semantically.

Consider the signaling behavior that we sometimes observe in the learned behavior of our house pets. If Lobo goes to the back door and barks, he signals me to let him out. Lobo is not necessarily talking to me. He is repeating behavior (ontogenetic coupling) that has worked for him in the past. In other words, Lobo and I have conditioned each other. (Still, I find it comforting to think that Lobo is talking to me.)

The activities within this linguistic domain are not yet language, but they are the basis of language. Learned social communication among animals is their way of adapting to conserve their social system. Our language serves the same purpose. Language, in this context, is best understood as an evolutionary adaptation built upon the linguistic domain of animals.

LANGUAGE, THE OBSERVER, THE "I"

As "languaging" animals, we put our linguistic domain into our linguistic domain; we "objectify" our objects; in other words, we describe them. In languaging, observers become aware of their linguistic domains and express them. it. Persons, thereby, co-create linguistic objects *and* are co-generated as observers. "With language arises also the observer as a languaging entity; by operating in language with other observers, this entity generates the self and its circumstances as linguistic distinctions of its participation in a linguistic domain" (1987, p. 211). Along with language and the observer, meaning also makes its appearance.

Languaging, observing, being a self, and creating meaning are the stuff of being human. We maintain our ontogenetic selves through our languaged meanings. "We are observers and exist in a semantic domain created by our operating in language where ontogenetic adaptation is conserved" (1987, p. 211). With our language, we create our peculiarly human societies and our peculiarly human individuality.

Our individuality and our language are closely linked phenomena. Self-consciousness is a social phenomenon dependent on language. Maturana and Varela put it this way. "There is no self-consciousness without language as a phenomenon of linguistic recursion. Self-consciousness, awareness, mind-these are phenomena that take place in language. Therefore, they take place only in the social domain" (1987, p. 230). The "*I*" is formed as a social singularity in the ongoing distinctions within our bodies that separate it from linguistically distinguished elements of its background. "In the network of linguistic interactions in which we move, we maintain an ongoing descriptive recursion which we call the '*I*'. It enables us to conserve our linguistic operational coherence and our adaptation in the domain of language" (1987, p. 231).

Humans create their "*I*'s" in societies as do all animals. The human creation is unique in its sophistication and complexity because of the language that elaborates and embellishes it. The "*I*" is like every other living thing in its metastability; the environment in which the *"I"* has its metabolism is the social world of language.

MODELS OF COGNITION

Francisco Varela, Evan Thompson, and Eleanor Rosch, in *The Embodied Mind* (1991), extend ideas that were expressed in *The Tree of Knowledge*. They survey and critique current theories of cognitive science. Then they propose a theory of cognition as enaction, that is, as embodied actions that create distinctions.

Cognitivist and Connectionist Models

Varela and his associates examine two principal models for artificial intelligence (AI), which they label cognitivist and connectionist. The cognitivist model is a computer programmed to solve problems by exhaustively examining possible outcomes of situations and picking the best option; it is programmed from the top down with algorithms. It deals with symbolic representations of reality rather than experienced reality. It is efficient at relatively simple tasks like playing chess, but it founders in trying to accomplish perceptually guided activities.

The connectionist model is programmed from the ground up; it consists of a web of weak information processors that organize themselves to accomplish their assigned tasks on the basis of heuristics. It is built upon the physics of non-linear thermodynamics. It can be modeled simply with Boolean algebra on a computer; and it finds analogues in numerous dissipative structures.

The theory behind this model is "connectionist, emergent, self-organizing, associationist, network dynamical" (Varela, 1993, p. 98). In this theory, cognition is "the emergence of global states in a network of simple components" (Varela, 1993, p. 99). It works "through local rules for individual operation and rules for changes in the connectivity among the elements" (p. 98). In this connectionist approach, "symbolic computations are replaced by numerical operations" (p. 98).

Artificial intelligence (AI) mechanisms built on this model consist of great numbers of weak computers hooked up in multiple parallel configurations. They build their own intelligence from the bottom, as little global states here and there connect with each other. They continue combining nodules with nodules in supercyclic fashion under the constraints imposed by their programs. Connectionist mechanisms seem to model the actual evolution of cognition. They mimic the ways in which nerves, ganglia, and larger nodules combine in supercycles.

Cognitivist theories openly embrace the mind-body dichotomy that is usually ascribed to Descartes. Connectionist theories push the envelope of "detached observer," but they fail to escape from its confines. Varela's own theory of enaction denies that there is any objective observer who could have an envelope.

Foundation of the Enaction Theory

The enaction theory grows out of the ideas of "knowing is doing" and structural coupling. It is also draws upon the work of Merleau-Ponty, the mindfulness tradition of Buddhism, and some advances of cognitive science.

Merleau-Ponty pondered the dual aspect of our lived physical bodies that are both the subjects (the experiencers) of our perceptions *and*, at times, also the objects (the experienced) of those perceptions. What does it tell me, for example, when I feel my left hand touching a table *and* then touch my left hand with by my right hand? It seems that my left hand changes from being the observer to being the observed. In this simple interaction of two hands and a table, my body bridges the supposed gulf between my consciousness and the external world.

The mindfulness tradition of Buddhism is a philosophical tradition of direct awareness. The goal of mindfulness meditation is to experience the fragility and contingency of our mental activity, and to go beyond words and categories to the boundlessness that lies behind them. Mindfulness is a long-tested discipline of introspection that avoids the pitfalls of the Cartesian-warped introspectionism of pre-Watson psychology.

The progress of cognitive science has prompted some theorists to stress the ecological embeddedness of cognition. Connectionist learning machines do not only learn their environments by interacting with them. They also shape their environments. (Organisms do the same.) This mutual adjustment is identified with the idea of *structural coupling* as it has been already presented.

A vivid example, of such structural coupling is provided by the vision of honeybees and the colors of flowers. On the one hand, honeybees need flower nectar to survive; they have developed sensitivity to variations in the ultraviolet spectrum in order to find flowers. On the other hand, flowers need honeybees for pollination; they have developed flashy colors in order to attract bees. "The colors of flowers appear to have *coevolved* with ultraviolet sensitive, trichromatic vision of bees" (Varela, 1993, p. 201).

An example of global structural coupling millions of years ago is given by the multiplication of anaerobic bacteria that produced oxygen in earth's atmosphere. Global examples from the present day are the developments of technology that are changing earth's ecology.

The evolutionary theories basic to the theory of enactive cognition are structural coupling and natural drift. Structural coupling is the spontaneous complexity that arises when organic structures react with their environments. In this general sense, it is the process of dissipative structuring as it is formulated by Prigogine, in which systems self-organize into higher states of complexity by metabolizing with their environments. In a more particular sense, it is the mutual structuring that takes place when two organic structures join forces, structures, and energies.

Natural drift is the organic proclivity to create changes that endure on the condition that the created changes (such as species) are able to survive and reproduce. It is a concept of evolution that downplays the competitive theory of natural selection that specifies the survival of the organism that *best* fits into its ecological niche. Instead, it stresses that mutant organisms survive if the mutation *satisfices* (that is, it does not kill its host or stop its reproduction). In other words, evolution creates its own path: any mutation is successful if it allows its host to live and to reproduce. Evolution has no formula; changes happen; some survive.

Cognition as Enaction

Using ideas derived from these theories, enactive cognition is: "a history of structural coupling that brings forth a world" (Varela, 1993, p. 206). Enactive cognition is a coupling of bodily activity with the environment that modifies both the lived body and the lived environment. This enaction brings forth an experienced world. Enaction (cognition) is also a history of past enactions (cognitions).

Enaction works "through a network consisting of multiple levels of interconnected, sensorimotor subnetworks" (p. 206). The networks spontaneously generate new complexity in the manner of the connectionist theory of AI. This spontaneity mimics the antics of genetic and ontogenetic changes in evolutionary theories.

In enactive theory, a cognitive system is functioning adequately "when it becomes part of an ongoing existing world (as the young of every species do) or shapes a new one (as happens in evolutionary history)" (Varela, 1993, p. 207). The constraint on new complexities is the satisficing one: are they able to survive?

In enactive cognition, then, one makes one's path by walking. One acts; and if one's actions are successful, one adds that successful action to the ones that have worked in the past. No matter how much strategy and logic go into a cognitive action, that action is basically a step onto unknown territory. A successful step lengthens the path in one direction; an unsuccessful one dictates a change of direction. In either case, one lengthens the path.

The paths of cognitive enaction are not primarily individual enactions; they are a shared world of significance that a group has forged in a wilderness. The main task of cognition is to enter into that world. Only when the communal cognitive system breaks down is enactive cognition focused intently on the environment.

The Enaction of a Theory

The theory of enactive cognition is the latest advance along a path that began with the idea of autopoiesis and biological phenomenology. The hypothesis, "Knowing is doing," was advanced in *The Tree of Knowledge*. The progress of this theory is itself an example of the embodied cognition that it describes; this theory has laid down its path by walking.

Marking the beginning of this path is the theory of autopoiesis that is a classic example of a cognitive pillar of Hermes. The pillars of Hermes in archaic Greece were mounds of stones that accumulated, as wanderers in the wilderness marked their paths by situating stones as landmarks. Over time, prominent places on well-traveled paths had virtual pillars of stones that were called "hermai" in honor of Hermes who was the patron god of merchants and thieves. The theory of autopoiesis identified one such pile of biological stones that dealt with the core essence of a living thing. It formulated in a precise way the self-generating essence of living things.

Beginning from the pillar of autopoiesis, Maturana and Varela moved to explain the development of cognition in an embodied way. They developed a theory of biological phenomenology that tried to describe evolution and learning from the viewpoint of the organism involved. They were guided by a system of logical accounting that enabled them to separate the topological planes of enactive explanation and behavioral description.

The principle "knowing is doing" constituted a signpost along this advancing trail; it too was a pillar of Hermes that gathered together lapidary ideas strewn around the idea of embodied cognition. The explanation of evolutionary history in terms of structural coupling and natural drift blazed a further trail that culminated in the appearance of an observer. The observer's appearance was explained through the concepts of communication, linguistic domains, and language. The emergence of language and the observer from embodiment constitutes another pillar of Hermes for the theoretical path that begins with autopoiesis.

REPRESENTATION VS. COMMUNICATION

In *Language and Species* (1990), Derek Bickerton opposes the usual conception that language is just a means of communication. He defends the theme that language is primarily "a system of representation, a means for sorting and manipulating the plethora of information that deluges us during our waking life" (Bickerton, 1990, p. 5). In taking this position, he opposes those "antiformalist" linguists who ignore and/or downplay the formal structures of language. He addresses the evolution of language from a formalist viewpoint.

The Continuity Paradox

Bickerton identifies the "Continuity Paradox" as the crux of linguistic evolution. If human language evolved, then there must be a continuity of linguistic development from animals to humans. Thus far, attempts to establish a direct line of development from animal communication to human language "have signally failed to produce a convincing 'origins' story" (Bickerton, 1990, p. 9). So there is a paradox: "language must have evolved out of some prior system, and yet there does not seem to be any such system out of which it could have evolved" (Bickerton, 1990, p. 8).

The Extent of Non-Human Communication

The linguistic attainments of the great apes are remarkable. Roger and Deborah Fouts, the long-time researchers with Washoe, summarize their findings of chimpanzee language as follows:

> Chimpanzees can pass this [sign] language on to the next generation . . . They can use it spontaneously to converse with each other as well as with humans . . . They can use their signs to think with, as evidenced by their private signing . . . They can have an imagination . . . They have good memories . . . They may even be able to perceive seasonal time (Fouts and Fouts, 1993, p. 39).

The same authors recorded over 5200 instances of chimpanzee-to-chimpanzee signing. They analyzed the content of that signing as follows:

> The majority of signing by the chimpanzees occurred in the three categories of 'play,' 'social interaction,' and 'reassurance'; these accounted for over 88 per cent of the chimpanzee-to-chimpanzee conversations. The remaining 12 per cent was spread across the categories of 'feeding,' 'grooming,' 'signing to self,' 'cleaning,' and 'discipline' (Fouts and Fouts, 1993, p. 33).

The findings of Fouts and Fouts document not only the achievements of great ape communication, but also its limitations.

The differences between animal communication and human language are qualitative, and their discontinuities immense. Even the most sophisticated systems of animal communication (such as those demonstrated by Washoe and her companions) are limited to a finite number of topics on which information is exchanged. Human communication, on the other hand, is practically infinite. Animal systems have very limited ways of combining message components, while human language has strict and flexible rules of unlimited combinations. Where other animals communicate only about things that have evolutionary significance for them, human beings communicate about anything. What is adaptive for humans, and qualitatively different for them, is their system of reference as a whole.

The continuity of human language with its forebears is not to be traced through communication, but through representations. The prior representational systems of animals culminate in human language. Moreover, human language is constituted by the species-specific formal structure of human representation.

> Formal properties of language do exist and do matter, and . . . without the very specific types of formal structure that language exhibits, it could not perform the

social and communicative functions that it does perform, and could not convey the wealth of peculiarly human meaning that it does convey (Bickerton, 1990, p. 9).

KINDS OF REPRESENTATION

Atlas and Itinerary

Language is a representational system in two ways: as an atlas or as a set of itineraries. An atlas represents those static items that are useful for a map-reader. It shows stable components within a fixed frame of reference. Language, as atlas, creates a cognitive map in which stable and useful elements of our experience are represented within a frame of meaning. The map provides the lexicon of language, its meaningful words.

Itineraries are charts that allow us to steer our course on the terrain of a map. Linguistic itineraries are the sentences that we hear and utter. These sentences are not arbitrary strings of words; they are constrained by their semantic terrain. Syntax, as a schematic book of itineraries, tells us how to navigate the terrain pictured on our cognitive map: it guides our use of words.

Levels of Representation

Bickerton uses the term 'representation' in a way that includes fleeting perceptions, memories, and fixed knowledge systems. At this level of abstraction, representation is simply our way of knowing the world. It includes "activities as disparate as seeing, hearing, and smelling, on the one hand, and believing, thinking, and knowing, on the other, under the same heading" (Bickerton, 1990, p. 76): the heading of representation.

Simple levels of representation are the foundations for more complex levels. Simple sensing mechanisms coalesce into sophisticated ones. Highly structured cognitive states and behaviors derive "through evolutionary processes, from far simpler ones" (Bickerton, 1990, p. 77).

When living things are disturbed, they react to stay alive. But they do not react to everything in their environments. To stay alive, organisms selectively react to differences that they perceive as differences (Bateson). Some differences, those it does not sense, make no perceptual difference for an organism; they are not information for it.

What is it, then, that makes a difference make a difference for an organism? The quality of an organism that leads it to sense a difference is its ability to react to that difference. Living things react only to the things they sense (information). Conversely, if an organism cannot react to something in its environment, then that something is not information for it.

Sensing and reacting are the poles of an organism's activity. To connect those poles, an organism must have some memory reference, some way of representing its activity. Reacting requires three things: sensing information, representing it in some way, and behaving on the basis of those representations.

Plant Protorepresentation

Plants have inferior representational powers, compared to animals, because they lack nervous systems. Plants, however, are not without representations. They remember their responses to night and day, to seasons, to humidity, etc. In their water and nutrient circulatory systems, they distribute and deposit molecular messengers (cf. Rossi, 1993), things such as the enzymes cAMP, ATP, and ADP.

The "endocrinal" communication system of a plant is rather slow. A plant's representational system is in large part global. An enzyme released into its circulatory system activates the DNA of each cell individually, of course; but it effectuates changes in the sprigs of a bush and its leaves in about the same way at about the same time.

Some plant behavior, however, is not global. The "sensitive plants," that Bickerton focuses on, react to touch in a localized manner and with some dispatch. These plants lack neural activity. Their sensoria are very narrow, being limited to those elements of the environment that affect them and that they can handle. Within such limits, sensitive plants, such as sundews, have cells and/or systems of cells that have ability to both discriminate touch and react to it.

The sundew is a sensitive plant that eats insects. It has leaves covered with sticky hairs. When a fly is stuck on those hairs, the leaf gradually closes and the plant consumes the fly at its leisure. Bickerton interprets this as an information-behavior transaction because the cells that gather the information about the fly are distinct from the cells that cause the leaf to close. He says, "Cell A . . . registers the fact that a fly (or something) is on it, and then transmits that information to cell B. Cell B, together with other cells of a similar type, then closes the leaves" (p.78). He asks, "Could we describe the reaction of cell A, a primitive sensory cell, as a representation of the fact that a fly (or something) has landed on it? (p. 78). He answers,

> We cannot really say that the reaction of cell A to the fly is the sundew's representation of a fly, or even of a fly-landing event. And yet . . . [cell A did distinguish] two states . . . So, to call the firing of cell A kind of protorepresentation of the state 'assumed presence of potential prey' should not seem altogether unreasonable (pp. 78-79).

Plant discrimination does not reach the level of perception, in Bickerton's view, because neural activity is absent; but it is 'protorepresentation' because it is the evolutionary basis of the representation found in nervous systems.

The ability of cell A to distinguish two states creates a distinction (prey present vs. prey absent) and a faculty of sensitive distinction that presage further evolutionary distinctions and more complex distinguishing faculties. The ability of cell B to react to the distinction made by cell A creates an adaptation that increases an organism's survival options. These increased options open more possibilities for more sensitive distinctions by cell A and its likely cohorts.

Animal Representation

Nervous systems enable animals to process environmental information with dispatch and specificity. They augment endocrinal processes with electrical ones. They send messages faster and with more specificity.

Messages can be sent along nerves for fairly long distances from one specific location to another in an animal's body. Those messages are recorded and encoded in patterns of interneural nerve and neuropeptide activity. They are "meaning" for animal sensoria. They are relayed back and forth, to and from the relevant cells in an animal's body to enable its sensing, recording, and reacting to information.

The connections between sensitivity and behavior are patterns of accumulated wisdom contained in an animal's interneural/neuropeptide system. Webs of neural and messenger molecule interaction are interposed between an animal's sensing and its behavior. The patterns in these interacting webs represent both stimuli and responses. These patterns match up and form an overall representation that is the animal's response to the stimulus. They are not necessarily representations of an external world; they are patterns of behavior that keep the animal alive.

Species-Specific Reality

Animals from the simplest to the most complex get the senses they need for the behaviors they are capable of, and they get their views of reality from their sensing systems. Their 'reality' is species-specific; it is what they sense and react to. Bickerton puts it this way: "What is presented to any species, not excluding our own, by its senses is not 'reality' but a species-specific view of reality-not 'what is out there,' but what it is useful for the species to know about what is out there" (p. 82). Bickerton makes this point even more forcefully: "It is meaningless to talk about 'a true view of the world.' To attain such a view, a minimum prerequisite would be for the viewer not to belong to any particular species" (p. 86).

THE PRIMARY REPRESENTATIONAL SYSTEM (PRS)

An animal's species-specific view of reality is what Bickerton calls its primary representational system (PRS). PRSs may be simple or complex, but they are all of the same type. They arise from simple representations. Those simple representations arise from the interaction of three types of neurons: sensing, intervening, and motor. Bickerton puts it this way:

> Simple representations arose naturally from the conjunction of three things: cells that could discriminate between two states, the distinction between sensory cells that gathered information and motor cells that acted on it, and motor cells that could perform more than one type of behavior in response to a given stimulus. What most sharply differentiates simple from complex PRSs is the degree of processing that outputs of sensory cells undergo (p. 82).

Operations within PRSs

Anemones shrink away from two predatory species of starfish and ignore five other non-predatory species. Such activities require a web of interneurons connecting sensory and motor cells because information from different sources (external, interoceptive, and proprioceptive) has to be coordinated. These neural webs encode internal and external states into neural codes. In Bickerton's words: "Nervous systems translate both internal and external states into quantitative terms, represented by variations in the firing rates of the relevant neurons" (p. 83).

Within a PRS, these representations correlate. They enable an animal to behave in ways that are appropriate to its life situation. The actual correlations involve multiple factors and a bewildering (as yet undeciphered) mechanism that matches frequencies, intensities, and sequences of neural and neuropeptide discharges. The process of correlation, however, is not conceptually difficult. It can be represented as a simple 2x2 matrix in which hunger and prey, presence and absence are the only variables that influence a predator's activity.

	Presence	Absence
Hunger	1	2
Prey	3	4

In case (1 & 3) the predatory would spring into action. In all other cases, (1 & 4), (2 & 3), and (2 & 4), it would not attack. The predator would attack only if hunger *and* prey were both present. If *either* hunger or prey was *not* present, it would not attack.

More complex animals have the ability to make varied responses to slightly different stimuli. They can effectively vary their responses because they incorporate articulated senses and an ability to collate information from numerous sources. A lizard, for example, alters its behavior as it stalks a fly on the basis of

spatial measurement and directional sensors and a sense of the fly's alertness.
Lizards have a degree of autonomy. Their behavior is no longer totally predictable;
it is co-determined by external factors and by activities within their brains.

The Paradox of Representation

As animals distinguish things, they become distant from them. In opening the gates
of consciousness, they progressively distance themselves from the actual world of
objects and events. A stone has no representations; it is part of nature. The lower
levels of organic life have simple representations consisting of the activities of
sensory cells. Higher invertebrates and vertebrates have information processing
(nervous) systems that interpose PRSs between sensory and motor cells. Some
animals add secondary representational systems (SRSs) in the form of
protolanguages. Human animals add syntactical speech.

All these levels of representation create species-specific distinctions in the
'real world.' They enable species and individual animals to utilize their worlds for
survival and enrichment. Paradoxically, they also distantiate species from the
environments in which they are organically embedded. The very process of
endowing the world with rich distinctions is the process of alienating animals from
their roots. As a result, animals react to species-specific representations of reality,
not to 'reality' itself. For species, of course, representations *are* reality because a
species' representations make its survival possible.

The very process that enables us to have rich and productive interactions
forces us to deal with representations and tricks us into believing that we know the
'real world.' One conclusion to be drawn from this paradoxical situation is the
following: "It is absurd to speak about 'a true view of the world' because it is not
true for any creature that what it perceives is the world itself. What constitutes any
creature's view is essentially a system of categories" (p. 87).

Categories

Bickerton's very broad definition of representation is, to repeat: "responding or
having a permanent propensity to respond to x, an entity or event in the external
world, in terms of y, a particular pattern of neural activity" (p. 76). In this definition
y corresponds to what Bickerton calls a *category*. Categories are "particular
patterns of neural activity." They are concepts or protoconcepts.

What is Perception?

Now, if we do not perceive the world directly, but only through species-specific
representations, what is perception? When you tell me, "I saw a golden eagle,"

what are you saying at the mechanical level of cognition? Bickerton describes what you are saying this way: "An object presented itself to my visual field that seemed to correspond to an internal representation labeled golden eagle" (p. 87). In this explanation, 'golden eagle' is a neural category (or concept), which is expressed by the words golden eagle that distinguishes a particular bird. [Note: in Bickerton's convention, words in single quotation marks denote neurological 'concepts' and underlined words denote words.]

When we perceive something, we assign a category to a stimulus. We select the appropriate category from the set of all categories available to us. The fact that categorical assignments usually work for us "does not alter the fact that no 'perception' is anything more than a category assignment, a best guess at what may be out there" (p. 88). These categories have no separate existence in the external world. They exist as specialized mechanisms for comprehending the world and acting in it.

Some categories are innate as is revealed in a baby monkey's instinctual fear and avoidance of a snake. Activation of the category 'snake' links with other neural connections that automatically produce fear and avoidance. 'Snake' is a fuzzy category in its innate form: waving a piece of rope or hose can excite a naive baby monkey. 'Snake' can, however, be sharpened by experience; then it becomes a learned category. As a learned category, 'snake' allows greater discrimination and plasticity of response.

Choosing Categories

On what basis does a species or an animal discriminate its categories and how does it relate those categories to its life choices? Consider a wildebeest that does not flee from every lion that it sees; it just becomes frightened and flees when a lion is a threat. How does it do that?

Certain things in an animal's environment are vital to its survival. Recognizing a charging lion is essential to a wildebeest's life. Feeling fear (or panic) in that situation is also necessary. The wildebeest's flight from a charging lion is a 'representation' in Bickerton's sense of the word. That representation contains several categories and subcategories especially 'charging lion,' 'fear/panic,' and 'run.' These categories or protoconcepts are neurological patterns that are integral to the wildebeest's 'getting away from the lion.'

How do animals choose what categories to make? How do they ensure that categories relate to their survival as they pile more and more of them between their sensing and their behavior? The simple answer is that animals create, maintain, and inherit those categories that promote their survival.

Categories, then, have an evolutionary history. They accumulate as patterns of activity in ganglia and ever-larger nervous systems. They create nodules of neural activity that link with other nodules and integrate into ever-more-complex PRSs. At each level of this process, responses to situations (Bickerton's

<u>representations</u>) with their corresponding categories are integrated into larger patterns of neural activity. Representations and categories that satisfice for survival at one level are co-opted and adapted for survival at subsequent levels. The totality of an animal's stored neuronal patterns constitutes its PRS.

How Do Categories Originate?

Even if categories are formed, accumulated, and integrated on the basis of their survival value, how do they come to be? One crucial factor in category formation is the need to group together representations that require similar behavioral responses. 'Bird', for example, would be 'something that lays round things you can eat if you can find them.' 'Tree' would be 'something you can climb if pursued,' and so on. In general,

> The set of categories that constitutes the PRS of any species is determined by the evolutionary requirements of that species: what it eats, what other things it needs, what it fears, what its preferred strategies for survival are, and how those strategies interact with the strategies of other species. The PRS as a whole constitutes, as it were, a model of reality that serves to guide behavior (p. 91).

Negations

With humans, negations are crucial to categorization. If we need a tree to climb to get away from a bear, we do not look for some object with the essence of 'treeness'. We simultaneously eliminate any possible alternatives to the first thing that we can climb. We are able to do this because our minds work "with large numbers of weak parallel processors rather than a single strong processor" (p. 93).

In addition, negations provide the basis for survival categories like 'own species/not own species', 'predator/non-predator', 'prey/non-prey', etc. Such categories would naturally create object trees of the following general constitution:

```
species
                    /-- predator
non-species    {                      /--prey
                    \-- non-predator    {
                                         \--non-prey, etc.
```

Using dichotomies something like these and fuzzy parallel processing, pigeons can identify categories like humans and tree. They can distinguish different kinds of people and trees, and even trees with just a few branches in a painted cityscape. Vital categories like 'trees' and 'humans' have been necessary for the survival of pigeons for millions of years. They are probably hardwired into pigeon neuro-circuitry. Other recognitions would be learned and held in ontogenetic memory.

Representations of Space, Action, Society

Animals do not develop categories of objects in a vacuum. They sense and categorize objects in relationships. They construct and memorize cognitive maps of their surroundings. They negotiate their survival activities in relation to those cognitive maps.

Animals inherit some aspects of such relationships like the three-dimensionality of experience and distance measurement. In their ontogeny, they memorize other important stable relationships like invariant physical constraints. In the moment, they concentrate on their survival activity. They incorporate these elements of their spatial cognitive maps into their larger cognitive maps, their PRSs.

Within an animal's PRS, there are also categories for different kinds of movement. There are cells that represent particular aspects of movement: some indicate movement to the right, others indicate approaching movements, and so on. Larger cell groupings indicate things like quickness of movement, danger of movement, and accessibility of prey. Among primates, complex patterns of neural activity represent things like 'hand grasping'.

Representations of objects in neural patterns can be considered not only as proto-concepts; they can also be considered as proto-nouns. Representations of embedded activities like 'cow grazing' or 'hand grasping' can be considered as forerunners of proto-verbs. All these proto-concepts become more numerous and distinctive in social relationships.

Social relationships are both catalysts and objects of category formation among primates. Continuous interactions among parents and siblings are, by definition, neural representations based upon categories (social, behavioral, spatial, and material). As an animal becomes more social, it articulates categories in its PRS. In a circular process, articulated categories enable complex social relations, which create better-articulated categories, and so on.

Social interactions lead to coordination of effort, competition, infighting, and a recognition of individual differences in terms of wiliness, persistence, strength, aggressiveness, and productivity. Patterns of interaction between individuals also develop. All these activities, individual differences, and patterns of relating are represented in the neural patterns of individual animals and social groups. These activities further enrich the PRSs of individuals, social groups, and species.

Summary

The infrastructure of language was laid down in the PRSs of primates. This infrastructure progressively built upon a foundation of primitive cells that were responsive to the environment. First, it differentiated sensor and motor cells that coordinated cell activities. Then it added interneurons that function as processors of information by merging and summating the outputs of different cells. With the proliferation and integration of neuronal networks, species-specific PRSs appeared. These PRSs gradually integrated the spatial matrices, physical behaviors, social activities, and personality features of conspecifics. These developments enabled primates to determine their activities with a degree of autonomy from outside stimuli. Our ancestor ape based her activities on the "outcomes of internal computations that merged information from the external world with information from the creature's current state and past experience" (Bickerton, 1990, p. 101).

Five million years ago, the PRSs of primates had the necessary infrastructure for the beginnings of secondary representation networks (SRSs), that is, protolanguages and language itself. Advanced primates at that time had protoconcepts that could serve as nouns and verbs. Language of some sort was an evolutionary certainty given selective pressures favoring more plastic responses and a viable way to embody an SRS. In Bickerton's words:

> All that was then required for the emergence of at least some primitive form of language was . . . some set of factors that would make the development of secondary representations advantageous, and some means through which such representations could be made both concrete and communally available (p. 101).

RECONCILING BICKERTON AND VARELA

There are similarities between Varela's theory of enactive cognition and Bickerton's theory of representation. Both are theories of embodiment. Both hold that evolution is better understood as a process of satisficing rather than as one of optimizing.

There are also differences. Varela discusses cognition from an enactive viewpoint. Bickerton discusses representation in terms of nervous systems. Varela says that representations have no part in pre-linguistic communication and cognition. Bickerton says that representations are the ways that animals know their worlds. Reconciling these positions involves consideration of phenomenological domains.

Varela does not need representations to explain enactive cognition, because his theory is presented from the interior viewpoint of the animal, from its phenomenological plane. He does not need 'representations' as operative elements in his theory because the concept 'representation' does not exist in the neuro-biology of pre-linguistic animals. He attributes the common misapprehension that representations are necessary to a logical confusion in which we equate a description, which we might make, with the operation of the organism.

What reason is there, however, to preclude a behavioral description of enactive cognition? Does enactive cognition have to be explained on the plane of phenomenological enaction? Can it not also be described from the phenomenological plane of an observer who views enaction in a systemic context?

Bickerton supplies such a systemic description of enaction. He does it on the basis of his connectionist theory of representation in which representations are embodied in the actions and the neural patterns of animals. These neural patterns co-evolve as their hosts incorporate information. Animals develop species-specific primary representation systems (PRSs) in which they map their views of reality.

All these progressions of representation are enactive. They embody the actions and neural patterns of animals as they negotiate their lived world. They self-organize in the manner of Csanyi's replicative model. PRSs evolve inevitably into SRSs (protolanguages and language).

We have, then, a two-tiered description of enactive cognition: Varela's on the enactive level and Bickerton's on the descriptive level. We have two ways, and not just one, to come to grips with enaction. On the enactive level, we deal with praxis; on the descriptive level, we deal with theory.

These two conceptions consider enactive cognition in complementary ways: non-representational and representational. Varela's theory of enactive cognition is non-representational because it tries to get behind the distinctions inherent in syntactic language. It tries to reunite nouns and verbs, like "cow" and "grazing," into their pre-syntactic designation, "cow-grazing." The theory of structural coupling expresses the blurred relationship that exists in nature between objects and their connections.

The nectar-gathering activities of honeybees, for example, might be represented as "bee-flower-drink-reproduce-live-see-color-etc." In the neural patterning of bees and the hormonal variations of flowers these protoconcepts exist as interrelated webs of existential patterning. The structure of language enables and compels us to express "nectar" in a syntactical structure with discrete components. We express the "nectar" situation by saying things like: "'Bees' 'need' 'nectar' to 'live.' 'Therefore', they 'drink' 'flowers'." "'Flowers' 'pollinate'; therefore, they need bees 'to reproduce'." "'As a result', bees 'evolve' 'discriminating' 'ultraviolet vision' and flowers evolve 'vivid colors'." Nothing in nature is discrete in this way; rather, everything in nature is interconnected in structural couplings.

Varela's expository challenge, in presenting cognition as structural coupling, is to express a kind of non-linguistic cognition to a culture and a species that thinks that all cognition is done in linguistic terms. There is an objectification intrinsic in language that obscures the fact that language arises in non-linguistic structural coupling-enactive cognition. This ever-present objectification in our discourse blinds us to the presence of underlying non-objectifying thought processes. The theory of enactive cognition tries to illuminate this not-objectifying enaction and remove our blind spot toward it.

There are, then, two different kinds of cognition or, at least, two tendencies of cognition: enaction and description. Enaction is embodied interaction. Description is representation by a somewhat disembodied observer. A dichotomy between enaction and description is very unlikely, however, because (1) 'a disembodied observer' is a semantic fabrication, and (2) every enactor of cognition represents experiences in its PRS.   Therefore, some synthesis of enaction and description is involved in every cognitive act.

                                       *   *   *

The preceding chapters on the history of systems thinking and theories of evolution establish the deep background for understanding how systems science can be applied to social situations.   The next chapter on the Habermas/Luhmann debate provides proximate background for several of the following chapters.

CHAPTER 4

THE HABERMAS[1]/LUHMANN[2] DEBATE

ABSTRACT

For the purposes of this work, the Habermas/Luhmann debate centers on two questions. One is theoretical, and one is ethical. The theoretical question is: Can all social processes be adequately explained in systemic terms? The ethical question is: How does reliance upon systems theory normatively affect an advanced industrial society? Since the original debate, Habermas has described an evolutionary process of the constitution of society that accords a role to systems theory. He continues, however, to voice objections and reservations about systems theory and about Luhmann in particular.

My consideration of this debate faces a major handicap because I am not at all proficient in German, and the debate has never been published in an English translation. This handicap is not, however, an insurmountable obstacle for the following reasons.

- In the past 25 years, Habermas and Luhmann have written profusely on the themes discussed in the debate, particularly those which are the topics of this volume. Their later works reformulate their positions on those themes.
- The debate sets the stage for this research but is not its principal focus; therefore, an accurate but brief summary of the debate is an adequate, though not optimally desirable, platform for launching this research.

---

1. The thinking of Habermas during this period is most accessible in English in *Knowledge and Human Interests* (1971a), *Toward a Rational Society* (1971b), and the article: "On systematically distorted communication" (1970).

2. Before and during these debates, Luhmann published the essays contained in *The Differentiation of Society* (1982). These essays can be read as an elaboration of his positions during the debate.

- Two admirable summaries of the debate are available in English, one from the Habermasian side by Friedrich Sixel (1976), and one from the Luhmannian side by Stephen Holmes and Charles Larmore 1982).

For these reasons, the debate will be synopsized from the reflections of the above-mentioned commentators.

Two questions in the debate are directly related to the theme of this book. One of the questions is theoretical: Can all social processes be explained in primarily systemic terms? Or alternatively, can social processes be completely explained without resort to systems thinking? In this debate, Habermas contends that social action requires consensual decision-making. Luhmann, on the other hand, contends that social activities are too complex for consensual bartering; they require impersonal systemic regulation.

The ethical question concerns the issue: How does reliance upon systems theory normatively affect an advanced industrial society? Habermas claims that applications of systems theory tend to repress free personal agency because they operate mechanically without recourse to common sense, democratic discourse, and social justice. Luhmann counters that the complexity of pluralistic societies precludes normative consensus in the particulars of contested situations. Moreover, impersonal, positive laws are the safeguards of individual and community rights. Finally, insistence on personal norms in social contexts is a remnant of dysfunctional metaphysical narrow-mindedness.

LUHMANN

Meaning

Luhmann defines social systems as organized patterns of behavior. As such, they include "actions, interactions, roles, symbolic meanings, choices, and so forth" (Habermas and Luhmann, 1971, p.36; translated in Holmes and Larmore, 1982, p. xx)-but not individual human beings. "Subsystems and organizations, including the political system, must be conceptualized as excluding men as concrete psycho-organic units. No man is completely contained inside them" (Habermas and Luhmann, 1971, p. 37). Concrete individuals in today's societies belong to multiple social groups, and they cannot be considered as "parts" of any one overarching society. As Luhmann points out, one possible exception is found in "the insane asylum."

Social systems and individuals, therefore, have a certain autonomy. They exist as environments for each other. They both exist in environments of complementary and competing physical, biological, psychological, and social influences. They are frequently forced to make choices on the basis of incomplete evidence and on the spur of the moment.

Social systems are centers of collective agency. As they adapt to novelties in their environments, these adaptations are its selections. A social system, then, has to select some way to proceed amid its complexity. In its selection, it is influenced by many factors but is not determined by them. It makes contingent selections (ones that could go one way or another). By making these selections, the system constitutes *meaning* for itself.

Meaning for Luhmann is both a process and a result. As a process it is the selection-making that has just been described. As a result, it is a system's accumulation of meanings produced by that process. It is a system's dynamic process for making its future *and* its memory of those selections.

As a system continues to reduce complexity with its selections, it builds up a unique backlog of selections made and selections negated. It uses this accumulation of selections, its meanings (in the second sense), as values for making future selections.

Communication

Social systems communicate among themselves by assimilating each other's meanings, that is, by sharing the ways that they have reduced their individual complexities. Systems use some of the meanings that they receive via communication to reinforce their accumulated sense. They use other communicated meanings to change or adapt it. In this way, systems make their arrangements of meaning more complex, more discerning of their environments, and more flexible for making future reductions of complexity.

Self-Reflection and Subsystems

As systems become more flexible because of their internal complexity, they become more aware of the contingency of their selections. This awareness of contingency is observable, for instance, in the behavior of rats that continue to pull a lever for a reward pellet even though the reward is provided only on a random basis. Their very ability to be programmed in such a manner demonstrates their awareness of contingency. This awareness is also observable in social systems, such as philosophical movements[1], that exist in universities, libraries, and even cyberspace.

Social systems have studied themselves as well as their environments. For example, the American Psychological Association studied the way its studies were

---

1   "Theoretically guided research . . . can be nothing other than a self-referential social system" (Luhmann, 1995, p. 487).

Social systems have a life of their own that gives direction to the thought and activity of individuals. This is easily observed in the history of cultural and intellectual fads or, in a more serious way, in multigenerational national and ethnic hatreds.

Social systems have studied themselves as well as their environments. For example, the American Psychological Association studied the way its studies were being presented and generated its well-known style manual. Over the course of time, social systems have become very interested in how they do things. They have studied learning, bought and sold money, influenced the exercise of political power, and valued art appreciation. In these cases of self-reflection, many individuals make contributions, but the self-reflection of the whole is more than the sum of individual self-reflections.

Systems use this self-examination ability to devise subsystems that regulate their dealings with their environments. They simplify their experienced complexity by devising internal mechanisms that specialize in handling various aspects. Capitalist economies, for example, generate bureaucratic procedures and systems of positive law. In this way, they make bureaucratic and legal aspects available for deeper consideration. Social systems also develop interactions between themselves and their subsystems. Eventually, they come to observe all of those interactions.

HABERMAS

The Ideal Speech Situation

Habermas, at the inception of the debate, had already developed the concept of the ideal speech situation in *Knowledge and Human Interests* (1971a). He defines it as a discussion in which participants express themselves freely, forthrightly, and truthfully; therefore, they put aside external power relationships and address each other on an equal footing. In such an ideal discussion, every viewpoint and argument is heard and decisions are made by the force of the better argument (cf. Thompson, 1984, pp. 271-272).

Arguing from this concept, Habermas maintains that certain unspoken premises underlie every attempt to communicate. He says that communicants (1) engage each other as persons in some understood relationship (claim of rightness), (2) claim that their assertions are true (claim of truth), and (3) profess truthfulness in their intentions (claim of truthfulness). It is through such communicative acts, which presuppose rightness, truth, and truthfulness, that they make decisions regarding meaning and value (Habermas, 1984, p. 99).

Meanings and Values

Habermas believes that this kind of interpersonal communication creates the meanings, norms, and values of a society. Such values give individuals and groups bases from which they can challenge oppressive social situations. In particular, they provide ramparts of truth, justice, beauty, honest and fidelity against a rampaging technocracy that regards functionality as its chief value. [These ideals are very similar to those proposed by Russell Ackoff for the corporate future (1981, pp. 37-42).]

ISSUES

Meaning

Habermas has a view of meaning that is radically different from that of Luhmann. Luhmann holds that meanings are a construction of past selections made in the course of a system's survival. Habermas holds that meanings are created through interpersonal communication; that they rest upon interpersonally accepted norms. Luhmann replies that the origin of the norms brought to any discussion lies in an accumulation of past selections.

Habermas claims that Luhmann's theory is a theory of *abstract* meaning because it ignores the corporeal reality of existing systems (societies and persons). He claims that Luhmann's theory does not recognize the *intersubjectivity* that arises in dialogue as real people acknowledge each other as beings of sense and body.

Luhmann grants that a person may recognize himself or some other person in a speech act but he insists that one does not, in that way, make oneself a meaning-endowed subject. Instead, one creates one's subjectivity by making interconnected choices among a range of possible actions and possible motives. One's actions have meaning because of a history of selections in complex situations. One's ability to make choices depends upon already constituted meanings that make further choices possible. Only on such a background, received from previous meanings, can one choose certain actions for particular reasons.

Technocracy

Habermas attacks Luhmann's position as being a very sophisticated defense of technocracy. He holds that Luhmann's concept of meaning reduces all values to the technocratic one: how to make the existing social system function better.

Such a technocratic bias is inherently conservative and resistive to change, according to Habermas. It fosters decisions made on the basis of expertise and precludes discussion of ethics and environmental concerns. It ignores the legitimate claims of people who are victims of technological systems. It makes it almost impossible for people to uphold their own values, meanings, and rights through the process of democratic debate.

Habermas notably objects that Luhmann creates an immediate relation between the theory of society and the praxis of society in which system complexity, and not human intersubjectivity, is the key mover of progress. Luhmann, on his part, points to the complexities of the modern world that inhibit extended discourse among multiple concerned parties. In this modern complexity, the ideal speech situation can never happen. It is not only improbable that equality, truth, and truthfulness would prevail in such a situation; it is impossible because of complexity and time constraints. Reason, in these circumstances, ceases to be a means of communication.

Luhmann says that a political order based on reason, truth, legitimation, and power is not up to the job of handling social system interactions in the modern world. An adequate political constitution has to abstract from such value questions in order to provide processes of greater selectivity for its subsystems. Thus, for example, law becomes positive-a mere system of operational rules with a memory of past decisions. Because of its alleged value-neutrality, positive law provides a framework in which persons of widely divergent value orientations can live together freely in peace and prosperity.

Norms

Habermas defends certain norms as quasi-transcendentally necessary in the Kantian sense; that is, such norms are prerequisites for human communication. He specifies three norms as basic claims that underpin all human communication: truth; rightfulness; and truthfulness. He asserts that human communication is impossible without a basic acceptance of these norms.

Luhmann contrasts normative and cognitive approaches to reality. He says that cognitive expectations are easily modified in the face of contrary experience. Normative (value) expectations, on the other hand, are not modified by contrary facts; rather, they demand conformity. Normative expectations tend to blind us to the differences existing in situations.

SINCE THE DEBATE

Both Habermas and Luhmann altered their positions as they attempted to understand and respond to each other. Habermas began to think almost like a systems theorist, at times, while maintaining certain reservations. Since the original debate, he has developed an evolutionary architectonic that accords a role to systems theory (*The Theory of Communicative Action, vol. 2*). In this work, he develops a social evolutionary theory to describe the increased rationalization of the lifeworld and to set the stage for the systemic steering media of money and bureaucracy. In so doing, he coordinates elements of the ideal speech situation and parts of Luhmann's systems analysis into a functioning theoretical unity in which societies are conceived as simultaneously systems and lifeworlds.

Luhmann has incorporated Maturana's biological concept of autopoiesis (self-generation) into his definition of social systems. He uses this concept in *Essays on Self-Reference* (1990) to explain things like the evolution of the state and world society, the self-reproduction of art, and the self-reproduction of law. In *Social Systems* (1995), he re-frames the whole discourse of sociology as it concerns meaning, communication, subjectivity, system/environment relationships, interpenetration between systems, structure and time, conflict, and rationality.

The Habermas/Luhmann debate is an admirable dialogue marked by both openness and critique. The dialogue sharpened and deepened the thought of both its protagonists. It reveals that difficulties of translation between points of view and disciplines can be large hurdles for even extraordinarily capable thinkers. To obviate some of these difficulties and ease others, the standards and methodology indicated in Chapter 19 can be extremely useful, chiefly for our debates of a less lofty nature.

The following chapters indicate the thoughts of various systems thinkers, including Habermas and Luhmann, which have been developed in the years since the original debate. The final chapters relate these thoughts to various social processes. Evaluations of the thoughts of all these thinkers are withheld until those final chapters, where contextual comparisons are made among all of these thinkers.

CHAPTER 5

HABERMAS SINCE THE DEBATE

ABSTRACT

In his thinking after the formal debates with Luhmann, Habermas has developed his thought in many ways. Three of those ways are explored in this chapter. First, he grounds the necessity and authority of rules on the three speech claims of truth, truthfulness, and rightness. Second, he shows the progression of those claims as they evolved into the structures of the lifeworld and eventually into systemic media of communication. He describes the ideal lifeworld/system relationship and the actual relationships between them. Third, he continues to voice fundamental objections to social applications of systems theory.

THE ORIGIN OF NORMS

From his analysis of speech acts, Habermas formulates the basic rules for communicative action. He traces the genesis of rules in three stages: from gesture to symbol; to grammatical speech; and to discourse. He develops a method for deriving the rules, shows how rules obtain their authority, and traces a historical process that he calls "the linguistification of the sacred."

Speech Acts

Habermas starts with basic questions about our talking to each other that were originally developed by John Austin and John Searle: What is a speech act? How is it more than just language? What are its formal elements? What are its criteria of validity?   Are   these   criteria   presupposed   in   every   speech   act?

Starting from the assumption that the goal of communicative action is reaching an understanding (1973, p. 17, 1984, p. 99), Habermas finds four essential elements in successful communicative action: comprehensible language, representation, expression, and legitimate interpersonal relation (1973, p. 18, 1979, p. 2).

When people are talking, their (1) shared language makes their communication possible. In their utterances, they (2) pass factual information, (3) reveal their inner thoughts, and (4) establish interpersonal relationships. Habermas says that anyone who reaches an understanding with someone else is:

a. *Uttering* something understandable;
b. Giving [the hearer] *something* to understand;
c. Making *himself* thereby understandable; and
d. Coming to an understanding *with another person* (1979, p. 2).

Uttering something comprehensible requires a common language.   Giving something to understand is based upon a supposition that what is given is true. Making oneself understandable presupposes that one is being truthful. Coming to an understanding with another person implies that one is striking a rightful relationship (for example, dialogue or assuming authority) with someone.

The fourth claim, expressing a goal of reaching consensus, requires rules for understanding and behavior.   In Habermas's (and Wittgenstein's) definition, rules lay down how someone produces something like a table or a sentence. They are propagated primarily through examples.   For instance, one teaches arithmetic progression by examples, such as, "1,3,5,7,..."   One comprehends the rule by continuing, "...9,11,..."   As Habermas says, "The application of the rule makes the universal in the particular apparent to us" (1989, p. 16).

As rules develop, they give authority to claims of comprehensibility, truth, and truthfulness by proclaiming their adherence to rules. It would seem that the reaching for rule understanding drives the evolution of all stages of communication. The urgency of rules is particularly obvious in the interpersonal stages of signals, grammatical utterances, and discourse. A rule is necessary to create a signal out of a gesture.   Mutually accepted rules (often unconscious) are necessary to create grammatical utterances from signals.   Rules are necessary for resolving misunderstandings and achieving consensus.

From Gesture To Signal

Habermas develops the thoughts of George Herbert Mead to explain how a signal could develop from a gesture.   For Mead, a gesture is a movement by some organism that influences another organism.   A wolf, for example, might yap at another wolf that yaps back. In a primal situation, in which this is a first meeting of a strange wolf for both wolves, there are, presumably, no mutual expectations. The wolves, then, share a *gesture-mediated* interaction in which "the gesture of the first organism takes on the meaning for the second organism that responds to it" (Habermas, 1989, p. 11). In other words, the second wolf finds some meaning in the yapping of the first wolf.

Now, wolves come to anticipate that one's yapping provokes another's yapping. How do they achieve this anticipation? Stimulus-response explanations merely beg this question. Mead and Habermas say that the first wolf "takes the attitude of the other." The wolves "internalize a segment of the objective meaning structure to such an extent that the interpretations they connect with the same symbol are in agreement, in the sense that each of them implicitly or explicitly responds to it in the same way" (1989, p. 14). Each wolf interprets that its yapping has a meaning *for it* that is like, but not yet the same as, the meaning that yapping has *for the second wolf.*

This initial internalizing of meaning structure is the first intermediate step between a simple gesture and a true signal. In the second intermediate step, the wolves learn how to employ a gesture with communicative intent; and enter into *a reciprocal relation* between speaker and hearer. Two wolves might, for example, yap at a bear to express their mutual sense of alarm. Implicitly, the wolves would recognize their roles as speakers and hearers of the alarm; they would look upon each other not merely as objects that react to gestures but as communicating counterparts. This second way of internalizing the attitude of the other leads to a third step in which the wolves might ascribe to the same gesture an *identical meaning.*

Having an identical meaning to at least two individuals is the essence of *symbolic signals.* Consider the case of wolves who interpret an adult wolf's gesture of lying on its back and exposing its belly as a signal of submission. This wolfish activity might not yet constitute a symbolic signal because it might not involve the idea of *should.* "There is an identical meaning when ego knows how alter *should* respond to a significant gesture; it is not sufficient to expect that alter *will* respond in a certain way" (1989, p. 14). If the attacking wolf does not heed the gesture of submission, he might kill his prey and thwart its expectations; but the idea that the ravaging wolf *should* have heeded the gesture might not arise. But Habermas does little to discuss whether "shoulds" arise for non-human animals. His focused methodical consideration is human social interaction.

In a human context, if ego's communicative expectations are unexpectedly thwarted, he or she might express disappointment. Alter might also express disappointment over this failure to communicate. In this situation, the mechanism of internalization can be applied for a third time-to these mutual expressions of disappointment. In this process, creating rules for using symbols obviates future failures. Habermas says,

> In adopting toward themselves the critical attitude of others when the interpretation of communicative acts goes wrong, they develop *rules for the use of symbols*. They can now consider in advance whether in a given situation they are using a significant gesture in such a way as to give the other no grounds for a critical response. In this manner, *meaning conventions* and symbols that can be employed with the same meaning take shape (1989, p. 15).

Yet, how can individuals come to such an agreement about identical meaning? Habermas answers this question by referring to Wittgenstein's analysis of *the concept of a rule*. Rules dictate how someone produces something, as we have already noted. Rules apply to symbol formation because "We can explain what we mean by the *sameness* of meaning in connection with the ability to follow a rule" (1989, p. 17). Consider the case at the presymbolic level of interaction where an individual, S, utters a call for help, q_0, to his mates, **T,U,V**..., and his call is not heeded. Let us say, that S is a young man yelling "Attack!" at the approach of an animal which **T,U,V**,... recognize as a pet wolf. In this situation where q_0 was used inappropriately to the context, S might make a decisive step where he "*internalizes this dismissive reaction by T,U,V... as a use of q0 that is out of place*" (p.21). By internalizing the negative positions that **T,U,V**... take toward him when he goes wrong "semantically," S, and subsequently the other members of his tribe, would come to

> *anticipate critical responses* where q_0 is used inappropriately to the context. And on the basis of this anticipation, expectations of a new type can take shape . . . based on the convention that a vocal gesture is to be understood as "**q**" only if it is uttered under specific contextual conditions. With this we have reached the state of symbolically mediated interaction in which the employment of symbols is fixed by meaning conventions (1989, p. 22).

Similarly, symbols arise as conventions that fix the sameness of meaning. They develop through rule formations among communicating individuals.

Signal languages are early stages of conventional abstraction that meet the survival needs of endangered organisms. They form the basis for symbolic interactions. Among primates, they were used and developed for centuries before grammatical utterances arose. Many more meaning conventions arose as primates continued taking the attitude of the other by internalizing parts of the objective meaning structures of their lives. Thing words, action words, modifiers, and rules of appropriateness were created. Pronouns (or pronominal endings) of the first, second, and third person were employed.

Grammatical Speech

The advent of grammatical speech signals a quantum leap of communicative ability. The new flexible rule in speech acts, "I tell you that . . . [for example, the bear is asleep]," provides a new power for reaching understanding and coordinating action. Within its flexible structure, grammatical speech can objectify almost anything within the speaker's internal and external world-including the very structure of speech.

The validity claims of speech acts have already been described. Within the context of grammatical speech, the formal components of those speech acts can be explicated. Habermas discusses these components of the speech act by discussing the activity of a "competent speaker." He says that a competent speaker is able:

1. To choose the propositional sentence in such a way that . . . the truth conditions of the proposition stated . . . are supposedly filled (so that the hearer can share the knowledge of the subject).
2. To express his intentions in such a way that the linguistic expression represents what is intended (so that the hearer can trust the speaker).
3. To perform the speech act in such a way that it conforms to recognized norms or to accepted self-images (so that the hearer can be in accord with the speaker in shared value orientations; 1979, p. 29).

Each of these three functions can be found, implicitly at least, in every utterance from one person to another. Usually, one function predominates, but the other functions are also present.

As an example, consider the sentence addressed from one person to another: "Peter smokes a pipe." This utterance contains more than its *propositional* content. In its full meaning, it says something like this:

> "I [hereby] assert to you that
> I believe that Peter smokes a pipe."

In this utterance, the clause, "I [hereby] assert to you," the speaker establishes contact with the hearer by stating the rule under which he or she is proposing a belief that Peter smokes a pipe. The force of this clause sets up an engagement between the parties according to mutually shared rules of communication. The speaker would propose a different rule if he or she would state, for example, "I deny that Peter smokes a pipe." This part of a speech act, that establishes a rule for understanding its propositional content, is its *illocutionary* aspect. The illocutionary parts of utterances set the rules for conversations, discourses, and conflict resolution; they encompass manners and commonsense ways of coming to understandings.

The "I believe that" clause in the above utterance expresses a claim of veracity that the speaker is sincerely expressing his or her subjective truth. The claim of veracity is the expressive aspect of speech acts. A different meaning would be given to the utterance about Peter's smoking if the "I believe that" clause were replaced by "I doubt that" or "I am sorry that."

The *propositional* content of an utterance, "Peter smokes a pipe," is a statement of objective fact expressed in the third person. The speaker claims that it is *true*. The validity of this claim is determined by whether it conforms to the objective facts or not. The *illocutionary* content of an utterance, "I [hereby] assert to you that," is expressed in the first and second person pronouns. It makes a rule bound connection between people about how the propositional content of the utterance is to be understood. The validity claim for this part of the utterance is *rightness*; it is validated if a correspondence of understanding between the first person (ego) and the second person (alter) is established. The *expressive* part of the utterance, "I believe that," is expressed with the first person pronoun; it expresses a correspondence between ego's inner convictions and his or her outer words. Its validity claim is *truthfulness*. This validity claim can be questioned if alter believes that the speaker is lying.

The structure of speech acts allows its illocutionary and expressive aspects to be expressed as propositions. For instance, one can say, "I tell you that I feel queasy" (self-expression expressed as a proposition); or "I think that I am telling you that . . ." (illocution expressed as a proposition). This ability of grammatical speech, to "propositionalize" its illocutionary and expressive components, gives it purchase on the multitudes of "I-You" relationships and the varieties of introspective experiences. Grammatical speech can talk about anything, bring it up for discussion, and organize it into schemata.

In the course of history, grammatical speech has come to express many phases of our lives and to structure our experiences with the logic of language that makes our lives rational. These rationalizations provide the structure of our social existence.

The grand theoretical project of Habermas is to chart these rationalizations and coherently express them within a theory of communicative action. He charts a progression in which social life develops into a profusion of rationally constituted institutions that he calls the "lifeworld." He proceeds to show how systems of capitalism and bureaucracy, steered by money and power, develop within that lifeworld, gain autonomy, and then create stress within the fabric of the lifeworld.

Discourse

In simple consensual communication, the validity claims of truth, rightness, and truthfulness are assumed. If disagreements arise about truth, ego has merely to give factual evidence and testimony to validate his or her assertion of facts. If ego's veracity is challenged, he or she proposes evidence of past truthfulness, consistency, and responsibility and, perhaps, proposes a trial period as guarantee of earnestness. If ego's right to state a claim, for instance, "I declare this marriage null and void," is questioned, then ego has to validate his or her claim by appealing to law, custom, or precedent. In consensual speech situations, validity claims are justified in a simple manner.

If a speech act's validity is not easily established, a contested situation arises that begs for some kind of resolution. The contestants could part company, fight, or engage in strategic (competitive) action with each other. They also might try to hash out their differences.

This hashing out of differences requires examining validity claims in an abstract manner that Habermas calls discourse. Although any validity claim can be considered in discourse, Habermas places special attention on the claims of rightness and truth (insofar as truth can be put to pragmatic tests). Rightness is the subject of practical discourse and truth is the subject of theoretical discourse. For Habermas, praxis and theory exist symbiotically: praxis tests theory and theory gives creativity to praxis. Discourse about truth leads to the development of philosophy and science. Discourse about rightness leads to customs, laws, and moral codes.

Discourse requires varying conventional rules for deciding what statements are true and what course of action is desirable. Discourse geared to understanding develops rules of logic and evidence. Discourse geared to action develops rules about who has the right to make decisions and command the allegiance of others, how rules are made, how they are to be obeyed, and how conflicts are settled.

Habermas holds a democratic ideal in which truth and rightness, meanings and values, are provisionally settled by the force of the better argument in an open discourse. He readily admits that such "an ideal speech situation" is seldom achieved in practice and that it is counterfactual. He claims, however, that its existence is a necessary fiction that underlies all of our structures of human relations. He holds that this counterfactual ideal is necessary as a goal to be approached asymptotically. He says we have to proceed according to this ideal because "on this unavoidable fiction rests the humanity of intercourse among men who are still men" (in McCarthy, 1978, p. 291).

Reconstruction

The way that Habermas arrives at rules is of particular interest because he uses partly empirical methods. He calls his theory of communicative action a "reconstructive science." By this, he means that he analyzes speech acts, signals, grammatical speech, and discourse as they are empirically presented in actual communication and reported by participants. Using phenomenological, hermeneutic, and historical methods, he draws out the key formal elements of those communicative acts and expresses them as hypothetical rules. He tests those hypotheses and pitches them to a critical audience for evaluation. Then, he revises those rules according to *a posteriori* experience and experiment.

Reconstruction, then, is an empirical science that tries to reconstruct, in language, the underlying structures of social existence that operate inarticulately. It is put together in a different model than the empirical-analytic model of the physical sciences. The reconstructive model offers testable hypotheses that apply to the formal structures, which underlie lived experience. Habermas includes other sciences, like linguistics and developmental psychology, among the reconstructive sciences. Such sciences recognize their empirical limitations. They recognize the structural interconnection between experience, action, and theory. They pay attention to the structures of communicative action. They account for the development of competencies in children (ontogenesis) and the historical development of those competencies in evolution (phylogenesis).

Habermas also describes one of his working methods as "reconstructive." In this typical way, he takes a theory apart (like Marx's historical materialism or Durkheim's phylogeny of speech) in order to put it back together again in a new form. With this reconstruction he helps the original theories achieve the goals that they set for themselves.

The Authority Of Rules

In his reconstruction of Mead's analysis of a signal, Habermas employs Wittgenstein's analysis of a "rule" to show how one gesture could come to have the *same* meaning for two organisms. Such rules often develop in society as ways of doing something. They have authority because they are the fabric of society. Formulated rules usually come later. They have their authority insofar as they accurately express that fabric.

The rule behind signal recognition is an *ontogenetic* necessity; rule conformity makes signaling possible. The ontogeny of propositions, illocutions, and expressions underlies communicative action in all its forms because those communicative thrusts are the stuff of grammatical speech. Similar formal rules are found by Habermas to underlie discourse, the lifeworld, and systems. All these rules have authority because they describe the underlying structures and workings of social existence.

Habermas uses the (ontogenetic) development of morality in the child as a bridge to a *phylogenetic* theory for the authority of rules. He reports the development of morality as Jean Piaget and Lawrence Kohlberg chronicle it. In their account, he emphasizes the stages of moral development that are characterized as preconventional, conventional, and postconventional.

At the preconventional level, the child knows good and bad as reward- and punishment-begetting activities. At the conventional level, he or she maintains group expectations and loyally defends its rules. At the postconventional, autonomous, or principled level, "there is a clear effort to define moral values and principles which have validity and application apart from the authority of the groups or persons holding these principles" (Kohlberg in McCarthy, p. 251).

This schema of Kohlberg shows a shifting of the authority of rules from physical intimidation, to group loyalty, to rational rightness. Habermas uses this rule consciousness development as a clue to the phylogenetic authority of rules-that is, how rules originated in an aura of sacredness and evolved to democratically derived norms.

To compose his scenario for the phylogenesis of rules, Habermas reconstructs Durkheim's theory of the sacred origin of rules, as follows. In the early days of human evolution, primates existed in a mostly continuous web with their surroundings. They had only a few signals to alert the group to united action and to facilitate their living together. In other words, they had a simple symbolic communication system that had not yet progressed to language (as grammatical speech with propositional content). In this state, they relied on symbols with their implicit meanings for their survival. They lacked the more explicit meanings that we possess via propositional speech.

The first symbols were used to point to awesome occurrences like the frightening appearance of a cave bear or the mysterious birth of a child. Eventually, the symbol for bear-appearing would encompass the bear itself, the fright experienced by the primates, the defenses used to fend off the bear, the sacred relics (claws, teeth, skin) of slain bears, the pride of the tribe in its mastery of its fear, its reverence for the might of the bear, and its ritual re-enactments of historic encounters with bears. In a similar way, the symbol for birthing would come to include the mother, the agony of the mother, her joy, her nurturing of the baby, the child, the magic of new life, the feelings of caring aroused by the child, the pride of producing new life, and the solidarity of the tribe.

In these ways, symbols for totems and natural events produced group solidarity through a social structure based on communication. Equiprimordially with the symbols, and practically equivalent to them, were the rituals that accompanied the use of symbols. These rituals re-enacted how the symbols of understanding developed in historical circumstances; they thereby showed how a symbol was to be used and understood. In this way, rituals encoded the rules for understanding and laid the groundwork for further communicative action.

"Primitive" societies, according to anthropologists, separate the sacred and profane arenas of communicative action. In the profane acts of everyday life, they govern their lives with propositional speech and make practical use of technology. In the sacred areas that comprise worldviews, customs, and morality, however, they guide their lives with ritual symbolism backed by the force of tradition.

Rituals, and their accompanying symbols, refer back to the foundational time of the tribe. They express how the tribe was put together. They constitute its lifeworld structure. They are sacred because they recall the very appearance of meaning in the tribe's life; they let the group re-enact that meaning and make it present and valuable for enriching human relationships. The rules of ritual have force because they express the foundations of the tribe's existence.

The Linguistification Of The Sacred

Habermas traces the evolution of rules from their ritual source through a process that he calls, "the linguistification of the sacred." By this, he indicates a process in which the rules of speech acts are progressively challenged, justified, challenged, and justified. In this way, a challenged taboo would be justified by an appeal to tradition; a challenged tradition would be justified by an appeal to theological myth, and so on. Eventually, taboos become justified by appeals to common sense, common agreement, and the common good. Throughout this process, the implicit symbolic ritual is increasingly explicated in language.

The linguistification of the sacred is a restatement of what Weber called "the rationalization of the lifeworld." This process is synopsized in the following passage:

> All the paths of religious rationalization branching through civilizations, from the beginnings in myth to the threshold of the modern understanding of the world, (a) start from the same problem, that of theodicy, and (b) point in the same direction, that of a disenchanted understanding of the world purified of magical ideas (1984, p. 196).

In their millennial progression from taboo to democratic laws, rules have gained an explicitness that enables nations of people to live together and thrive. Over time, however, the sacred force of rules has become obscured. Multicultural diversity and the progress of scientific rationality have eviscerated the force of commandments and their theological justifications. Like many other things in modern society, the authority of moral rules has sometimes sunk to a lowest common denominator: conformity to avoid reprisal (Kohlberg's first stage of moral development).

In Habermas's project, rules are conventional but not arbitrary. They are developed conventionally over millennia to meet the needs of communicative (social) action. They have been tested and revised to account for environmental and historical challenges. In their deep structures, they express how communication and society more generally are put together. These deep structures are the essence of civilization and evolved humanity.

The authority of all rules, sacred rituals and democratic laws alike, lies in their expression of some common good. Laws are valid because they express, even if counterfactually, what is necessary for successful communicative action.

Habermas distinguishes between the *logic* of development and the *dynamics* of historical processes. The logic of development is expressed in the statement of Weber's position that has just been quoted. A dynamic explanation of history requires the identification of the external historical determinants that influence the development of worldviews. Some questions, which a historical explanation must answer, are the following:

1. What the conflicts that overload the structurally limited interpretive capacity of and existing worldview look like, and how they can be identified;
2. In which historically caused conflict situations a theodicy problematic typically arises;
3. Who the social carriers are that establish or rationalize a new worldview;
4. In which social strata a new worldview is adopted, in which sectors and how broadly it affects the orientation of everyday conduct;

5. To what extent new worldviews have to be institutionalized in order to make legitimate orders possible-merely in elites or in entire populations;
6. Finally, how the interests of the carrier strata guide the selection of the contents of worldviews? (1984, p. 197).

In tracing the linguistification of the sacred, Habermas finds two main threads of social evolution in the rationalizations of communicative and strategic action. Communicative action comprises those activities that make possible mutual understanding, communal living, conflict resolution, and culture. In its earliest stages (of signal exchange, grammatical language, and basic discourse) communicative action is intensely interpersonal. Even in its earliest stages, however, communicative action is societal because it is embedded in family and tribal mores. These mores expand in complexity to generate the legislative, judicial, and executive forms of governance that eventually become governmental bureaucracy.

Strategic Action

As communicative action (the illocutionary aspect of speech) develops the lifeworld, it simultaneously enables strategic action. Strategic action comprises social activities that try to get something done. When acting in this mode, people are trying to force other things and people to do their will whether the activity is lifting a beam onto a wall or forcing slaves to lift it for you. Strategic action involves the *perlocutionary* (getting something done *through* locution) aspect of speech.

In the early stages of strategic action, individuals wield strategic power because of their personal characteristics: strength; wisdom; brutality; charisma; or mystical pretensions. Later, they rule because of accumulated power, heredity, and wealth. Today, they prevail because of their positions in the monetary and bureaucratic hierarchies that regulate the exponentially expanding strategic actions of modern societies.

Strategic action develops originally within a lifeworld where decisions are made and goals are accomplished within the confines of personal and group decisions. A pinnacle of this kind of lifeworld was the democracy of the Greek city-states. Later lifeworlds increased in size and required some kind of bureaucracy and economic system, like those of the Roman Empire or those of late medieval monarchies with their merchant and guild economies.

In Europe, the size and complexity of the lifeworld became unmanageable on the basis of only interpersonal and *ad hoc* decision-making. As strategic actions multiplied and became more complex, they placed impossible demands upon the lifeworld processes of negotiating and bartering. They called for more expeditious means of facilitating conflict resolution and economic activity. As a result, systems of law and money grew up to handle strategic complexity in more or less automatic ways. The system of laws and judgment make conflict resolution more or less automatic. The monetary system makes the exchange of goods and services automatic.

Law and money are media of communicative action that regulate the activity of the bureaucratic and economic arenas without requiring interpersonal negotiating and bartering. They are at the core of systems activity, as understood by Habermas. They coordinate activity without requiring consensus. As systems activity, they stand in contrast to lifeworld action that develops consensus on what activities are to be undertaken.

In Habermas's view, the twin systems of bureaucracy and capitalism have waxed so strong in the Westernized world that they severely limit operations that rightly belong to the lifeworld. In his words, these systems "colonize the lifeworld." An example would be the plight of middle-class families where both parents work. Instead of deciding what mutual behaviors would be best for their marriage and the welfare of their children, parents work ever-longer hours to maintain their position in the economic system. In other words, keeping their place in the economic system . has demeaned their desires to improve their family lifeworld.

The rules for communicative action provide the organic structure upon which gestures, symbols, and grammatical speech develop. They also lie at the basis of the lifeworld and system environments that are the topics of the following section.

LIFEWORLD AND SYSTEM

In *The Theory of Communicative Action, Volume 2*, Habermas integrates his ideas of democratic interaction and social integration with some of Luhmann's ideas about activity coordination and the systemic integration of effects. He says:

> We conceive of societies simultaneously as systems and lifeworlds. This concept proves itself in a theory of social evolution that separates the rationalization of the lifeworld from the growing complexity of societal systems so as to make the connection . . . between forms of social integration and stages of systems differentiation tangible (Habermas, 1989, p. 118).

Social Integration and System Differentiation

Habermas contrasts social integration with systems differentiation. The first (lifeworld action) develops consensus on what activities are to be undertaken. The second (systems activity) coordinates activities without requiring consensus. Habermas shows how lifeworlds develop systems as ways to manage their affairs, and he defends the necessity of consensus even in highly differentiated societies.

The lifeworld (action) theory of society is an explanation of society in terms of its conditions of communication. Lifeworlds nurture the evolution of communication, norms, speech acts, and other communal developments. They function within "the Weberian theory of action, which is tailored to purposive activity and purposive rationality" (Habermas, 1989, p. 2)-a theory that is, according to Habermas, hermeneutic and idealistic. The lifeworld approach can explain the symbolic reproduction of a group's lifeworld from an internal perspective, but it cannot adequately explain the reproduction of society as a whole.

In the course of time, lifeworlds increase their internal complexity to a stage where they need to regulate their interactions in an impersonal way that does not require barter and consensus. In the course of time, they spawn steering systems (principally money and power) that make such exchanges. The systems theory of society explains how those interactions are accomplished without a meeting of minds but by a simple regulating of effects. Putting a quarter in a parking meter, for example, abridges innumerable discussions about the morality, efficacy, and legality of the procedure, and allows society to attend to more important concerns.

Habermas finds the origin of systemic sociological thinking in Durkheim who said that the division of labor is not a sociocultural manifestation but a "phenomenon of general biology whose conditions must be sought in the essential properties of organized matter" (quoted in Habermas, 1989, p. 114). Durkheim viewed society from the outside, objectively, as an observer. He looked *not* at how society functioned in its internal evolution, *but* how it developed its structures through the interplay of general forces. In this way, Durkheim arrived "at an analytical level of 'norm-free sociality' which can be separated from the level of ...communicative action, the lifeworld, and the changing forms of social solidarity" (Habermas, 1989, p. 114). In Durkheim's view, "the typical social relation would be the economic, stripped of all regulation and resulting from the entirely free initiative of the parties" (quoted in Habermas, 1989, p. 115).

This external viewpoint on social change is called functionalism. Since the inception of general systems theory and cybernetics, functionalism has been called systems functionalism or just systems. [Systems theory, however, in a sense not used by Habermas, is not inherently functionalist; it can describe and enable operations in the lifeworld.]

While *lifeworld* mechanisms coordinate actions that harmonize the *action orientations* of participants, *systems* mechanisms stabilize the unintended interconnections of actions by way of functionally intermeshing *action sequences*. One set of mechanisms effectuates *social integration*; the other, *systemic integration*. In one case, society is viewed from the inside as the *lifeworld of a social group*. In the other case, it is viewed from an uninvolved perspective as a *system of actions* in which actions have functional significance for maintaining the system (cf. Habermas, 1989, p. 117).

THE LIFEWORLD (ACTION) THEORY OF SOCIETY

Habermas defines progressively more comprehensive ideas of what a lifeworld is. He starts with the phenomenological idea of a social horizon and articulates the relationships between language and culture. He then considers the lifeworld as its ordered institutions and personality structures. Finally, he considers the everyday concept of the lifeworld that is the stories that people tell about their world.

The Lifeworld of the Social Horizon

Habermas first defines the lifeworld "as the horizon within which communicative actions are 'always already' moving" (Habermas, 1989, p. 119). He builds upon Husserl's idea of a horizon of consciousness "that shifts according to one's position and that can expand and shrink as one moves through rough countryside" (Habermas, 1989, p. 123). In transposing this notion to the world of social communication, Habermas says that we have to consider a social activity's "context of relevance" that is its "culturally transmitted and linguistically organized stock of communicative patterns" (Habermas, 1989, p. 124). This "context of relevance" is the social horizon.

In short, he says that "language and culture are constitutive for the lifeworld itself" (Habermas, 1989, p. 125). Language and culture are the media of mutual understanding; they are not a formal frames of reference or objective things in the social world. They are invisible and unrecognized by us when we are in our normal performative (non-self-reflective) attitude. They are "at our backs" and have a peculiar form of "half-transcendence."

Language and Culture

The semantic contents of language (its meanings) contain and transmit cultural patterns of interpretation, valuation, and expression. They supply people with common background convictions upon which they make selections, reach mutual understandings, and negotiate new situations. These meanings "preinterpret" new situations so that people do not "step in a void" when they go beyond the horizon of a given situation. Instead, "every new situation appears in a lifeworld composed of a cultural stock of knowledge that is 'always already' familiar" (Habermas, 1989, p. 125).

Habermas describes the power of language and culture that are the structure of the lifeworld as follows:

> The structures of the lifeworld lay down the forms of the intersubjectivity of possible understanding. The lifeworld is, so to speak, the transcendental site where speaker and hearer meet, where they can reciprocally raise claims that their

utterances fit the world (objective, social, and subjective), and where they can criticize and confirm those validity claims, settle their disagreements, and arrive at agreements (Habermas, 1989, p. 126).

As language and culture, the lifeworld is both beyond question, always constant, and always changing. It is beyond question because it is the presupposition of every selection and every discussion. It is constant because it accompanies us in our every shift of horizon, making the new vistas already intuitively familiar. It is always changing because it incorporates new vistas into itself as new cultural knowledge. Also, as language and culture, a lifeworld "cannot become controversial . . . at most it can fall apart" (Habermas, 1989, p. 131).

Institutions and Personality Structures

The lifeworld, however, is more than just language and culture; it is also ordered institutions and personality structures. These latter elements *do* confront an actor in a lifeworld as elements of a situation because they restrict his or her initiatives. Institutions and personalities have a double status-"as elements of a social or subjective world, on the one hand, as structural components of the lifeworld on the other" (Habermas, 1989, pp. 134-135). Such institutions and personalities are not always recognized; they often work *a tergo* (from the rear) as do language and culture.

Human action in the lifeworld, accordingly, has an iterative aspect: It is conditioned by lifeworld structures; it alters the lifeworld; it continues to be conditioned by this altered lifeworld, and so on. Habermas points out the advantages of this action/lifeworld symbiosis:

> Action, or mastery of situations, presents itself as a circular process in which the actor is at once both the *initiator* of his accountable actions and the *product* of the traditions in which he stands . . . Whereas *a fronte* the segment of the lifeworld relevant to a situation presses upon the actor as a problem he has to resolve on his own, *a tergo* he is sustained by the background of a lifeworld that does not consist only of cultural certainties. This background comprises individual skills as well-the intuitive knowledge of *how* one deals with situations-and socially customary practices too-the intuitive knowledge of *what* one can count on in situations-no less than background convictions in a trivial sense. Society and personality operate not only as restrictions; they also serve as resources (Habermas, 1989, p. 135).

In this quotation, Habermas contends that the lifeworld should not be excessively abstracted into an "autopoietic" system. Lifeworlds do not produce all the elements that constitute them. They incorporate bodies, intuitions, material/social institutions, and personalities with all their activities and accomplishments.

Through this enmeshment, lifeworlds produce webs of psycho-socio-material communication that imbed, restrict, and nurture growth. They, therefore, have an unquestionable quality that provides their members with well-established solidarities and proven competencies. In other words, lifeworlds provide basic material, social, and psychological certainties for their members.

The Everyday Concept

Habermas calls the concept of the lifeworld, which we have discussed so far, a description from the *perspective of participants.* He describes a further elaboration of the lifeworld, what he calls the *everyday concept of the lifeworld.* The everyday concept includes the stories that people tell.

Persons do not only encounter one another in the attitude of participants; they also give narrative presentations of events that take place in the context of their lifeworlds. Actors base their narrative presentations on a lay concept of the "world," in the sense of the everyday world or lifeworld, which defines the totality of states of affairs that can be reported in true stories. "This everyday concept carves out of the objective world the region of narratable events or historical facts" (Habermas, 1989, p. 136).

This everyday concept of the lifeworld, as stories, comprises the totality of a society's culture. It is the stuff of which social theory is made. It, in fact, includes the social theories that are constructed about it because those social theories are, of course, stories that are told by a society about itself. Thus the stories of sociology become additional stories that sociology has to study. The everyday concept of the lifeworld, in other words, provides a cognitive reference system based on the structure of language. Therefore, according to Habermas, a theory that explains communicative action is a theory that explains the lifeworld.

THE SYSTEM (ACTIVITY) THEORY OF SOCIETY

There is more involved in society and its activities than the background and foreground of the lifeworld. From their beginnings in human protolanguage, lifeworlds developed through evolving stages of evolution. They developed sophisticated means for settling disputes. Eventually, however, they generated a society with such speed and complexity that personal mechanisms of consensus-formation were inadequate. At that time of crisis, they developed impersonal mechanisms of exchange.

Systems mechanisms, like money in the economy and bureaucratic power in business and politics, handle relations impersonally. They functionally intermesh action-consequences instead of harmonizing lifeworld action-orientations. They relieve human purposive (lifeworld) activity of a burden that would stall the development of society as a whole. They transform purposive behavior into routine, functional behavior.

The origins of capitalism with its steering mechanism of money can be found in the Western tradition. The rich soil of capitalism was a society that separated religion from politics, had a tradition of (canon) law and a mercantile use of money, and valued individual achievement. That soil had been prepared during the late Middle Ages, the Renaissance, the Reformation, and centuries in which the scientific ideal was formulated.

Steering Media

The steering media in society (principally money and power) function as alternative languages to coordinate economic and political transactions. As such, they displace the consensual lifeworld haggling in the trading of commodities and in the execution of organized activities. Habermas explains how this happens in the following sentence:

> Media such as money and power attach to empirical ties; they encode a purposive-rational attitude toward calculable amounts of value and make it possible to exert generalized, strategic influence on the decisions of other participants while *bypassing* processes of consensus-oriented communication (Habermas, 1989, p. 183).

Habermas examines the various media that Parsons had divined as operative in steering different subsystems of society. He evaluates them on the basis of their relative autonomy from lifeworld interaction. He concludes that money is the strongest steering mechanism in the realm of property and contract, and that power is also a strong steering mechanism in the realm of bureaucratic organization. He finds that other proposed steering mechanisms, like influence in the realm of leadership and value commitments in the realm of morals, function weakly if at all to steer activity without communicative interventions.

The Integration Of Lifeworlds And Systems

Habermas conceives the proper integration of lifeworlds and systems as follows: "a system . . . has to fulfill conditions for the maintenance of sociocultural lifeworlds" (1989, pp. 151-152). He offers the following schematic formula: "societies are *systemically stabilized* complexes of action of *socially integrated groups*" (1989, p. 152).

> [Society is] an entity that, in the course of evolution, gets differentiated both as a system and as a lifeworld. Systemic evolution is measured by the increase in a society's steering capacity, whereas the state of development of a symbolically structured lifeworld is indicated by the separation of culture, society, and personality (Habermas, 1989, pp. 151-152).

Colonization Of The Lifeworld

Habermas concludes his volume on lifeworld and system by echoing Weber's musings about the loss of meaning and freedom in the modern world. "Weber saw the noncoercive, unifying power of collectively shared convictions disappearing along with religion and metaphysics" (Habermas, 1989, p. 301). These shared convictions have waned as we have relied upon a subjective reason of self-assertion and neglected objective reason with its reliance upon truth. In this process, freedom has succumbed to the domination of technocracy.

> [Society is becoming a] shell of bondage which men will perhaps be forced to inhabit someday, as powerless as the fellahs of ancient Egypt. This could happen if a technically superior administration were to be the ultimate and sole value in the ordering of their affairs, and that means: a rational bureaucratic administration with the corresponding welfare benefits (Weber quoted in Habermas, 1989, p. 302).

Weber's foreboding applies to the entire technocratic economy/society of our modern world, where the value of self-survival, expressed as efficiency for growth and profit, is supreme. Efficiency overrides all other values, such as justice, honesty, fairness, and mutual consent. The single-minded quest for efficiency imposes an external standard of system conformity upon decision-making and scorns the internal intentions of living people. This technocratic purgatory has been brought about, according to Habermas, through a process of "mediatization," in which interpersonal activities are progressively objectified and regulated by the media of money and/or power.

How such mediatization occurs can be seen on a large scale in the history of the welfare state. In the history of modern Western society, capitalist states have avoided disruptions to the social order by creating welfare. In an overall situation of radical social inequality, they grant a degree of need gratification to capitalism's underprivileged in order to maintain social stability. As time goes on, the welfare state grows increasingly complex as new human needs are identified and brought into the political arena. The welfare state generates bureaucracies to administer its social regulations that become systems unto themselves; they generate subsystems and subsystems of subsystems.

Similar processes take hold in economic, communal, and familial arenas because the economy and the bureaucratic state propel this process. These processes eventually enfold our lives; they objectify and control all aspects of social living. Structures of lifeworld integration, such as families, theoretic research, arguments over principle, commonsense conflict resolution, art, and even human personalities, are objectified and made objects of manipulation. These lifeworld structures are no longer integrated through the mechanism of human understanding; instead, they have become *formally organized domains of action*. Within the bureaucratic/systematic context, lifeworld structures become elements of system environments and they congeal into "a kind of norm-free sociality."

As increasing portions of the lifeworld are placed within bureaucratic/systematic contexts, areas of life lose their connection to human values and human decision-making. People in organizations are faced with a situation in which their value orientations and personal opinions lack validity because organizations do not have to achieve consensus by communicative means. In such a lackluster lifeworld, people lack vitality; they lose contact with cultural traditions; they lack a sense of personal and social meaning; they become "specialists without spirit," and "sensualists without heart." "In the deformations of everyday practice, symptoms of rigidification combine with symptoms of desolation" (Habermas, 1989, p. 327).

Habermas deplores the rape of the lifeworld by the forces of economic and bureaucratic efficiency. He calls the process in which lifeworld activities are subjugated to systems' imperatives, the *colonization of the lifeworld.* "When stripped of their ideological veils, the imperatives of autonomous subsystems make their way into the lifeworld from the outside-like colonial masters coming into a tribal society-and force a process of assimilation upon it" (Habermas, 1989, p. 355).

HABERMASIAN OBJECTIONS TO SYSTEMS THEORY

Habermas's conception of systems is a narrow one. It derives from Durkheim, Parsons, and Luhmann almost exclusively. It neglects the General Systems Theory of Von Bertalanffy, complex systems theory, and autopoietic systems. It does not consider the theory of general systems of Churchman and Ackoff. This narrow conception serves his purposes well in the context of his argument. The narrowness of this conception demands an effort of translation from us, however, when we try to apply his criticism of systems theory to the various schools of systems theory.

Habermas's deepest objection to systems theory, particularly the theory of Luhmann, is contained in his view of the tension between democracy and capitalism. He says,

Between capitalism and democracy there is an *indissoluble* tension; in them, two opposed principles of societal integration compete for primacy. If we look at the self-understanding expressed in the basic principles of democratic constitutions, modern societies assert the primacy of a lifeworld in relation to the subsystems separated out of its institutional orders. The normative meaning of democracy can be rendered in social- theoretical terms by the formula that the fulfillment of the functional necessities of *systemically* integrated domains of action shall find its limits in the integrity of the lifeworld . . . On the other hand, the internal dynamics of the capitalist economic system can be preserved only insofar as the accumulation process is uncoupled from orientations to use value. The propelling mechanism of the economic system has to be kept as free as possible from lifeworld restrictions as well as from the demands for legitimation directed to the administrative system. The internal systemic logic of capitalism can be rendered in social-theoretical terms by the formula that the functional necessities of systemically integrated domains of action shall be met, if need be, even at the cost of technicizing the lifeworld. Systems theory of the Luhmannian sort transforms this practical postulate into a theoretical one and thus makes its normative content unrecognizable (Habermas, 1989, p. 345).

Habermas sees "the technicizing of the lifeworld" as an imperative foisted upon democracy by capitalism; he berates Luhmann for cloaking capitalism's assertion of power as a theoretic necessity.

Habermas's further objections and reservations about systems theory are gathered and numbered in the following paragraphs.

1. Societies cannot be smoothly conceptualized as organic systems because their structural patterns are not accessible to [purely external] observation; they have to be gotten at hermeneutically, that is, from the internal perspective of participants. The entities that are to be subsumed under system-theoretical concepts from the external perspective of an observer must be identified beforehand as the lifeworlds of social groups. They must be understood in their symbolic structures (Habermas, 1989, p. 151).
2. It is evident on methodological grounds that a systems theory of society cannot be self-sufficient. Not only do the structures of the lifeworld, with their own inner logic placing internal constraints on systems maintenance, have to be gotten at by a hermeneutic approach, [but] the objective conditions [also] under which the systems-theoretical objectification of the lifeworld becomes necessary have themselves only arisen in the course of social evolution. And this calls for a type of explanation that does not already move within the system perspective (Habermas, 1989, p. 153).
3. Sociopathological phenomena cannot be reduced to systemic disequilibria, because what is specific to social crises gets lost in the process. From his perspective as an observer, the systems analyst can judge whether these disequilibria reach a critical point only if he can refer to clearly identifiable survival limits, as he can with organisms. There is no comparably clear-cut problem of death in the case of social systems. The social scientist can speak of crises only when relevant social groups *experience* systematically induced structural changes as critical to their continued existence and feel their identities threatened (Habermas, 1989, p. 292).

Habermas pays special attention to the systemic theory of Niklas Luhmann who has been his long-time debating partner. He holds Luhmann in high regard and calls his theory "incomparable when it comes to its power of conceptualization, its theoretical imaginativeness, and its capacity for processing information" (Habermas, 1987b, p. 354). Habermas, nevertheless, has severe reservations about Luhmann's systems theory of society. Some of these reservations are:

4. Luhmann eviscerates the lifeworld and hypostatizes it into "society." As such, the lifeworld merely forms the background for formally organized interactions; it is no longer directly connected to action situations (Habermas, 1989, p. 155). Further, Luhmann simply presupposes that the structures of intersubjectivity have collapsed

and that individuals have become disengaged from their life worlds-that personal and social systems form environments for each other (Habermas, 1987, p. 353).

5. This is an oversimplification. Luhmann's "gains in abstraction" make his theory a "tireless shredding machine of reconceptualization" which separates out as indigestible residue those lifeworld phenomena that most interest traditional sociology" (Habermas, 1987, p. 353).

6. Luhmann overlooks "the mutual interpenetration and opposition of system and lifeworld imperatives" (Habermas, 1987, p. 355). He overlooks, for example "the fact that media such as money and power, via which functional systems set themselves off from the lifeworld, have in turn to be institutionalized in the lifeworld" (Habermas, 1987, p. 355).

7. Luhmann portrays modern societies as acentric, "without central organs." For him systemic monads, such as economy, state, education, and science, have "replaced withered intersubjective relationships with functional connections" (Habermas, 1987, p. 358). These monads balance themselves out functionally; none of them can "occupy the top of the hierarchy and represent the whole." As a result, "modern societies no longer have at their disposal an authoritative center for self-reflection and steering" (Habermas, 1987, p. 358).

Habermas holds that, in his theory of *linguistic* socialization, there is an effective center for everyday communicative action even in modern, largely de-centered societies. He says that the different lifeworlds that collide with one another do not stand *next to each other* without mutual understanding. As totalities, they follow the pull of their claims to universality and work out their differences until their horizons of understanding "fuse" with one another, as Gadamer puts it (Habermas, 1987, p. 359). This fusing of horizons is a virtual center of self-understanding, projected though it might be. It gives societies an intuitive grasp of even functionally specified systems of action, "as long as they do not outgrow the horizon of the lifeworld" (Habermas, 1987, p. 359).

8. Luhmann's systems theory is a successor to the abandoned philosophy of the subject. It seeks to inherit the basic terms and problematics of the philosophy of the subject, while at the same time surpassing it in its capacity for solving problems (Habermas, 1987, p. 368). Luhmann inserts the concept of system into the slot traditionally held by the concept of the knowing subject that was developed from Descartes to Kant. Because of its abstraction, Luhmann's concept of society as a system relates to its environment in functional analogy to the way that the Cartesian subject relates to the material world. As a result, Luhmann's systems theory flounders on the same theoretical rocks as does the philosophy of the Cartesian subject.

Thomas McCarthy, in his commentary on Habermas's dual conception of lifeworld and system (1991), re-expresses some Habermasian objections to systems theory that Habermas himself no longer makes.

9. The notions of self-maintenance, boundaries, goal states, and structures are fuzzy in the socio-cultural context. In earlier writings, Habermas made clear the problems

that the system paradigm poses for empirical social analysis. Such analysis depends importantly on our ability to identify boundaries, goal states and structures essential for continued existence (McCarthy, 1991, p. 122).

In *Legitimation Crisis* (1975), Habermas argues that social systems have neither clear boundaries nor identifiable life-spans. Therefore, meaning (as a system's memory of past self-maintenance selections), rests upon problematic foundations. As McCarthy says, "The whole notion of self-maintenance becomes fuzzy, even metaphorical at the socio-cultural level" (McCarthy, 1991, p. 122). Habermas no longer stresses this point; instead, he stresses the dependence of systems theory upon action theory to obtain its definitions of limits, goals, and structures. In other words, "If it is not connected with action theory, systems theory becomes empirically questionable, a play of cybernetic words that only serves to produce reformulations of problems that it does not really help to resolve" (McCarthy, 1991, p. 123).

10. System complexity is only one point of view to judge progress in the realm of natural evolution. The degree of complexity is not a sufficient criterion for level of development. McCarthy expresses it this way: "Increasing complexity in physical organization or mode of life often proves to be an evolutionary dead-end" (McCarthy, 1991, p. 133, quoting Habermas).
11. Systems theory is still the mortal enemy of democracy. Social systems theory represents today, no less than it did in 1971, 'the *Hochform* of a technocratic consciousness' that enters the lists against any tendencies toward democratization and promotes a depoliticization of the public sphere by defining practical questions from the start as technical questions (McCarthy, 1991, pp. 133-134).
12. McCarthy reiterates the Habermasian objection that systems theories lack the empirical determinacy to predict empirical results. They show a "neglect of internal factors-structures, processes, problems" (McCarthy, 1991, p. 134). They neglect the nitty-gritty and lack effectiveness on the practical level. Systems theories are *normative-analytic* rather than *empirical-analytic*. In other words, they merely posit the goals against which they measure change, and do not establish those goals by empirical processes.

As a normative-analytic process, systems analysis can be used as an aid in planning; as such, "It provides what Habermas once called 'second-order technical knowledge'" (McCarthy, 1991, p. 134). Systems theories, then, have a strictly limited importance.

13. Systems theory is imprecise regarding feedback and control mechanisms. "One might point, in particular, to the unsatisfied requirement of specifying empirically and with some precision the feedback and control mechanisms which are supposedly at work to keep the system directed toward it supposed goals. Without this social systems theory can "achieve little more than a translation of old ideas into new jargon" (McCarthy, 1991, p. 136).
14. Social science does not need systems theory. Causal interconnections "whether stabilizing or destabilizing, functional, or dysfunctional . . . can be investigated

with comparatively meager theoretical means" (McCarthy, 1991, p. 137). What need is there for systems theory?

15. We do not need the paraphernalia of social systems theory to identify unintended consequences. Nor do we need them to study the 'functions' that an established social practice fulfills for the other parts of the social network, for these are simply the recurrent consequences of this recurrent pattern of social action for those other parts (McCarthy, 1991, p. 137).

CONCLUSION

Habermas's derivations of rules and his integration of lifeworld and systems perspectives are major accomplishments. His objections to systems theory provide a goad and a measuring rod for theoreticians and practitioners alike.

*   *   *

Chapters 2 to 5 supply historical and evolutionary background for understanding current social systems thinking. They explain trends in sociological and design science. They help us to understand Luhmann's later thinking and that of Kampis and Goertzel. They show the need for critical thinking about metaphors and cognitive maps. Finally, they provide integral components of the integrating syntheses that are developed in Chapters 18 – 22.

PART TWO

**INCORPORATING
HUMAN PARTICIPATION
INTO SYSTEMS THEORY AND
DESIGN**

INCORPORATING PARTICIPATION

Many Habermasian objections have been faced by systems theorists. Luhmann faces some of them from an adversarial position. Differentiation theory has modified functionalism to accommodate criticisms similar to Habermas's. System designers have incorporated lifeworld concerns into their methodologies.

Cybernetics faces some of these concerns following its own logic without reference to Habermas and critical theory. Evolutionary systems theory has new sciences and radically altered systems discourse in the past twenty years. Of special note are the thoughts of Kampis on information and those of Goertzel on the cognitive equation. Lakoff shows how metaphors ground our lofty ideas in our bodies. The General Evolution Research Group proposes that cultural cognitive maps are the reproductive core of social evolution. These various strands of social theory, design, and practical philosophy offer us opportunities for dialectical interaction.

The semantics for such dialogues are quite confusing. For example, critical and soft systems theorists deal explicitly with lifeworld concerns. They oppose many positions that Habermas calls "systems" because they consider those positions to be inadequate. The semantic confusion bred by such anomalies needs to be addressed. Nevertheless, systems science, if it is loyal to Churchman's ideal of comprehensiveness, must work out some engagement with Habermas and Luhmann. Many branches of systems science are not at all engaged with Habermas, Luhmann, and critical theory. In return, Habermas and Luhmann give scant recognition to any systems theories other than their own.

How do the ideas of Habermas, Luhmann, soft and critical systems, cybernetics, and general evolutionary theory influence one another? How can they cross-catalyze efficiently and in a humane manner? It would seem that forums for bringing these ideas into explicit conflict and possible synthesis are in order.

Chapters 6 to 17 explore the thoughts of prominent systems thinkers in their own contexts. This section, comprising chapters 6 to 10, brings together systemic themes that employ individual and lifeworld considerations. Chapter 6 details how sociologists have recursively modified the basic concepts of functionalism and differentiation theory to meet the objections of critical theorists such as Habermas. Chapter 7 describes the philosophical standpoint behind soft systems theory and its practical application in the soft systems methodology. Chapter 8 describes how critical systems theory advances soft systems by focusing on the power imbalances that skew the designing processes, and by devising procedures to counteract them. Chapter 9 explicates major themes of Banathy's work that deal with designing social systems. Chapter 10 presents Warfield's science of generic design and methodology of Interactive Management.

The third section, chapters 11 to 17, presents themes integral to evolutionary systemic thought. The final section, chapters 18 to 23, synthesizes themes from the preceding chapters under the topical headings of the Practice and Ethics of Design, the Structure of the Social World, Communication, Cognition, and Epistemology. In this way, we create a modest forum for cross-disciplinary interaction.

CHAPTER 6

DIFFERENTIATION THEORY

ABSTRACT

Sociology is a forum in which dialogue between systems theory and critical theory is ongoing. Sociologists now elaborate the main systems theory of sociology, differentiation theory, in ways that satisfy salient points in the critiques made against them. They situate differentiation as the master trend of social change, but they describe patterned departures from that trend and describe several countervailing forces that work against uniform differentiation.

FUNCTIONALISM AND CRITICAL THEORY

The historical context of the Habermas/Luhmann debate mirrors the historic conflict between functionalism and critical theory: Habermas represents critical theory, and Luhmann represents functionalism.

Functionalism, a dominant school of sociology, describes societies as natural and biological processes. Durkheim, for example, saw a parallel between social and biological evolution. He said, "The law of the division of labor applies to organisms as well as societies. It may even be stated that the more exalted a place in the animal hierarchy an organism occupies, the more specialized its functions are" (Durkheim, 1984, pp. 2-3). Functionalism, then, explains social processes as natural ones and sees society functioning as an organism. It models societal function on the biology and evolution of organisms.

The origin and nature of functionalism are explained by Giddens as follows:

> A naturalistic perspective converged with an advocacy of functionalism in social science, informed by the general notion that the mechanics of biological systems have close affinities with the operation of social systems . . . The relation between naturalism and functionalism for a considerable period defined what was regarded by proponents and critics alike as the 'mainstream' version of social science (Giddens, 1987, p. 55).

Functionalism is still the dominant underlying paradigm of empirical social researchers. It no longer enjoys universal high theoretical repute, however, because of the seminal criticisms of Wittgenstein (1953), Habermas (1989), Kuhn (1970), and various post-modernists.

Reaction To Criticism

Theoretical sociologists had embraced functionalism as their "research programme" (Lakatos). How were they to react to philosophical challenges to their main theory? Jeffrey Alexander broaches this question (Alexander and Colomy, 1990) in a general formulation: "What does a general theory do when it is challenged?" (p. 9). He answers, "They [its defenders] differentiate between core notions, which are positions considered essential to the theory's identity, and others that are more peripheral" (p. 10).

In this way, the theory of social functionalism has twice transformed itself in the past forty years. These transformations have been synopsized by Paul Colomy (Alexander and Colomy, 1990). In the mid-1950's, functionalism was attacked within the sociological community. It was accused of being conservative, resistive to change, oblivious to internal conflict, and exceedingly abstract.

> These critiques claimed that structural functionalism, premised on value consensus and the internalization of norms, was inherently conservative and could not account for either social change or conflict. They also asserted that action theory was excessively abstract and could not be applied empirically (p. 466).

Stung by these criticisms, some functionalist scholars refined their systematic theorizing and their empirical research.

Differentiation Theory (I)

"Differentiation theory was the product of this collective intellectual endeavor" (p. 466). In its early formulation, differentiation theory rested on three fundamental tenets: (1) differentiation is the master trend of societal evolution; (2) it is directed by societal needs; and (3) it increases the adaptation, generality, and inclusivity of society.

First, this approach described a "master trend" of social change. The master trend of differentiation identified the replacement of multifunctional institutions and roles by more specialized units as one of the most theoretically and empirically significant aspects of modern social change . . .

Second . . . it typically invoked a societal need explanatory framework to account for [this] transition . . . Often that societal need model was tied to the idea of structural strain.

[Third,] the institutionalization of more specialized structures increases the adaptive capacity of a social system of subsystem. In addition . . . high levels of differentiation are correlated with value generalization, greater inclusion, and the emergence of specialized integrative institutions (pp. 466-467).

With this formulation, theorists like Parsons (1966, 1971) and Smelser (1959, 1968) met the criticism that functionalism could not analyze change.

Criticism Of Differentiation Theory

The critics of functionalism were not pleased, however, with its reincarnation as differentiation theory. They attacked differentiation theory on four fronts: vagueness, idealism, unfounded assumptions, and conservative bias.

This initial version of differentiation theory instigated, in turn, a second round of criticism which indicted the theory on four grounds. First . . . the functionalist theory of change lacked historical and empirical specificity and presumed, unjustifiably, that modernity is associated with a radically complete break from the past . . . Second . . . the functionalist model of differentiation suffered from three maladies: an idealist bias which undermined the explanation of transitions between evolutionary stages; a failure to examine the impact of concrete social groups; and a neglect of power and conflict . . . Third . . . differentiation theory falsely assumed that structural change invariably increases systemic adaptiveness, was unable to describe specific integrative mechanisms, and tended to treat the absence of integration in highly differentiated systems in a residual way . . . Finally, the ideological charge that functionalism and its theoretical progeny, including differentiation theory, were inherently conservative was more forcefully articulated (p. 468).

ADDING PARAMETERS OF CHANGE

Sociologists have responded to this new round of criticism by (1) increasing the empirical scope of differentiation theory, (2) attending to concrete groups and power relationships, (3) recognizing the variable consequences of differentiation, and (4) formulating bases for a critical differentiation theory. The authors in this volume examine degrees, rates, and sequencings of structural change. They describe not only differentiation but also patterned deviations from it. They regard differentiation "as one element in a larger pattern of change" (p. 472). Colomy describes some of their advances below:

> First, they elaborate the empirical scope of the original model, supplementing description of the master trend with the identification of patterned departures from that trend. Second, they open up the explanatory framework of differentiation theory, specifying the cultural and structural parameters conditioning structural differentiation in a more multidimensional manner, and giving greater attention both to how concrete groups affect the course of change and to the role of power, conflict, and contingency in structural differentiation. Third, the original characterization of the effects of differentiation is modified, with reintegration and greater efficiency now being treated as a subset of a much larger array of possible consequences. Finally, the still widely accepted ideological charge of conservatism is challenged and the rudiments of a "critical modernism" are outlined (pp. 468-469).

In their reports and analyses of detailed empirical research, the authors in this volume significantly modify earlier differentiation theory.

Smelser

> Smelser's discussion of English mass education in the first three-quarters of the nineteenth century provides a rich historical-sociological account of what can be called "blunted differentiation." During this period, the establishment and institutionalization of primary education for children of working-class and poor parents were impeded by a concatenation of social forces that pulled mass education in opposing directions (p. 470).

In this case, educational reformers promoted education, but they met a backlash fueled by religious rivalries, suspicion of government intervention, and patterns of economic restraints. Universal primary education did not become a reality until time changed the economic and political climate.

Colomy

> In my [Colomy's] study of political change in the antebellum United States, I introduce the term "uneven differentiation" to designate the varying rate and degree of differentiation of a single institutional sector or role structure within a society (p.471).

In this case, political parties developed unevenly in different states of the antebellum United States. New York had the most developed political parties, while South Carolina retained "a diffuse gentry ruling class, a minimal degree of mass party development" (p. 471), and classical norms of governance.

Alexander

> Alexander . . . describes the contrasting developmental paths of France, Germany, and the United States toward a more differentiated news media (p. 471).

In this case, different paths of differentiation were followed, leading, ironically, to different forms of structural integration.

Other Observations

In addition to these case studies, Colomy identifies a "double-movement" of change-consolidation by new institutions and backlash by existing power centers. Lechner "demonstrates how value generalization . . . spawns "reductionist" movements designed to restore meaningful order on the basis of an absolute, substantive value principle" (p. 473). The authors in this volume do *not* find "a perfect inverse relation between advancing differentiation and eroding traditionalism . . . Vestiges or enclaves of traditionalism persist even in quintessentially modern societies" (pp. 473-474).

These essays locate intractable parameters that "precipitate change and establish broad limits on the types of possible transformations" (p. 474) that are available to a society. They identify voluntaristic elements that affect the course of differentiation. In brief, they show that considerable variation is in play within the parameters of overall differentiation.

These studies also determine that "integration is not the only or even the necessary outcome of differentiation" and that "integration and conflict . . . are not mutually exclusive" (p. 486). In addition, they present incisive specifications for how integration works.

POWER RELATIONSHIPS

What, then, "determines which particular option, from the delimited set of alternatives, is selected in response to structural contradictions and strains?" (p. 477). In this situation, concrete groups mobilize, form coalitions, obtain resources, and control conflict. Eisenstadt uses the term "institutional entrepreneurs" to describe "small groups of individuals who crystallize broad symbolic orientations, articulate specific and innovative goals, establish new normative and organizational frameworks . . . and mobilize . . . resources" (p. 476). He maintains that "the very occurrence of structural change and the specific direction it takes are shaped by the activities of institutional entrepreneurs" (p. 477). Colomy expands on Eisenstadt's idea by identifying "strategic groups [which are] collectivities whose members assume leadership roles in directing the course of institutional development" (p. 478).

When such leadership groups encounter opposition groups, the interplay of conflicting interests helps to decide the details of how structural change comes about. The presence of such "political" interests suggests that the "problem-solving explanatory model" of structural change needs to be supplemented with an "interest model" because "differentiation is partially contingent on the relative strength and position of contending groups" (p. 480). The struggle between such interest groups "is multidimensional and involves a battle waged over both conflicting material and ideal interests" (p. 481).

CRITICAL DIFFERENTIATION THEORY

Differentiation theory in its latest incarnation establishes bases for constructing a "critical differentiation theory." It locates an "Archimedian point" that is formed by the juncture of "institutionalized individualism" and collegial decision-making. From this point, differentiation theory can support movements that foster individual and collegial growth. It can also oppose movements that do not.

CONCLUSIONS

The progress of differentiation theory is the result of a fruitful dialogue with critical theorists. This progress bodes well for future effort to expand dialogue between other branches of systems theory and social theorizing. Differentiation theory now upholds standards for the valid application of systems (differentiation) theory to society that satisfy many of the demands of critical theory. It proposes that analyses of structural change must:

1. situate differentiation as one element in a larger pattern of any society's structural change;
2. acknowledge that differentiation is the master trend of social change, but that there exist patterned departures from that trend;
3. discern the societal needs and stresses that drive structural change;
4. specify the cultural and structural influences that control structural change;
5. recognize the influence of individuals and concrete groups in influencing the course of differentiation;
6. recognize the roles of power, conflict, and contingency in shaping differentiation;
7. recognize that differentiation is often only partially accomplished;
8. recognize that there are other possible results of differentiation than reintegration and greater efficiency;
9. identify the mechanisms that enable the integration of social change and acknowledge that integration is often accompanied by ongoing conflict; and
10. foster social movements that further institutionalize individuality and collegiality.

CHAPTER 7

SOFT SYSTEMS THEORY

ABSTRACT

This chapter considers Soft Systems Theory as it is expressed in the works of Churchman, Checkland, and Fuenmayor. Soft systems developed from roots in engineering approaches to systems problems. It deals with the "messes" of vague uneasiness in a social organization that escape the net of "hard systems" approaches. It considers ethical, heuristic, and epistemological aspects of consensual decision-making.

Churchman argues for a science of values; explicates the nature of design; and defends the systems approach against the approaches of politicians, moralists, religionists, and aesthetes. Checkland develops the Soft Systems Methodology for dealing expeditiously with complicated human organizational situations that do not yield to the hard systems approaches of organizational design. Fuenmayor spells out the epistemological and ontological rigor that underlies soft systems theory,

HISTORICAL BACKGROUND

The original application of systems to organizations took the form of "hard systems." With hard systems, one identifies the problem, brainstorms methods to correct it, evaluates those methods, picks the best one, and implements it. Systems, in this sense, is the long-used methodology of engineers. Hard systems approaches require accurate descriptions of problems. They work well to optimize results when a problem can be defined but poorly when the "problem" is an ill-definable sense of something wrong.

In the 1950's and 60's, Systems theorists attacked such ill-defined uneasiness in organizations with the tools they had, while innovatively trying new approaches. System consultants and theorists like C. West Churchman, Russell Ackoff, and Peter Checkland eventually designed a new systems method for these ill-defined situations: soft systems.

Soft systems is a methodology that organizations use in order to learn. In using soft systems, organizations open themselves to a variety of viewpoints. They encourage diverging views in order to clarify goals, objectives, and procedures through cycles of examining practice, theorizing, applying theory, examining results, retheorizing, and reapplying.

CHURCHMAN

In *Prediction and Optimal Decision* (1961), Churchman considers the philosophical issues of a science of values. He starts by trying to assess the role of management scientists in the context of democratic decision-making. These scientists are often called upon to assess complicated problems and recommend solutions, but they realize that they are not the decision-makers. How can they perform their role without subverting the democratic process? He poses the problem as follows:

> We seem driven to accept these apparently inconsistent consequences of democratic decision-making:
> a. Ultimately all interested persons should be partners in social decisions,
> b. Good decision-making depends on the selection of relevant information and its proper utilization,
> c. Only a few people have the time and means to obtain relevant information and to ascertain its proper use (p. 7).

He suggests that democratic societies usually delegate to scientists the jobs of obtaining information and ascertaining its proper use.

Forming the Prudential Ought

How, then, are management scientists able to perform their tasks without infringing on the authority of democratically elected decision-makers? They must go beyond fact-gathering and stratagem-forming. They are obliged to find out what the decision-makers want. They must find out not only how to accomplish results but, also, what results are desired. To this end, they must have a standard for determining how much the decision-makers want the results.

To measure how much a person values an outcome, a standard can be derived from that person's past behavior. If the scientist can discern what decisions have been made in the past, then he or she can propose consistent behavior in novel and problematic situations. In this way, the scientist can devise a standard of the *prudential ought*. This standard for value measurement is: "When we say 'X ought to do A in this environment' we mean 'X would do A if he were undisturbed, could think properly, and were not subject to external forces'" (p. 18). And, " 'X ought to do A' means 'X would do A if he were aware of all alternative actions and knew the probabilities of the possible outcomes'" (p. 19).

In accord with this prudential ought, the obligation of the management scientist to the decision-maker has two parts. "He must try to improve the decision-maker's state of knowledge, and he must try to find out how the decision-maker would act if his ignorance were removed" (p. 19).

From the Prudential to the Moral Ought

Using this prudential ought, Churchman derives one likely standard of a *moral ought*. He bases it on his impression "that ethical judgments are really judgments made for the sake of future generations" (p. 22). He does not, however, base the standard on the simple idea of doing what future persons might want, nor does he derive it from the past. He bases his standard on the idea of converging, yet dialectical, interests.

At one time, Churchman had advanced a simple standard of convergence in which one predicts "changes in men's interests, but expects a convergence of these interests toward a common set of ideals" (p. 24). At a later time, he finds this standard to be simplistic because "it ignored the reasonable thesis that men do not want to agree" (p. 25). He finds that the ideals of agreement and disagreement are contrasting but complementary standards for making decisions. He puts it this way:

> The rationalist will argue that men disagree in order-in the long run-to find ways of agreeing. The individualist will argue that men agree-in the long run-to find ways of disagreeing. The rationalist looks at human history as a trend toward less conflict, more consolidation, but realizes that such a theory of history can only be sustained if men disagree both with others and with themselves. The individualist believes that men must agree on methods of survival and more profitable living in order the better to express their disagreements.
>
> The viewpoints are not contradictory because both theories see no end to the story. Within this view of eternity, there is no distinction between means and ends. Science can interpret human history as a trend toward peace in which conflict is a necessary sustaining force or a trend toward conflict in which peace is a sine qua non of survival (p. 25).

Within this ethical universe, one honors individuality *and* consensual decision-making. One respects both conflict and agreement.

The Nature of Design

In *The Design of Inquiring Systems* (1971), Churchman first defines "design" and then turns to the project of designing a theoretical computer that would know reality as we do. To determine how we know reality, he considers the thought of the following seminal thinkers: Gottfried Leibniz, John Locke, Immanuel Kant, George Hegel, and Edgar Singer. He starts by asserting:

Design belongs to the category of behavior called teleological, that is, "goal-seeking" behavior.   More specifically, design is thinking behavior which conceptually selects among a set of alternatives in order to figure out which alternative leads to the desired goal or set of goals (p. 5).

He then states, as a first approximation, the following characteristics of design:

1) It attempts to distinguish in thought between different sets of behavior patterns.
2) It tries to estimate in thought how well each alternative set of behavior patterns will serve a specified set of goals.
3) Its aim is to communicate its thoughts to other minds in such a manner that they can convert the thoughts into corresponding actions which in fact serve the goals in the same manner as the design said they would (p. 5).   Design requires communicating with another designing mind such as, a computer, a philosophical system, another human being, or a research group. To this end, design has another characteristic:
4) It strives for generality by making clear its methodology.   For this reason, the designer tries to delineate the steps he or she uses in producing designs. "Once the designer has had some success in this fourth effort, he can say that he can tell *why* a design is good, in addition to telling the *what, when,* and *how,* which the first three efforts attempt to accomplish" (p. 6).

In addition, one can design a system only by working within the context of other systems. One has to decide what is the largest environmental system one is going to consider, and what are the smallest components that will not be further analyzed into systems. This leads to design characteristic number

5) "The systems designer attempts to identify the whole relevant system and its components" (p. 8).

Epistemology

Inquiry, like design, is thinking behavior. In inquiring, we seek knowledge either as individuals or as groups. In designing an inquiring system, then, we are trying to design a system (perhaps a computer) that thinks as we do. To do this, we need to determine how we think and know.

Churchman considers how Leibniz, Locke, Kant, Hegel, and Singer theorize about how we know. He describes how these thinkers understand us as inquiring systems.   He believes that we can best understand our thinking by critically combining their perspectives and methodologies.

He characterizes the epistemology of **Leibniz** as a system of fact nets that include streams of sentences (or charts), some of which may be true, others false, others irrelevant. The citizen's problem is to converge on one story that seems to hold them together in the best manner (p. 176).

According to **Locke,** we are born with a mind that is a tabula rasa. We receive our original ideas as they are written on our slate by reality. Then we develop more complex ideas by ourselves.

The rightness or wrongness of our ideas is then determined by consensus among thinking people. The careful recording of the steps one has taken to acquire a certain knowledge and the careful examination and/or re-enactment of those steps provides the consensus in a scientific community. "The Lockean inquirer displays the 'fundamental' data that all experts agree are accurate and relevant, and then builds a consistent story out of these" (p. 177).

Kant said that we impose reality upon our worlds through the activities of our minds. We construe things in terms of space and time; we organize them in terms of innate categories that exist in our minds. We grasp objects through the representations that we have of them. "The Kantian inquirer displays the same story from different points of view, emphasizing thereby that what is put into the story by the internal mode of representation is not given from the outside (p. 177).

Hegel introduces passion, drama, history, and progress into epistemology. He says that societies pull themselves together through coherent worldviews that contain their main action/ideas-their theses. A dominant thesis spawns opposition in the form of its "deadliest enemy." This deadly enemy observes the dominant society from a self-consciously superior position. It produces antitheses that attack dominant society on all possible fronts. Conflict ensues.

> The revolutionary looks down upon the reactionary. The reactionary in his conviction can only think that the revolutionary is crazy or criminal; he must utterly reject him as an unnatural evil or a meaningless mind. But the revolutionary understands the nature of the reactionary full well; for him the reactionary's conviction is based on a natural selfish greed and hypocrisy (p. 173).

Next comes another act of one-upmanship. Another observer focuses not on the conviction of the antagonists but on their very opposition. He or she passes judgment on the conflict and settles it by building

> a new world view in which the nature of the conflict is understandable, but which shows that at a higher level the conflict is merely one aspect of reality and not the critical aspect. The conflict in fact is devoured by the higher-level worldview (p. 176).

In the context of the Hegelian dialectic, then, an inquirer interested in discerning truth uses the same data as other inquirers but constructs a dialectical model.

> [He or she] tells two stories, one supporting the most prominent policy on one side, the other supporting the most prominent policy on the other side. The teleological issue is: Which method of telling the story will produce the optimally informed citizen when each is constrained by the same cost and time resources? (p. 177).

In this way, Hegelian storytelling forces us to go beyond routine rationality. It challenges us to give up explicit knowledge for implicit knowledge. It values passion and conflict because opposition produces ideas. The underlying life of the dialectic

> is its drama, not its accuracy . . . Drama is the interplay of the tragic and the comic; its blood is conviction, and its blood pressure is antagonism. It prohibits sterile classification. It is above all implicit; it uses the explicit only to emphasize the implicit (p. 178).

Singer was a philosopher and metrologist, a student of measurement, during the first half of this century; and he was Churchman's teacher. He proposed ideal standards of comprehensiveness and a method of "sweeping in" descriptions from divergent viewpoints so that a representation of something would ever-more-closely approximate the ideal.

Singer starts by describing scientific methods of measurement that allow for patterned changes in both measuring standards and the things that they measure. He describes, for example, the everyday process of measuring something with a yardstick. He recounts the ways that measuring rods have been made more accurate. He then describes how to "sweep in" other factors like the effects of temperature, humidity, motion, the phases of the moon, an so on. This sweeping-in process increases the accuracy and reliability of the measuring process.

In designing standards for measuring the truth of observations, Churchman works in a similar way. He gradually introduces Leibnizian, Lockean, Kantian, Hegelian, and numerous other observations about inquiring systems. In this way, he honors a realization "that human knowledge does not come in pieces: to understand an aspect of nature is to see it through 'all' the ways of imagery" (Churchman, 1971, p. 298).

According to Singer, when all is going well in the realm of theory and practice, "then is the time to rock the boat, upset the apple cart, encourage revolution and dissent" (p. 199). Singer does not advocate the resolution of the ensuing philosophical dispute, but its intensification.

Singer's theory of progress is dialectical, not linear. It sees nineteenth century optimism as only half of the process.

> On the one side . . . is production-science-cooperation, the trilogy of nineteenth-century optimism. The progress toward this trilogy is toward a world of enlightenment where men have the means to live out their individual lives in their unique ways, without having to disrupt the lives of others (p. 202).

The other side is alarmed by the harm done to the environment, human families, and human psyches. It ridicules "fat science proclaiming it will save the world while it odiferously defecates in public" (p. 203).

In the implicit world that is captured between thesis and antithesis, there are no explicit rules that guarantee success and/or esteem. The implicit world of intuition and perilous action is the realm of heroes and heroines. In this world, people stumble, fall, and find themselves. Sometimes they are notably successful for the good of their societies.

The Systems-Approach Hero

In *The Systems Approach and Its Enemies* (1979), Churchman defines his efforts to design an ideal inquiring system as "a theory of general systems." He contrasts this approach with the approach of General Systems Theory, which he defines as "a general theory of systems" (p. 51). Following Churchman's lead, soft systems theorists are interested in how we know, plan, make decisions, and ensure that those decisions are ethical. Their interest in how systems interact is grounded in their experience with organizations.

In his continuing quest to define an ideal inquiring system, Churchman adopts a semi-autobiographical approach. He identifies himself with the hero, described by Singer, "who wants the world to act sanely . . . [and he is angry because] his enemies are righteously bent on destroying the peaceful world of humanity" (p. 5).

His enemies are people who attempt to "solve" systems problems head-on by relying upon politics, morality, religion, and aesthetics. These enemies commit the "environmental fallacy." They fasten their attention on one thing and prescribe simple solutions like "outlaw the use of marijuana," and neglect the historic lesson called "prohibition." Even environmentalists and ecologists are his sometime enemies.

> The fallacy of the environment, says our hero, is far more serious than pollution, depletion of resources, and the like. These so-called ecological phenomena are easy enough to see, but the insidious pervasiveness of the fallacy of the environment seems not to be seen at all. Nothing the prophets of doom predict about population growth can equal the disasters that lie ahead as a result of our repeatedly basing large-scale social policies on the environmental fallacy (p. 6).

The corrective to the environmental fallacy, described by our hero, is: "the rather bland prescription to look at matters in as broad a way as possible . . . [This] bland prescription questions the whole existing enterprise of science, which is based on the publication of findings arrived at by 'acceptable' research methods" (p. 6).

History of the Systems Approach

Churchman says that the systems approach is "based on the fundamental principle that all aspects of the human world should be tied together in one grand rational scheme" (p. 8). He traces the history of this approach as a learning journey, starting with the *I Ching*, the *Bhagavad-Gita*, and the pre-Socratics.

From Aristotle, who "attempted to put all conceptualization in teleological form," he derives the lesson: "In order to conceptualize social systems, one has to think of conceptualization in general, and therefore the ways in which we think about the psychological and social worlds" (p. 39). If we adopt the "Aristotelian notion that the two domains of nature, the physical and the social-psychological, are non-separable . . . [then] we can link the findings of all the disciplines in order to explain what we mean" (pp. 39-40).

The heroic truth-seeker finds an ethical dimension in the Kantian concept of logic as "the basic justification of what we take to be true or false" (p. 54). He concludes that some arguments can and should be justified, while others can and should not. In his systems approach of comprehensiveness, he assumes that the facts of nature comprise a unified whole and, therefore, that those facts are to be "explained by what can be called the 'fact net' of nature's world" (pp. 54-55). The meaning of logic, then, "is roughly displayed by the idea of an explanatory fact net" (p. 55).

The Guarantor of Destiny (GOD)

In many traditional systems approaches, ultimate logic lies in the mind of God that contains the explanation of the whole system. God, in this instance, is conceived as the guarantor of systems designs in general and the inquiring system in particular. Churchman identifies three approaches to this question of God as guarantor that he identifies with Augustine, Spinoza, and today's systems thinkers. For Augustine, "God is the designer of the real system, as well as its decision-maker" (p. 41). For Spinoza, "God *is* the whole system: He is the most general system" (p. 41). For many of today's systems thinkers, the idea of god is only implicit in their ideas of progressive systems designs. For them, "god" is just "the best systems design we humans can create at any period of time" (p. 43).

> Such a god is incapable of finding a perfect design (to say the least). And god is not static; if we are hopeful, we can say that he learns or evolves, and that later on in his life he'll know a lot more than he does now (p. 43).

The logic of the "scientific laboratory," which was formulated by Galileo and others, replaces God's design as the guarantor of truthfulness with its methodical testing of hypotheses. This laboratory rests on the metaphysical assumptions that "nature is neither capricious or secretive" (p. 57). It also rests upon an epistemological assumption that its laboratory work can be shielded from social environmental influences. Upon reflection, however, it is obvious that "the planning laboratory cannot be decoupled from the rest of society" (p. 59). In particular, the bias of the scientific researcher, who collects "just those data that appear relevant, and can be obtained objectively . . . even though these data are not 'basic' in terms of human lives," (p. 62) must be confronted.

The Systems Approach

Churchman wants to develop "a methodology in which human bias is an essential aspect" (p. 62). He asks whether such a methodology is "scientific." He answers, "No, if we doggedly stick to the assumption that the classical laboratory *is* the basis of science. Yes, if 'science' means the creation of relevant knowledge of the human condition" (p. 62).

The methodology of an inquiring system that explicitly recognizes the bias of researchers

> is based on a schema that can be modified-for example, by expansion-as the planner learns more about reality, and especially about the reality of the human being . . . The schema . . . pictures the world of the decision maker as stages in time. First there is an action: at the next stage are the "outcomes" of the action [some good "ends," some unforeseen "side effects]" (p. 63).

Planners can take different postures in this scenario. They may accept the decision-maker's goals and simply engage in problem solving. They may accept those goals but put some constraints upon their proposed solutions (legality, feasibility, and so on). Or they may refuse to limit themselves to narrow considerations.

They might ask the basic question: "Who should be served by the social system and in what way?" (p. 64). In other words: who are its appropriate clients? The answer "may (probably should) include future and past generations" (p. 65). If that is the case, the questions: "who should design and how? and, who should plan and how? follow inevitably" (p. 65).

When such broad questions are involved, the ends of an inquiry are no longer identified "in terms of the intentions of the decision-makers who pay, but rather in terms of a general ethic" (p. 65). In this situation, "the 'inquiring system' becomes a 'systems approach,' and planning is 'ideal planning'" (p. 65).

Churchman points out that "the ambitions of the systems approach are enormous" (p. 65). They go beyond the ambitions of the classical laboratory. "Its adherents see no reason why all matters of human concern should not be tied together in one grand imagery of purposive behavior" (p. 67). Systems theorists, then, seek to tie all our goals together in a meaningful schema, and they try to accomplish that task without resorting "to mysticism or some deep inexplicable human essence" (p. 67).

Ethics

If an ideals approach to systems is to come about, it needs to confront issues of ethics. Churchman does this by considering the work on ethics done by Kant and Bentham during the 1780's, and the work of Singer and Jung who dealt with ethical matters during the 1940's and 1950's.

Bentham proposed that actions were good if they were useful in preventing pain and enabling pleasure. He proposed the principle that governments existed to minimize the pain of their citizens and maximize their pleasure. To discern the appropriateness of a piece of legislation, then, one simply totals the pleasures that might derive to people and subtracts the pains.

Kant radically dissents from Bentham. He says that "power, riches, honor, even health, and . . . happiness" can be perverted. Only Good Will "can be called good without qualification" (quoted by Churchman, 1979, p. 122). Good Will does not operate by conditionals; it is categorical. Kant's categorical imperative expresses the big idea that what your action expresses deserves to be considered as a universally valid law. Churchman paraphrases this imperative as follows: "If you can tolerate the burden of having your big idea become a universal law, then your big idea and the action are moral. If you cannot tolerate it, then the big idea and the action are immoral" (p. 123).

Jung elevates "the process of individuation to the pinnacle of the Good" (p. 130). He says, "individuation is indispensable for certain people, not only as a therapeutic necessity, but as a high ideal, an idea of the best we can do" (quoted by Churchman, p. 132). Jung's version of the categorical imperative, therefore, is "Thou shalt never interfere with the process of individuation of thyself or another" (p. 132). He makes individual morality the keystone of social morality.

Singer espouses a morality based on social ends. For him, "if we are to search in a rational manner for the good, it must be found in our purposes" (p. 135). He proposes a standard of ethics based upon the scientific notion of an "ideal." For him, "an ideal in principle is unattainable but in principle can be approximated within any prescribed limit" (p. 135). According to Singer, "the statements we make in response to questions about the natural world are always approximations" (p. 135). The actual *answer* to any question is beyond precise statement and is always an ideal. For that reason, "the real is always an ideal" (p. 135).

Now, since we humans are goal-seeking creatures, we also seek the power to obtain those goals. On the basis, of this principle, an "enabling" value theory can be constructed by categorizing the essential things "that increase an individual's chances of gaining what he wants" (p.137). The three necessary things are:

(1) a richness of means at his disposal-that is, "plenty"; (2) an awareness of the appropriate means to select-that is, "knowledge"; and (3) a desire for goals that are consistent with the goals of others-that is, "cooperation" (p. 137).

The need for cooperation is problematical for a person whose only interest is self-interest. "Why should I pursue these ideals to any extent beyond what serves me best? Specifically, why should I seek the betterment of the lives of future generations?" (p. 139).

If one desires the contentment that comes with a progressive life, one understands this need. One experiences a dissatisfaction with oneself and one's surroundings, and an urge to renew vital processes. Because of this dissatisfaction, a person may become actively involved in systematically designing a future worthy of our past history and our descendants. That person becomes a hero. In the words of Singer, "to win contentment, one needs all the qualities of the hero" (quoted in Churchman, 1979, p. 139).

Churchman suggests that "the quality of mind that, for Jung, makes a person seek his own individuation is the same quality that, for Singer, makes him a 'hero'" (p. 139). The psychic force that drives the hero to do ideal planning "is the very same psychic force that drives one through the 'stages' of individuation" (p. 139). In summary,

> Jung's basic exhortation is toward the psyche as a reality and toward the individual psyche, while Singer's is toward the collective "mankind" and toward the most general psyche. Jung's hero seeks the completeness of the Self, Singer's the completeness of all Selves (p. 141).

The Story Thus Far

Our disgruntled hero has organized his life around designing an inquiring system for himself that is as comprehensive as he can make it. He has developed a clear idea of his own ethics, but he has to admit that he does not have an overwhelmingly persuasive case for his concern for future generations.

> He may be overwhelmed by the image of the possible future of humanity going to its living hell, disease-ridden, malnourished, energy-less. But why should such an image disturb him? There may be psychological causes why it *does* disturb him, either archetypal or in his own past history. But why should it? (p. 143).

In addition, the morality of the heroic mood is not universal. The ideal planner finds himself in a we/they situation. "The hero dearly wishes to be at one with humanity, just one of the boys/girls, but can't help observing that a vast number of people are different from him/her in their moral concerns" (p. 144).

In his strangely alienated position, the hero can still comfort himself or herself with an image of human solidarity.

> He may take some comfort in the reflection that his vision, though not universal, is essential for the survival and the betterment of humanity, and that the vast majority of humans, taken collectively over the ages, agree that if betterment is possible, it should take place. [Still the hero has doubts. He] wants humans to accept some version of holism in their lives. But why? Why should humans obey the prescriptions of rationality? And there is another "Why?" Why are the rational types urging more rationality on the others?   What is there about these people that makes them continuously fight battles to overcome perceived stupidity? (p. 155).

A key to an accommodation to these "Whys," if not an answer, is provided by the Socratic "Know thyself." "The rational planner cannot exclude himself from the whole system, else his holism is unholy." (p. 155). Our hero must immerse himself in the reality of his enemies and recognize the limits of his rationality.

As a first step, he tries to incorporate the insights of groups that he calls "enemies of the systems approach" into that approach. He enters this discussion with the conviction that he has to cast as wide a net as possible. In the midst of the widely diverging information that he gathers with his net he intends to use three closely related strategies: "Bringing different observers more closely together is one strategy. Another is to use the divergences to suggest another area to investigate. A third is to decide that the divergences show the irrelevance of the information" (p. 147). He is convinced that "an observer is not 'objective' if he sticks to but one of these strategies" (p. 147).

The Enemies

The enemies of the systems approach are people who do not use rationality in the manner devised by our hero. For convenience, Churchman gathers them under the categories of politics, morality, religion, and aesthetics. People who think in these categories do not think holistically; they think and feel *issues*. They become emotionally engaged and think principally in a mode of defending preconceived convictions.

Churchman builds his idea of *politics* around the *vita activa* in the Greek *polis* (city-state) as Hannah Arendt described in *The Human Condition.*. "I've taken this idea and called it 'making polis' around an issue" (p. 157). A family becomes polis, for example, if it gathers together around a common concern like getting a child through school. Groups and counter groups become polis over issues like pollution, anti-development, gay rights, and so on.

There are several aspects of polis-making that differ from the rational systems approach. First, polis making takes shape around an issue. If it succeeds, it may go out of existence with a celebration. If it does not succeed, it may go on for a long time. Second, success is measured by feelings of satisfaction. Third, "a definite, often hard attitude toward information occurs in the life of polis" (p. 157). Because of their issue-commitment, polis-formers take an ideological stance and become incapable of seeing facts that might undermine their sense of rectitude.

Churchman dramatizes the plight of the hero by considering "the tragicomedy of the systems hero as he encounters polis at work" (p. 158).

> He wants to tell the conservationist that conservation by itself means very little: whether we conserve or use depends on aspects of the larger system. But the conservationist's joy comes from saving a redwood forest from destruction, and *not* from planning the whole future of the state or nation. The issue dissipates when it's put in the larger context . . . To a keen conservationist, the real enemy is in the counterpolis, the guy who says, "Once you've seen one redwood, you've seen them

all." He cannot even hear the voice that tells him to put conservation in the proper context (p. 158).

Because of their ideological stance, polis-makers consider our hero an enemy. They think that he is trying to co-opt them.

Moralists espouse values that our hero holds dear: "fairness, equity, treating humanity as an end withal" (p. 198), but they resist any attempt to balance values by cost-benefit analysis or some other means. They insist that there are no trade-offs when it comes to robbery, suppression, or murder because those acts are sins. Moralists repudiate any attempts to rationalize or make dialectical decisions. For moralists, morality is a universal feeling and rationality is largely irrelevant.

Religion demands that the systems approach recognize that we are not the only designers of systems. The reality of God needs to be recognized. God needs to be worshipped by humans, who are basically non-rational beings, that obtain their direction from revelation and follow emotional urges. In fact, "the hero's persistent urge to improve the human condition is simply the hero's mode of worship" (p. 199).

The *aesthetic* person demands that our hero look toward the quality and not just the content of his planning. One should consider the psychological impact of proposed changes. In particular, our hero should take into account the uniqueness of every individual. At this juncture, the hero perceives the trap. "Everyone's uniqueness is a world in itself, incomparable with any other uniqueness" (p. 199). Therefore, "Gone is tradeoff. Gone is adding up values. Gone is any sensible way of assessing change" (p. 199).

The Retreat to a Vision

In his encounters with these enemies, the hero tries to sustain his vision of comprehensive rationality, but he is done in at every turn by appeals to comprehensiveness that undermine rationality. In these straits, he turns to visions of progress.

> Kant's vision included the perspective of humanity gradually creating a world in which virtue and happiness begin to coincide-that is, where virtue produces happiness. For this to happen, the vision needed a guarantor, God, the possibility of free choice, Freedom, and an opportunity for the endless struggle, Immortality. But is there any reality in such a vision? Kant calls the three conditions "postulates," as though in effect there was a demand for them, else the whole plot collapses (p. 201).

Singer improves on Kant's vision in two ways. First, he enriches his vision with a "heroic mood," which "stirs us out of our humdrum life of satisfaction to search for new ways to follow the pathway of progress" (p. 201). He finds evidence for this mood in tragedy, comedy, and heroic music. Second, he likens his vision to ideal images that are employed in physical science to approximate measurements.

Just as ever-more-precise measurements reduce the error-factor in a measurement so that they converge on the ideal of zero-difference without ever attaining it, so by using the technique of sweeping in alien perspectives, one converges on an answer. "Hence the real becomes the ideal, because ideals are desirable ends (signposts constantly beckoning us on) which can be approximated but never attained" (p. 202).

Still, can one observe that progress is really taking place? Can systems visionaries integrate their vision with reality? Not necessarily. Some visionaries escape the need for integration by retreating into the realm of abstract models, and proclaiming that their models have no practical consequences. Others cut off a piece of their vision and do something practical by using politics, persuasion, and tricks of the trade-and lose the vision. The visionary finds himself espousing a faith in propositions that fails in every instance-but is somehow true, in general.

Be Your Enemy

As our hero pushes himself toward a greater and greater comprehensiveness, he finds himself in a realm of incomprehension. He is now faced with a paradox: his own reflection on the deep puzzles of systemic reasoning leads him to results that run counter to that reasoning. He grins. He knows that paradoxes are a driving force of rational advances. He feels that "there is much to be said . . . for purposefully designing paradox" (p. 204).

To deal with the quandary, he offers himself the prescription: "Be your enemy" (p. 204). With this prescription, he opens up a "new teaching of rational planning-that is, a new and more general meaning of rationality, which goes beyond dialectical reasoning" (p. 204). This new teaching says,

> Rational humans need to leave the body of rationality and to place the self in another body, the "enemy," so that the reality of the social system can unfold in a radically different manner. From this vantage point he/she can observe the rational spirit and begin to realize not only what has been left out of it, but also what the spirit is like, especially its quality of being human (pp. 204-205).

By releasing the bonds of hard rationality, one can see that rationality is a tool, an expression (among other expressions) of what it is to be human.

In this stepping out of the body of rationality into the bodies of politics, morality, religion, and aesthetics, our hero does not lose his identity; he continues to operate as a deeply involved rational planner. He embraces the kind of "sane schizophrenia" that Otto Rank (1932) offered to visionaries: "at one and the same time . . . [to live] visions and the reality of the collective consciousness" (p. 213).

In stepping out, our hero gains objectivity about his rational self and leavens it with some humor. In looking at himself *as his enemy*, he begins to see how foolishly he pushes "one point of view, of model building, statistical analysis, game theory, ethics, or holism" (p. 214). Churchman expresses the satisfaction that comes with this recognition as follows:

> Once you are your enemy, you at last see yourself as you really are: a human being, wise and foolish, who has a quirk about the destiny and the improvement of the human condition, just as all the rest of humanity has its quirks (p. 214).

CHECKLAND

Checkland was working as a university-based management consultant during the 1970's. He experimented with systemic approaches to problem-solving following the logic of his problems and the leads of other scientists, notably Churchman and Ackoff. He eventually devised the Soft Systems Method (SSM) as a generic approach to the kind of problems he was facing.

The SSM is a progressive learning cycle with five stages (in its simplest form). The cycle can begin at any of those stages. Three of the stages concern events in real time, and two involve abstract systemic definitions and models. The first (real world) stage is the gathering information and viewpoints about a problem situation.

The second (thinking) stage is the formulating of critical core definitions. The formulation of these definitions requires abstract thinking about delivery systems, coordination systems, and so on. At this stage, one develops several definitions. Each of these definitions explicitly identifies six elements about the problem situation.

These six elements are identified by the CATWOE acronym. They are: Customers, Actors, Transformation process, Worldview (Weltanschauung), Owner, and Environmental constraints. Customers are those "who would be victims or beneficiaries of this system were it to exist." Actors are those "who would carry out the activities of this system." Transformation process answers the question: "What input is transformed into what output by this system?" Worldview answers "What image of the world makes this meaningful?" The Owner is the one "who could abolish this system." Environmental constraints answer the question: "What external constraints does this system take as given?" (Checkland, 1991, p. 69). Each of these elements is carefully integrated within its worldview. Each Worldview is assessed for its theoretical and ethical relevance.

The third (thinking) stage is also critical. It develops conceptual models based on these root definitions. These models explore the theoretic and practical consequences of the root definitions. They abstractly point out elements, viewpoints, and consequences that are not part of real-world awareness.

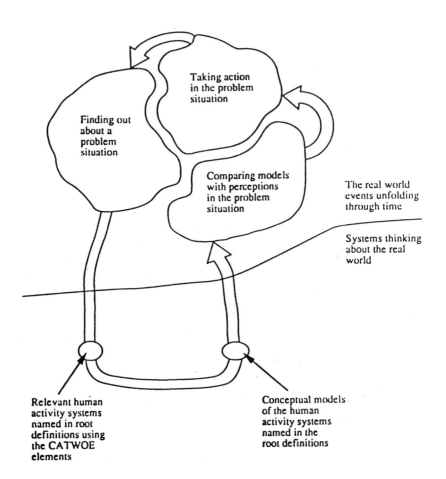

CATWOE

C ("customers")   Who would be victims or beneficiaries of this system were it to exist?
A ("actors")   Who would carry out the activities of this system?
T ("transformation process") What input is transformed into what output by this system?
W ("Weltanschauung")   What image of the world makes this meaningful?
O ("owner")   Who could abolish this system?
E ("environmental constraints") What external constraints does this system take as given?

The nature of soft systems methodology (SSM)

From Checkland in Flood and Jackson (1991) p. 69,
Critical Systems Thinking: Directed Readings. Reproduced by permission of John Wiley & Sons Limited.

The fourth (real world) step "compares" these systemic models with the existing perceptions within the problem situation. This step brings together real life perceptions with abstract perspectives and novel recommendations. It leads to a real life decision about what will be done to address the organization's problems.

The fifth (real world) step is the taking of some action in the problem situation and the monitoring of the resulting patterns of behavior. Hereupon the stage is set for renewing the process.

Clearly, SSM is a learning system. Organizations use it to clarify their viewpoints, examine theoretic consequences, confront their perceptions with novel schematic possibilities, and use those insights to intervene into their problem situations. They then monitor results and reiterate this process in an upward spiral of organizational communicative improvement.

Checkland also sketched the phenomenological and social bases of SSM. He thought SSM was applicable by anyone to any organizational problem in a value-free manner. He saw compatibility between SSM and critical theory, and noted the similarities between SSM and Habermas's "ideal speech situation" (Checkland, 1981, p. 283). SSM achieved rapid acceptance as a management tool in the 1980's because it offered systematic approaches to real-life situations.

FUENMAYOR

Fuenmayor explores the cross-implications of holism and worldview as they are used in soft systems thinking. He interprets soft systems as interpretive thinking because it recognizes that social facts depend upon the point of view of their interpreters. SSM, for example, emphasizes that different worldviews produce different "facts" for the participants of soft systems problematics (different customers, for example, and different external restraints). These differing facts relate to an existing reality that soft systems tries to see as a whole, that is, from all (relevant) viewpoints.

Fuenmayor argues that soft systems is untrue to its genius if it stays only on a pragmatic-regulative level geared to interpretive management. At this level, soft systems is useful, but it effectively limits the acceptable worldviews to those of the managers and the regulators. At this level, moreover, soft systems is not self-reflective; it fails to consider its own worldview. A self-reflective system would recognize its own partiality of viewpoint in its quest for holistic understanding.

Interpretive Systemology

Fuenmayor undertakes to present the onto-epistemological context for a philosophically self-aware soft systems approach to reality. He uses a quotation of Checkland as his springboard:

[SSM] as a whole clearly articulates a phenomenological investigation into the meanings which actors in a situation attribute to the reality they perceive . . . In contrast to [the] paradigm of *optimizing* (or satisficing), soft systems methodology embodies a paradigm of *learning*. The notion of "a solution," whether it optimizes or satisfices, is inappropriate in a methodology which orchestrates a process of learning which as a process is never-ending (Checkland, 1981, pp. 278-279).

Fuenmayor considers learning itself to be the focus of systems thinking. In his theory of *"Interpretive Systemology,"* the organizations that utilize SSM learn during the course of their cycles of planning and implementation. This learning is not just incremental. It develops crises and removes organizational blindness towards unfamiliar aspects of reality.

An organization's worldview provides an interpretative context within which it interprets events. By making explicit different worldviews in the formulation of root definitions, SSM opens new vistas for organizations and enables them to conceptualize their situation in the context of different "facts." In this way, SSM articulates the complexity of the planning situation.

What is complex in interpretive complexity is the difficulty of seeing the variety of interpretations rather than the variety itself. The source of that difficulty and, at the same time, its most immediate consequence, lies in the difficulty of seeing our own states of mind (Fuenmayor, 1991, p. 234).

In doing SSM, interpretive systemologists model various contexts of meaning in an effort to espy their own perspective from afar and, thereby, understand phenomena in a more rounded, holistic fashion. For them, critical thinking is "a thinking that has to think itself in order to interpretively understand others' actions" (1991, p. 237). In order to see many sides of a thing, they admit that their distinctions of it express only a few of many possible differences.

The "facts" of interpretative systemology do not exist in themselves; they are not things in themselves; they are things that cannot be without an interpretive context. The "customers" and "environmental restraints," for example, are different facts depending on the worldview of their root definitions. Fuenmayor says: "An 'interpretation' is something which 'is-not-in-itself'. The mode of Being of an interpretation is that of a possibility, since that which it may be depends upon different 'interpretive contexts'" (1991, p. 236).

To view phenomena as interpretations flies in the face of empiricism that considers experiences facts. Where empiricists try to model stable properties and explain the relations among them, interpretative systemologists model contexts of meaning and derive their facts from those contexts. They probe insistently for deeper contexts of meaning in order to uncover richer understandings of situations and more fruitful manners of addressing them. "The deeper the contexts of meaning are theoretically explored, the more critical and liberating the discussion"(1991, p. 236).

The kind of interpretive thinking advocated by Fuenmayor uses the phenomenal act of making a distinction in the continuum of experience as its bedrock. Independent of our verbal distinctions, that is, as "whatever-is-the-case," reality is both seamless and incredibly individual. Our distinctions keep referring to this "whatever-is- the-case" without ever capturing anything but partial, fleeting aspects. An appreciation of the observer's limitations is the fundamental source of holism. This appreciation also reveals the need for an interpretive approach to "facts," especially "social facts."

A House of Iterations

Interpretive systems thinking constructs an onto-epistemological building out of "facts" that are meanings and interpretations. It builds this philosophical house in never-ending iterations, always constructing it anew to systemically accommodate its every addition. Fuenmayor describes this process as follows:

> The methodical search for knowledge is characterized by the modeling of various contexts of meaning; by explicitly interpreting the phenomenon with regard to such contexts of meaning and by discussing the various interpretations in the light of their respective contexts of meaning (1991, p. 236).

This model house of truth can never be completed because "truth is an essentially dynamic process which cannot be finished or stopped because it would turn into the opposite of truth" (1991, p. 240).

Human life enfolds a tension of truth and practicality. Truth is never objectively achieved, but practical living involves constant interactions with objective things. Life, therefore, connects wonder about the ineffability of things with practical distinctions and decisions in everyday and political life. One finds oneself, therefore, constantly acting on approximations of an unattainable truth, acting out of uncertainty. This tension between truth and practicality forms a dialectic. "Authentic interpretive systems thinking and intentional action are the opposite parties of a dialectic that cannot dissolve in a non-dialectical synthesis" (1991, p. 241).

Fuenmayor does not think that the soft systems approach needs to import ideas from Habermas or any other critical theorist because it is rigorously critical in its own right. "Interpretive Systemology does not need the eclectic addition of other approaches in order to be critical, for it is critical in its very marrow" (1991, pp. 242-243).

SUMMARY

Soft systems theory expresses a conscious attempt to craft a theory of organization and design based upon humanistic and ethical principles. Churchman laid its philosophical foundation. Checkland and others moved beyond hard systems to deal with problems that defied precise definition. They defied efficient methodologies that embodied democratic principles. Fuenmayor and others have explicated soft systems epistemological and ontological bases.

SOFT SYSTEMS STANDARDS

Churchman, Checkland, and Fuenmayor propose certain standards for the responsible application of systems theory to social processes. Some of their imperatives are listed below in the form of shoulds and should nots. Systems, systems theorists, and systems practitioners should:

Churchman

1. find out what decision-makers want
2. make plans and decisions that serve the future of humanity
3. avoid the environmental fallacy
4. take human bias into account
5. tie all matters of human concern together in one grand imagery
6. be as comprehensive as one can be
7. not exclude themselves from the whole system
8. ultimately, "be the enemy," and observe their rational selves with some humor.

Checkland

1. identify the Customers, Actors, Transformation process, *Worldview*, Owners, and Environmental constraints of their plans
2. create several root definitions in accord with divergent *Worldviews*
3. carry on all discussions and decision- making openly and without coercion
4. understand the systems process as learning process.

Fuenmayor

1. recognize their own partiality of viewpoint
2. recognize that phenomena are interpretations because they are perceived in a context that is formed by a *Worldview*
3. acknowledge that no truth can be completely expressed. Any claim to expressed truth is a lie
4. see that truth and life exist in an essential, dynamic tension.

CHAPTER 8

CRITICAL SYSTEMS THEORY

ABSTRACT

Critical systems theory has roots in soft systems theory. It addresses theoretical, ethical, and practical issues in systems practice with an eye towards balance and equity. Jackson points out the skewing that power imbalances inflict upon the soft systems method. He proposes three processes for remedying this situation. Ulrich describes the normative impact of break-offs in systems applications. He proposes a program of systems research based upon the social rationalities identified by Habermas as: instrumental, strategic, and communicative. He advocates a critically normative approach that operates dialogically with all the parties who are affected by planning decisions. Flood undertakes the liberation of systems theory from its self- imposed insularity, its objectivist and subjectivist illusions, and its internalized subjugations of discourse. He introduces, along with Jackson, a complementary approach to the use of different systems modalities.

JACKSON

Power Imbalances

Jackson perceives a weakness in soft systems methodology--its inadequacy in dealing with situations where there is an imbalance of power relations. SSM functions appropriately only when all the stakeholders in a decision enter into free and open discussion about proposed changes. This condition of untrammeled communication is unfulfilled in many applications of SSM. Furthermore, soft systems is unrealistic in any expectation that this condition will be met often, because persons in power are not likely to undercut their dominant positions. Jackson puts it this way:

> Privileged stakeholders (in terms of wealth, status or power) are unlikely to risk their dominant position and submit their privileges to the vagaries of idealized design or whatever. So soft systems thinkers, if they take their own criterion of validation seriously, will have to steer clear of the very many social systems where full and effective participation cannot be established (1991, p. 129).

If soft systems methodology is to be applied in contested situations, it must safeguard itself from these power inequalities. These inequalities produce domination of one view over another by forces other than that of the better argument. If it does not, SSM will ensure the imposition of privileged ideologies upon the unprivileged who lack the means to recognize their own true interests. Moreover, it must overcome the built-in bias that arises in the hiring of soft systems consultants. Consultants are usually hired by managers and regulators to secure their ongoing dominance over the unrepresented victims of a policy. Because of these factors, soft systems practice frequently advances a managerial and regulatory bias into social situations.

Hard systems do not treat conflict any better than soft systems. They tend to ignore conflict. They "reduce the problem (and, in the process, distort it) so that it can be tackled using a methodology based upon the functionalist/engineering model" (Jackson, 1991, p. 123). They tend to split human beings into two classes: "the social engineers and the inmates of closed institutions" (Jackson, 1991, p. 124, quoting Habermas, 1974). Hard systems approaches often fall back upon the criterion of applicative control. But this criterion is invalid because of the close link between social theory and social practice. Social coercion, for example, can easily make a social theory seem rational (cf. 1991, p. 121).

Corrective Measures

Jackson recommends corrective measures in three stages to improve the probity of systems theory. He advocates forming and extending critical social theories, the organizing processes of enlightenment, and selecting appropriate situation-specific strategies.

First, he holds, with Habermas, that the construction of explicit social theories is an essential part of any social systems science. To meet their critical task, social systems theorists have to escape the functionalist paradigm in order to cope with conflict, contradiction, power, coercion, and change. They must recognize the part that cognitive systems (paradigms) play in producing knowledge and controlling activity. In a society full of contradictions, they also must make room for conflict and uncertainty. Otherwise, these theories will fall into "the most fatal contradiction of all, namely that of existing between its own structure and that of its object" (Adorno; quoted by Jackson, 1991, p. 120). Systemic models do not have to be functionalist; they can also be organismic, morphogenic, factional, or catastrophic. Jackson suggests these alternate models "for systems theorists who wish to follow Habermas" (1991, p. 133).

The second stage, the organization of enlightenment, goes beyond the first-stage work of professional scientists. In this stage, designers authenticate this knowledge in the lives of ordinary people, "the social actors at which it is aimed" (1991, p. 133). In this stage with its validations, designers strive "enlighten those to whom it [the design] is addressed about the position they occupy in an antagonistic social system and about the interest of which they must become conscious in this situation as being objectively theirs" (Habermas, quoted by Jackson, 1991, p. 134).

The enlightening of stakeholders enables the third function: the selection of appropriate strategies. Because they now "possess a social theory which enables them fully to comprehend their position in the social world and the possibilities for action that this affords" (1991, p. 134), the group can make enlightened choices. They can embark deliberately upon a soft-systems methodology with their opponents or they can choose to struggle with them.

Every group, in interaction with other groups, has to choose between broadening its circle of enlightenment or struggling. Even in struggle, however, opposing groups eventually resume communication.

ULRICH

Ulrich focuses on applied science as "the study of contexts of application" (1991, p. 105). He finds that the key difficulty of applied science lies in its surreptitious moralizing because of "the normative content that its propositions gain in the context of application" (1991, p. 105). By *normative content* he does not mean just the value judgments encapsulated in recommendations for action, design models, planning standards, or evaluative judgments. He also means the "life-practical consequences and side effects of the 'scientific' propositions in question for those who may be affected by their implementation" (1991, p. 104). He is concerned, in other words, about the coercive effects of pseudo-science narrowly focused on operational and strategic ends.

Critical Heuristics

Ulrich recognizes the cogency of Habermas's "ideal speech situation" for theoretically addressing this problem, but he emphasizes its "ideal," that is, impractical character. He finds, in particular, that the ideal speech situation takes no account of "the inevitability of argumentation break-offs. In practical discourse . . .

every justification attempt must start with some material premises and end with some conclusions that it cannot question and justify any further" (1991, p. 104).

Ulrich sees this inevitability of argumentation break-offs as a wedge that people can use to defend their positions against hurtful "scientific" judgments. He proposes a kind of guerrilla warfare in which affected citizens question the break-off points of inquiry and demand justification for them. He calls this method the "critical heuristics of social systems design."

Critical heuristics posits three requirements: (1) "a clear understanding of . . . the unavoidability and critical significance of justification break-offs"; (2) "a conceptual framework . . . to identify effective break-offs of argumentation"; and (3) "a tool of cogent argumentation that would be available . . . to 'ordinary' citizens" (1991, p. 105).

The first requirement refers to boundary judgments about the size of the whole system and its appropriate break-off points. It requires that planners make explicit reckoning of their boundary judgments. Such reckoning is deemed practically "adequate if it makes explicit its own normative content" (1991, p. 107).

Because boundary judgments are made arbitrarily on the basis of (often unconscious) normative considerations, systems practitioners cannot apply objective solutions to social situations. They can, however, apply "a critical solution to the problem of boundary judgments" (1991, p. 107) by specifying the norms they use in creating their boundaries. A social system, therefore, can be called rational when it "renders explicit the underlying justification break-offs and thus enables both those involved and those affected by the design to reflect and discourse on the validity and legitimacy of these break-offs" (1991, p. 107).

Ulrich proposes twelve boundary questions for ascertaining the "*a priori* concepts of practical reason" (1991, p. 108), that is, the norms that are in use in a particular situation. He asks of the design in question, what are its sources of motivation, control, expertise, and legitimation?

By employing these questions, ordinary citizens can make explicit the power relations and agendas that are contained in a systems design. They can use these inquiries to form cogent argumentation against noxious system designs. They need not claim any expert status; They only need to demand that the "experts" avoid cynical or dogmatic pronouncements about the "scientific" nature of their proposals.

Instrumental, Strategic, and Communicative Approaches

Ulrich also proposes a *program of research* in the areas common to systems thinking, systems practice, and practical philosophy. He states the need for such a program as follows:

> The systems approach, *because* it strives for comprehensiveness, must learn to live with its own unavoidable incomprehensiveness and must draw the necessary conclusions from this insight. It must bother to take into account that which is not

> systemic in its nature and hence cannot be rationalized in the terms of systems rationality. Otherwise its quest for comprehensiveness . . . is bound to lead into new kinds of reductionism; for example, by reducing everything to "nothing but" functional systems aspects (pp. 245-246).

To accomplish this task Ulrich relates the differing strands of systems rationality to the social rationalities identified by Habermas as: instrumental, strategic, and communicative. Instrumental rationality is non-social and functional, geared toward the efficient use of scarce resources for given ends. Strategic rationality is social and functional, geared to the effective steering of complex systems. Communicative rationality is social and consensual, geared to the social integration of conflicting interests.

Operational systems management works on the level of non-social instrumental action. In this category are operational research, systems analysis, and systems engineering. These methods "decompose" problems and strive for technical control. They are particularly valuable for solving problems concerning the management of scarcity.

Strategic systems management relates to strategic action. Like operational action, strategic action works towards maximizing one's own self-interest, but it takes into account the activities of other actors. If there are conflicts of interest, strategic action works to maintain a social system and optimize its performance. It generally does not consider the ethics of a situation or the plight of "victims" who are not part of the decision-making process.

Strategic action deals with the management of complexity in a fashion that is not required in instrumental action. It deals with

> "*strategic potentials* of success"-i.e., systems capabilities of self-regulation, resilience, and innovative adaptation in the face of turbulent environments . . . The focus on problem decomposition and control is replaced by a focus on problem identification and understanding with a view to basic policy decisions that will minimize surprise and lost opportunities (1991, p. 259-260).

To this end, an impressive array of simulation techniques, cybernetic modeling, game theory, "key factors of success," and portfolio management have been developed. Some of the specific tools and concepts are Forrester's "system dynamics," Ashby's "law of requisite variety," Beer's "managerial cybernetics," and Vester's "cybernetic sensitivity model."

Despite its great practical success and its holistic perspective, strategic systems management has a utilitarian orientation. It does not address questions of communicative rationality that are practical but not functional. It deals with situations of conflict monologically without dialoging with affected parties who are not decision-makers. Because it does not enhance communicative rationality, but expands only the reach of functional systems rationality, strategic systems management tends to be self-involved.

> [It] threatens to pervert the critical heuristic purpose of systems thinking . . . into a mere heuristics of systems purposes. This means that it is no longer "the system" and the boundary judgments constitutive of it that are considered as the problem; instead, the problems of the system are now investigated (p. 261).

In other words, a purely cybernetic or organic (lacking a communicative aspect) approach to problems is not holistic; it is parochial and imperialist.

A systems rationality that corresponds to communicative action, on the other hand, must be oriented toward mutual understanding. It must be *dialogical*. Therefore, it inevitably concerns the norms upon which decisions are made. In other words, it takes into account the values of the victims of social systems as well as the values of their architects. In so doing, it expands the realm of rationality to truly human concerns.

The Critically Normative Approach

Ulrich begins his development of a "critically normative approach" for management by stating that no standpoint, not even the most comprehensive systems approach, is ever sufficient in itself to validate its own implications. Hence,

> A definition . . . of a system can be called 'objective" only in as much as it makes explicit its own normative content; whether or not it does so cannot be established "monologically," by reference to the expertise of the involved, but only communicatively, by reference to the free consent of those affected. Thus a critically normative systems approach will, of necessity, be a communicative approach (p. 262).

Ulrich says that such a critically normative approach faces three challenges: (1) It cannot simply add normative considerations to instrumental and strategic ones to get the job done. It demands philosophical competence of systems theorists because they have to incorporate normative considerations into the essential body of systems rationality. (2) Not just any debate fulfills the requirements of Habermas's ideal speech situation. In order to achieve rationally motivated consensus, one has to arrange for, and cultivate, processes of maximum undistorted communication-that is, communication in which the force of the better argument gets a chance to prevail over other forces (p. 262).

Because we rarely can achieve Habermas's ideal, a lack of complete rationality is inevitable. Therefore, we should not try to force the impossible. Instead, we should sharpen our abilities to deal critically with conditions that are less than ideal. Ulrich puts it this way: "The systems movement will make a real contribution toward communicative systems rationalization if it puts the systems idea to work on the job of dealing critically with conditions of imperfect rationality" (p. 263).

(3) Systems methodology can never supersede democratic legitimation when decision-making affects people who are not privy to the systematic process. Therefore, "systems practice should not misunderstand itself as a guarantor of socially rational decision making; it cannot, and need not, 'monologically' justify the social acceptability of its designs" (p. 263).

Conclusion

In his discourse on critical heuristics, Ulrich seeks to lay down epistemological foundations for a critically normative systems approach that takes account of incomplete rationality. He points out that critical heuristics, as such, does not seek objective solutions to the problems of practical reason; rather, "it aims at a merely critical solution. A critical solution does not yield any 'objective' justification of normative validity claims. It prevents us, rather, from submitting to an objectivist illusion regarding such claims" (p. 263).

Normative systems management is still a fledgling compared to instrumental and cybernetic approaches. Some of its principles are embodied in soft-systems methodology with some borrowed from Habermas and Foucault. Critical and liberating systems theorists have recently outlined its rationality, proposing new tools and concepts.

Ulrich provides a list of concepts and tools that are available to critically normative systems practice: (1) Kant's original presentation of the systems idea as an "unavoidable" critical idea of reason; (2) Churchman's "ethics of the whole system"; (3) Jantsch's concept of "vertical integration"; (4) Checkland's concept of "root definitions"; (5) Mason and Mitroff's "assumptional analysis"; (6) Ackoff's and Churchman's concept of "ideal planning" and the "process of unfolding." Also of value are some of the concepts of critical heuristics: (7) the *a priori* judgments of practical reason; (8) argumentation break-offs; (9) the polemical employment of boundary conditions; (10) the purposeful systems assessment of designs with regard to their normative implications; and (11) the critically heuristic training of citizens.

Ulrich concludes his "program of research" as follows:

> I do not say that the task of integrating the communicative dimension of rationality will be easy. But if eventually we succeed, this will considerably ease the quest for rational systems practice, for it will free the systems approach from the impossible (and elitist) pretension of securing a "monological" justification of rational practice (p. 265).

FLOOD

Flood and Jackson (1991) develop Ulrich's core ideas to assign instrumental, strategic, and communicative approaches to practical situations with their program of Total Systems Intervention (TSI). They base TSI on a philosophy of "complementarism," "sociological awareness," and the promotion of "human well-being and emancipation."

Flood, in his *Liberating Systems Theory* (1990), undertakes to liberate systems theory from (1) its self-imposed insularity, (2) its objectivist or subjectivist delusions, and (3) its internalized subjugations of discourse. He also presents a theory that (4) addresses emancipation in response to domination and subjugation in work and social situations, and (5) opens up systems theory by introducing significant philosophical perspectives (cf. p. 13).

Insularity

The insularity of the systems movement derives from its tendency to explain everything in terms of a closed set of key concepts. Flood counters this insularity with two lessons:

> Lesson 1: Systems thinking has a natural tendency to be conceptually reflexive . . . and would benefit by looking beyond its own horizons.

And,

> Lesson 2: The study of (systems) concepts requires a 'historical and developmental' investigation that attempts to deal with the subjectively intended meaning of authors (pp. 23-24).

Flood looks beyond the horizons of systems science into social, ideological, ontological, and epistemological concerns. On the basis of his improved viewpoint, he examines the historical and developmental forces behind systems practice, its blind spots, and its conservative bias.

Objectivist and Subjectivist Delusions

The biases of dualistic positivism make the subject/object split absolute and take scientific ideas to be the essences of reality. Flood challenges those biases by examining critical, phenomenological, and hermeneutic thought since Kant. He portrays this thought as either foundationalist or anti-foundationalist.

Foundationalism attempts to provide knowledge with a justification when possible and a critique when none is possible. The aim is to ensure that our knowledge is on firm, indubitable and unshakable foundations. Anti-foundationalism proposes a theory about subjectively shared practices of language, against the notion of indubitable ideas of the individual thinking subject. It "attempts to establish that there can be no absolutely neutral standpoint for inquiry outside ongoing interpretations, values, and interests of actual communities of inquirers at work in current social practices" (pp. 17- 18).

In this context, Flood portrays foundationalism as Structuralism and anti-foundationalism as Deconstruction. He favors an interpretivist stance that bridges this dispute by acknowledging the actuality of exterior reality while maintaining that verbal interpretations express necessarily only partial conceptions of that reality. He offers Habermas and Foucault as "scholars of union" who help him build his interpretivist bridge. He uses Habermas's ideas of knowledge- constitutive interests and critical hermeneutics as the bases for his ideas of complementarity and objectivity. From Foucault's interpretive analytics, he draws the idea of liberating repressed and neglected points of view.

According to Flood, the belief that systems thinking expresses the status (or essence) of social reality is a fallacy. That kind of thinking surreptitiously assigns normative parameters to a situation. In other words, "questions of status must be considered as normative questions in disguise" (p. 24). Therefore, it is important to ascertain how "contrasting paradigmatic viewpoints influence our use of systems concepts" (p. 24).

Paradigms enter every definition of "system" because they provide the underlying metaphors upon which particular systems are built. "There is no single metaphorical/analogical term which is 'system.' Such singularity would clearly be nonsense" (p. 24). Instead, the idea of what a system is has continuously been redefined through analogical reasoning drawn from the organic world (feedback, autopoiesis, feedforward), but also from mechanisms, culture, politics, and psychic prisons. Each of these underlying metaphors carries a different paradigmatic (unreflective normative) understanding of what a system is and there are several metaphors that comprise a paradigmatic system.

Each underlying metaphor of systems thinking rests upon a basic normative theory of how the social world is put together: mechanically, organically, communicatively. Each metaphor also reflects an objectivist or subjectivist viewpoint, and an attitude about the possibility and desirability of social change. Therefore, mixing metaphors of the natural and social worlds can promote "problem solving as a means of contributing to appreciation of systems" (1990, pp. 80-81). The paradigmatic divergence of these metaphors reveals the utter abstractness of the category "system." It also releases our organizational ability for cross-disciplinary understanding of our world.

Flood provides an expanded systems semantics. He takes terms that are ordinarily defined in positivist terms and redefines them in an interpretivistic sense. In doing this, he hopes to remove from these concepts the "value constructions of politics, religion, morality, and aesthetics" with which they have become encrusted. He wants to reconceptualize them as

> abstract organizing structures through which we can conceive of a world around us and which can help us to critically organize our thought on matters of action in the world of natural and social dynamics (1990, p. 65).

He believes that this change in systems underpinnings from positivism to interpretivism will free the concept of "system" for the abstract integration of science.

Flood repudiates the idea that systems exist in the real world and embraces the idea that systems are the result of our thought processes as we try to make sense of our world. He does this in conscious opposition to cybernetic theory that claims to uncover the natural laws of organization. He thereby sets up a dialectic on this issue within the systems community. He adds a polemic edge to this dialectic with the following:

> The positivist theoretical position leads to describing the world as if it were a complex of . . . 'open systems,' both natural and social . . . This proposition is obviously misconceived. Consider the idea of a boundary. In reality it is a non-entity. A boundary is the ultimate expression of systemic abstraction . . . It will not be surprising, therefore, to find that a main concern of traditional systems thinkers is an expansionary difficulty in boundary identification (1990, p. 91).

He shows his predilection for complementarism, however, in another quotation:

> Since . . . the world is not a complex of systems, the exercise of system identification in messy social contexts (qualitative or quantitative) becomes all but redundant . . . [Considering the world a complex of systems] is not all, but is one viewpoint (1990, p. 92).

Finally, he summarizes his case against cybernetics as follows:

> Cybernetics as such deals with natural laws . . . Nevertheless, with interpretive reasoning, the current argument rejects the use of the word or concept 'system' to describe any 'real world' things, stating clearly that it should be saved exclusively for systems thinking. The argument is that if we are able to agree upon the structure and process, and it is the notion of structure and process that makes the label 'neurocybernetic system' meaningful, then why not abandon the word or concept 'system' for what is essential to our understanding, namely 'structure and process' (or perhaps 'organization' in some contexts). This saves the reader from the task of making inference and, in fact, adds meaning to an otherwise nonexpressive label" (1990, p. 98).

Flood's viewpoint is radically subjective but not radically relativistic. While every viewpoint and every systems approach may be partial and relative, not every viewpoint and approach is equally valid. One can reflect on the sources of knowledge *and deception* that are contained in a viewpoint or approach.

In particular, the concept of false consciousness can be raised concerning any viewpoint or approach. False consciousness "can be understood as [the] freezing of people's *worldview*" (1990, p. 103). It occurs in at least two interrelated ways. First, in the creation of forced visions (invisible to the dominated) about the value of work and other social issues. Second, in terms of a super-science, or a dominant rationality, that shapes the way we treat 'the' external natural world, 'our' social world and 'my' internal world" (1990, p. 103).

By promoting critical awareness of false consciousness and other sources of deception, Flood points out our theoretical and ideological subjectivities. He helps us realize that we are subjective in our judgments both individually and socially. He holds that "No systems approach for handling real world complexity should seriously be considered rational unless it allows for and encourages (self) critical reflection of free citizens" (1990, p. 104).

Subjugated Discourse

General Systems Theory (GST), with its hubris and genius, has undergone severe rebuke from positivists, neo-positivists, and interpretivists. Positivists pointed out that GST could not provide "conclusive verifiability" for its theses. Neo-positivists attacked GST's use of isomorphisms across disciplinary lines because statements about them are not falsifiable and are, therefore, meaningless as scientific statements. Within philosophical sociology, interpretivists were "paradigmatically unable to accept concepts of nomic isomorphism, in particular for social contexts" (p. 127). So, they rejected GST out of hand. As a result of these attacks, GST is held in almost universal disfavor.

Flood shows that this disfavor is unwarranted by discrediting the arguments against GST. He shows, in addition, that GST has been repressed paradigmatically by powerful webs of Foulcauldian "truth." He does this by, first, demolishing the positivist, neo-positivist, and interpretivist criticism and, then, by looking at the politics of science.

Positivism and neo-positivism are refuted by the simple *ad hominem* argument that they do not live up to the very standards that they demand of systems science. Positivism cannot provide "conclusive verifiability" for any general statement. Neo-positivists, as Popper has shown, suffer from a naturalist fallacy: they cannot demarcate sense from nonsense and hence the meaningful from meaningless in philosophy.

Interpretivist attacks maintain their force only within a nomic context in which GST is "conceptualized as referring to ontological matters of fact" (p. 127). If GST simply accommodates to the fact that "no general proposition referring to so-called matter of fact can ever be shown to be necessarily and universally true" (p. 127), then why should it not apply to perceptual or communicative systems? Flood even points out that "An awareness that there might be such concepts as issue-based contextual, rather than nomic, isomorphisms might prompt interesting research" (p. 127).

The history of GST gives ample evidence that political repression is alive and well in the academic world. It demonstrates the Foucauldian truth that "Power is knowledge." GST was subjugated in a turf war. Its history confirms Foucault's observation that "truth" arises "because of domination at local discursivity levels imposed by nondiscursive subjugators" (1990, p. 119).

Flood proposes to correct GST on the basis of improved social awareness, ontology, and epistemology. He also proposes to expand GST into social contexts by the techniques of complementarity. With complementarity, novel and conflicting viewpoints are liberated (pluralism) and the appropriate viewpoint is allotted to the appropriate context. In this area, he generally adopts the schema laid out by Ulrich that he and Jackson later developed into Total Systems Intervention (cf. Flood and Jackson, 1991a; also 1991b, pp. 321-337).

Emancipation

Through complementarism, Flood hopes to emancipate the work of systems practitioners from blind allegiance to one particular systems model, be it hard, cybernetic, soft, or critical. By expanding the vocabulary and scope of critical systems, he hopes to expand the relevance of systems to conflictual social situations.

Liberating Systems Theory

Flood hopes to illuminate the whole world of systems thinking so that readers, researchers, and practitioners will possess a better view of their worlds, their work, and their lives. On the philosophical level, Flood uses Foucault as his opener of discourse and Habermas as his closer. Foucault is his icon of stubborn openness. Habermas is his model for critically open discourse that arrives at practical solutions. Flood wants to both liberate and critique, to widen discourse but narrow it down when necessary to make it practical, to have both resolute openness and epistemological rigor. He considers the tensions between Foucault and Habermas to be typical of his complementarist position.

Flood describes a way to integrate theory and ideology. He holds that no systems approach is ideologically neutral; that every approach is biased on the basis of its fundamental metaphors. He sees no escape from ideology. On the contrary, he urges a critical awareness of ideology as it is conceived and used in social situations. In particular, he believes that critical systems thinking ensures that all angles of a problem are considered by pointing the conservatism of both hard and soft systems.

Flood sums up the relationship between the systems ideal of comprehensiveness and the critical theory ideal of a moral imperative in the following:

The systems epistemological ideal—a critically motivated quest for comprehensiveness—in practice too easily lends itself to uncritical claims of comprehensive rationality, neglecting the fact that we never know and understand 'the whole system' (the totality of relevant conditions). On the other hand, the ideal of critique just as well lends itself to an uncritical absolutism of one's critical standpoint, for it is an impossible imperative to permanently question all one's presuppositions, including one's standards of critique; but presupposition-free critique is impossible. It seems that the two requirements mutually complement each other in a useful way: 'Think systems, but don't ever assume to grasp the whole!' implies the systems inquirer's need for critical self-reflection and 'Think critically, but don't ever allow your standards of critique to become absolute!' implies the critical scientist's need to think beyond his particular standpoint and to look for comprehensiveness in his understanding (1990, p. 177).

SUMMARY

Critical systems theory emphasizes how power imbalances subvert the democratic process even in methodologies that are as well-designed as soft systems. Jackson urges the formation of explicit social theories that supply the context for design decisions, in-the-field testing of those theories/designs, and the development of appropriate strategies that advance participatory democracy. Ulrich advocates a designing process that utilizes a communicative approach in preference to instrumental and strategic approaches. He also advocates an approach that makes explicit its normative assumptions. Flood, with Jackson, offers criteria for deciding whether to use instrumental, strategic, or communicative approaches in particular situations. Flood combats a perceived insularity of systems science by highlighting its dependence on metaphor. He utilizes the concept of complementarity to recognize instances of subjugated discourse, to free systems practitioners from blind allegiance to any one particular systems model, and to reconcile the systems ideal of comprehensiveness with the critical theory idea of a moral imperative. The following two chapters expand the circle of authors who describe how to make the process of design more democratic and more scientific.

CRITICAL SYSTEMS STANDARDS

Jackson, Ulrich, and Flood offer standards for a responsible application of systems theory to social processes. These standards can be stated in a list of shoulds and should nots. Systems, systems theorists, and/or systems practitioners need to recognize certain parameters, avoid certain assumptions and behaviors, and adjust for certain deviations. Some of those shoulds and should nots are presented below:

Jackson

1. Should recognize and compensate for imbalances in power relationships (cf. 1991, p.130).
2. Should recognize the part that cognitive systems play in the production of knowledge and the control of activity (cf. 1991, p. 119).
3. Should recognize that society is full of contradictions and allow for conflict and uncertainty in their activities (cf. Jackson, 1991, p. 120).
4. Should not judge the success of a social theory by criteria of social control (cf. 1991, p.121).
5. Should embody critique if they are to be adequate to social situations (cf. 1991, pp. 130-131).

Ulrich

1. Should bother to take into account that which is not systemic in its nature (Ulrich, 1991, pp. 245-246).
2. Should adopt a dialogical approach to social reality (Ulrich, 1991, p. 262).
3. Should look beyond the horizons of systems science into social, ideological, ontological, and epistemological concerns (cf. 1991, p. 247).
4. Should recognize built-in biases behind systems thinking because of the value-laden metaphors that give form to the abstract ideal of systems (cf. 1991, p. 260).
5. Should justify the inevitable break-offs in systems theories on normative grounds (cf. 1991, pp. 106-107).
6. Should make explicit their normative content (cf. Ulrich, 1991, p. 107).
7. Should forego aiming at an objective solution to the problem of boundary judgments, but aim at a critical solution instead (cf. 1991, p. 107).
8. Should build bridges between systems philosophy and practical philosophy. (cf. 1991, p. 247).
9. Should not confuse non-technical rationality with irrationality (cf. 1991, p. 251).
10. Should complement the "monological" utilitarian dimension with the communicative dimension of rational practice (cf. 1991, p. 251).
11. Should go beyond the actual state of practical philosophy and develop practical ways of mediation between the divergent requirements of cogent argumentation (on the part of the involved) and democratic participation (on the part of the affected; 1991, p. 253).
12. Should dare to be utopian. Should change from today's biological inspired systems concept . . . to a systems concept that would be inspired by practical philosophy's emancipatory utopia of a community of autonomous and responsible citizens (cf. 1991, p. 260).
13. Should not think that its standpoint is ever sufficient to validate its own implications (1991, p. 262).
14. Should not misunderstand itself as a guarantor of socially rational decision making (cf. 1991, p. 263).

Flood

1. Should recognize how non-sovereign power relations (Foucault) subjugate knowledges and prescribe which knowledges are operative in a society (cf. 1990, p. 42).
2. Should not be conceptually reflexive (cf. 1990, p. 23).
3. Should investigate its historical and developmental roots (cf. 1990, p. 24).
4. Should mix metaphors to obtain a rounded appreciation of situations (cf. 1990, pp. 80- 81).
5. Should consider status questions normative questions in disguise (cf. 1990, p. 82).
6. Should not use the concept "system" to describe real world things but should save it for systems thinking (cf. 1990, p. 98).
7. Should use metaphors critically (cf. 1990, p. 100).
8. Should acknowledge that "system" is not a concrete thing but an abstract concept that constitutes particular relationship that can be actualized in a number of ways (1990, p. 128).
9. Should remain self-reflective with respect to particular and all positions or approaches (1990, p. 165).
10. Should "Think systems, but don't ever assume to grasp the whole!" (1990, p. 177).

CHAPTER 9

BANATHY

ABSTRACT

Bela H. Banathy has a philosophical and idealistic bent. In our new information age, he envisages a culture in which people use design technology to benefit themselves, their societies and their ecology. He pulls together multiple strands of design wisdom, methodology, and practical experience and enlivens them with his fervor for ethics, transcendence, and design culture.

Banathy expresses outrage over the way that institutions, such as science and education, stunt human potential because of their adherence to a mechanistic operational paradigm. He documents how this mechanistic paradigm fails the human spirit and shortchanges the future. Being very much in the tradition of Churchman's "hero," he is an apostle of the systems vision. He asks the evolutionary research community: "Is the improvement of the human condition OUR field? Is the issue of human betterment our business?" (1993b, p. 17). His answer is YES.

In *Designing Social Systems in a Changing World* (1996), Banathy presents a compendium of four decades of research into systems design. He presents an in-depth knowledge base of design and related fields, and encourages his readers to generate their own personalized theories of design. He demonstrates how mechanistic systems are destructive in our information age. He urges a design culture in which people use design technology to replace dysfunctional structures with systems that enable fulfilled living.

Along the way, because of the encyclopedic character of this book, Banathy presents the theories, insights, and phrases of the giants of systems theory. He gathers decades of research into one volume and recapitulates the progress that has been made in systems design during the past 40 years.

The following discussion of Banathy's work deals with (a) the new world of organizational culture, (b) the idea of design, (c) the process of idealized design, the subordinate processes of (d) transcending the existing system and (e) imaging the new system. In addition, he summarizes philosophical reflections on the nature of systems thinking. He also shows the need for a design culture.

CHARACTERISTICS OF OUR INFORMATION AGE

We live in a world of new realities. In our new era, according to Bell (1976) and Banathy, society's creative focus has shifted "from producing goods to generating information and knowledge" (p. 43). Our society is characterized by: "(1) the centrality of knowledge, (2) the creation of new intellectual technologies, (3) the spread of a knowledge class, (4) a massive change in the character of work, (5) a focus on cooperative strategy, and (6) the central role of systems science" (p. 43). Banathy constructs the following table of our changing organizational culture:

The old story	The new story
Fixed, bureaucratic structure	Flexible and dynamic
Status-laden and rigid	Functional and evolutionary
Power resides at top	Power shared by empowerment
Motivate, manipulate people	Inspire, care for each other
Compliance is valued	Value creative contribution
Focus on problems	Focus on creating opportunities
Blame people for failure	Support learning from failure
Short-term focus	Long-term focus
Past regimen reinforced	Innovation and novelty nurtured
Work within constraint	Seek the ideal
Progress by increments	Progress by leaps
Technology and capital based	People and knowledge based
Linear/logical/reductionist	Dynamic/intuitive/expanding
Emphasis on high volume	Emphasis on high value
Insisting on "the right way	Encouraging learning/exploring
Driven by survival needs	Desire to develop, fulfill self
Motivated by production	Personal/collective satisfaction
Need external acknowledgment	Acknowledgment from self
Adversarial and competitive	Cooperative and supportive
Goals are to succeed, and to get ahead	Aim at having individual and collective integrity

Banathy (1996) *Designing Social Systems in a Changing World*, p.45.
Reproduced by permission of Kluwer Academic/ Plenum Press.

Because of the massive changes wrought by this change from industrial to information technology, the institutions inherited from the industrial age are in major ways dysfunctional. The old institutional machinery requires more than a major retooling, it needs to be scrapped and redesigned. In our new situation, social innovation and its new technology-social systems design-come to the fore as essential ingredients of survival because old technologies are confronting their ecological limits. We need to shift "from a trial-and-error, piecemeal tinkering with our systems to their radical transformation by design" (p. 42).

Banathy formulates a new story that combines realistic appraisal of what is going on in the world, critical evaluation of the organizational characteristics that are feasible in our new age, and a systemic idealism that emphasizes holistic values. He offers images of what our new age can be like. He invites us, at this moment of society's bifurcation, to work to make a desirable future real. He asks us to engage in the process of design.

THE IDEA OF DESIGN

We have known since the time of Plato that social systems are purposeful and that they can be guided by creative design. In our age, we have come to see those systems as evolving, as composed of bewildering systemic complexity, and as requiring enormous information management for their functioning. The worker of our new age is the knowledge worker, and knowledge is the capital of the age. "Therefore, design becomes the central activity in social systems, and competence in design becomes a capability of the highest value" (p. 15).

Social systems design is more complicated than the design of material systems because of the fluidity of social contexts, forms, actors, and boundaries. Also there exist multiple and often unchartable interactions within and among social systems. For that reason, social systems design is systemic and not merely systematic; in other words, it is a "creative, disciplined, and decision-oriented inquiry, carried on in iterative cycles" (p. 16); it is not a definitive once-and-for-all-time solution. Iteration in social design is necessary because it allows the testing of alternative solutions, integration of insights, the formulation of viable strategies, and a conscious attention to shifting parameters.

Design and science both employ disciplined inquiry. Where science is concerned with what is, "design is a process of creating things and systems that do not yet exist" (p. 17). Science is principally an academic discipline that is concerned with matters of fact. Design is a practical discipline that deals with technology in the broadest sense, as strategies and techniques for satisfying human needs. Design is the core element of professional activities.

Planning, problem-solving, and design share several characteristics: disciplined thinking, purposefulness, creativity, collection and evaluation of information, decision-making, and synthesis. Design differs from planning in that it deals with situations of greater complexity, ones that cannot be broken down into manageable pieces and which require iterative reformulations. Designers are not interested in merely "getting rid of what is not wanted," they design "for interdependence and internal consistency across all systems levels by integration" (p. 19). Design differs from problem-solving by its orientation towards the future of the whole system. It creates an image of the desired state, selects approaches, and models the most promising alternative.

Design confronts interrelated complexes of problems. It confronts "a system of problems rather than a collection of problems" (p. 29). These kinds of problems are embedded in uncertainty; they are "ill-structured and defy straightforward analysis" (p. 29). Such problems are the "wicked" ones that are discussed by Rittel and Webber (1984) in the following terms:

- It is not possible to provide a definitive formulation of a wicked problem . . .
- There is no stopping rule for wicked problems. The designer can always try to seek a "better" solution . . .
- Solutions to design problems are good or bad. They are not true or false. Judgments of "goodness of fit" of the solution may vary . . .
- There is no ultimate test of the solution of a wicked problem . . .
- There is no exhaustively definable set of solutions to wicked design problems . . .
- Every design problem can be considered to be a symptom of another problem.
- In the design world of wicked problems, the aim is not to find the truth but to design systems that enhance human betterment and improve human quality (p. 31).

A complex of these kinds of problems is sometimes called a "problematique," which Peccei (1977, p. 61) discusses as follows:

Within the problematique, it is difficult to pinpoint individual problems and propose individual solutions. Each problem is related to every other problem; each apparent solution to a problem may aggravate or interfere with others; and none of these problems or their combinations can be tackled using the linear and sequential methods of the past (in Banathy, p. 29).

In our age of uncertainty, we live in rapidly evolving processes in which our solutions do not stay solved.

There are four ways that people and systems react to situations of tumultuous change. Some try to go "back to the future" in a *reactive* mode that tries to find comfort in old verities. Others lean to an *inactive* mode in which they do nothing and go for the ride without rocking the boat. Still others anticipate change, prepare for it, and exploit its possibilities and thus ride the tide in a *preactivist* style. Finally, some people *interact* with a desired future and help bring it about by participating in its design and implementation in a style that Banathy calls "shooting the rapids."

The interactive style is the one that best conforms to the model of evolutionary progress and it suits our age of accelerating change. The interactive style is the only one that learns from the past, values what is good in the present, and takes responsibility for the future. It focuses on harnessing individual and collective aspirations, creativity, and intelligence for the purpose of seeking the ideal. On the basis of that ideal, it gives direction to change and shapes the future by design (p. 42).

Designers synthesize their viewpoints by creating models. With models, they are able to work on the beginning and ending of problematic situations, to analyze and synthesize them in overlapping timeframes.

The first step toward building a model is to assess the present complicated state of the problematic situation, what Ackoff calls "formulating the mess." The mess is "a set of interdependent problems that emerges, the problems being identifiable only in interaction" (p. 58). In making this assessment, the designer examines what would happen if nothing changes in the system and what would happen if individual parts were to change. After gaining some systemic insight into the current state of affairs, the designer selects the appropriate ideals for the system and formulates its idealized design.

IDEALIZED SYSTEMS DESIGN

For Ackoff (1981), ideals are basic to social systems design because designers conceptualize the systems that they want; that is, they design idealized systems. To be successful, their idealized designs must be technologically feasible, operationally viable, and capable of rapid learning and development. Idealistic designs are not utopian; they are "the most effective ideal-seeking system of which designers can conceive" (Ackoff, p. 107).

As technology evolves, some idealized designs become feasible that were once utopian. In the field of the design of design, for instance, computer assisted inquiry (like the CogniScope™) makes possible mathematical rigors that were previously infeasible and unimaginable. Still, no design should incorporate unknown or unusable technologies.

Designed systems must work in the field not just on paper. Also they should function after the design team has departed, not just during their presence. These requirements express a design's operational viability. For this reason, the stakeholders in the design must be participants in the designing.

Several conditions should be part of the designed process to facilitate its ability to rapidly learn and adapt. They are:

1. empowerment of the stakeholders to modify the design whenever they wish;
2. introduction of design experimentation in order to resolve issues that emerge in the course of the design;
3. use of processes that enable the system to learn from its experience and improve its design over time; and
4. the establishment of evaluative/monitoring processes in the course of the design inquiry that introduce corrections in the design whenever indicated (Banathy, p. 192).

The idealized systems designer does not aim to create an ideal system; rather he or she aims to create an effective ideal-seeking system.

To accomplish its goals, idealized system design cannot be a top-down operation nor can it be expert driven. It must actively involve the stakeholders of the design in shaping a shared vision that represents their ideas, aspirations, values, and ideals. It draws vitality and viability from that vision. It rewards participants not just with a product that approximates their vision, but with a process of dialogue and interaction that embodies their desires for community and effectiveness.

In the process of design, people put their vision "out there" on the horizon of their aspirations. As they do their design work, they move toward that horizon. As a result, their horizon changes; that is, they see their ideal model from a different perspective. Therefore, they need to continually review their ideal models with an eye toward revising and improving them. This is the "law of the moving horizon" (Banathy, p. 195).

Idealized systems design is an imperative of our age and it requires user designers with a certain amount of expertise and confidence. Unfortunately, social science departments, professional schools, and even R & D agencies are largely ignorant of its techniques and oblivious to its importance. Our information age, therefore, challenges these institutions:

> (1) to become experts in the intellectual technology of ideal systems design (ISD); (2) to develop professional preparation programs in ISD; (3) to prepare resources and programs that people in our communities can use to "get ready" and to engage in ISD; and (4) to assist our communities in their design learning and in carrying out their design program (p. 196).

THE PROCESS OF TRANSCENDING

One must conceptually leap out of an existing situation if one is to   envisage a new system or radically redesign an existing system. Such leaping is fraught with emotional and conceptual peril. To lessen the peril, Banathy offers an option field framework that allows designers to map the parameters of their task before they plunge into it. "Using the framework, designers can create an option field, consider major design options, and explore the implications of their use" (p. 63).

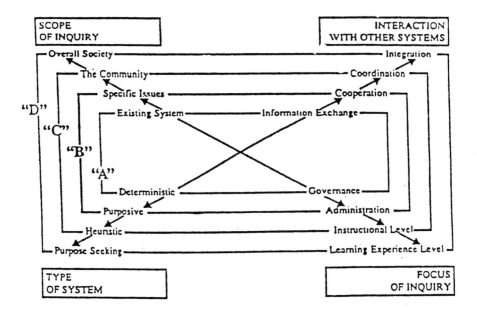

Banathy (1996) *Designing Social Systems in a Changing World.*
Reproduced by permission of Kluwer Academic/ Plenum Press.

Option fields are customized to the demands of specific situations, but they have common characteristics. These common characteristics can be found in a sample option field for education. This representation [confer insert] consists of four concentric areas in which the inner area is the most constricted. If designers work in this inner area, they consider only the existing educational system. In this inner area, participants merely share information with other agencies and systems; they are interested only in governance; and they develop in only deterministic ways. It is different for designers who work in the outermost area. They consider how education can integrate the learning experience with other vital experiences in the society for the purposes sought by the learner.

The four concentric areas are defined along four axes: scope of inquiry, interaction with other systems, focus of inquiry, and type of system. The diagram shows that each of these axes allows four degrees of generality. The scope of inquiry can be the existing system, specific issues, the community, or the overall society. The degree of interaction with other systems can be information exchange, cooperation, coordination, or integration. The focus of inquiry can be governance, administration, the instructional level, or the learning experience level. The type of system can be deterministic, purposive, heuristic, or purpose-seeking.

By choosing on these axes the degree of generality in which design will take place, designers specify the boundaries of their design and the degrees of feedforward and feedback that they will accept from the design's environment. As they expand the boundaries toward their limits, they approximate an ideal educational system while simultaneously making its implementation more problematical. To use the option field properly, designers must make professional judgments about what level of generality and idealism they are going to achieve.

> [The option framework] creates a place where designers can land as they leap out from their existing system. The framework creates a "comfort zone" in which the anxiety of leaping out melts away and the excitement of entering into the space of envisioning, imaging, and creating replaces troubling uncertainty and the fear of the unknown (p. 64).

IMAGING THE NEW SYSTEM

The basic bond of any societal group is its cultural cognitive map or image of itself. Two propositions of Boulding describe the importance of the image. The first proposition says, "Behavior depends on the image. The behavior of designers of social systems depends on their image of their society and the image they have about the societal function of the system they wish to create" (p. 69). The second proposition says, "Our experiences provide us with messages that might produce changes in our image" (p. 69). Most of our experiences do not produce major changes in our image, because images are resistant to change. Sometimes however the message of an experience hits the nucleus of the image, then "the whole thing changes in quite a radical way" (Boulding, 1956, p. 8). If the message persists, we begin to doubt our image. "Then one day we receive a message which overthrows our previous image and we revise it completely" (Boulding, p. 9). In our current society, old images are in doubt; new messages do not resonate within them. If we are to avoid putting new wine in old wineskins, we need to revise our images of societal institutions.

To create an image of the new system, the various stakeholders dip into their sense of justice, rightness, and feasibility. They also tap their own self-interests. They express their ideals in one of the various formats that have been developed in systems science and reduce them to an agreed upon list. The image then becomes the touchstone for evaluating subsequent design possibilities.

TRANSFORMATION BY DESIGN

Imaging the new system is one cycle of the dynamic of divergence and convergence that characterizes design. Designers diverge in the imaging stage to encompass a variety of ideals, then they converge to fix the image of the future system. Consequently, they diverge to consider possible structures and design features that might bring that image to reality, then they converge to agree upon the model of the future system. Further on, they diverge to consider ways of implementing the system, and converge to implement specific techniques and strategies.

SYSTEMS PHILOSOPHY

Two important systems philosophers, Vickers and Jantsch, have been slighted in this book thus far. What follows is a synopsis of some of their thinking as Banathy presents it.

Geoffrey Vickers (1981) was interested in "the effect of systems thinking on our outlook on life and our philosophy of life" (p. 19). According to him, our world is one of active and dynamic reactions "in which stability, not change, requires explanation" (p. 19). Inner relations are the essence of what things are. "We are accustomed to regard 'relating' as something which entities do, rather than something which they are. Should we rather view all entities as systems, created by their relations which sustain them?" (p. 20). A practical question suggests itself: "When does a system retain its identity and continuity through change and when does it itself vanish or become something new?" (Banathy,   p. 161). Social systems must meet demands for both stability and success (which involves change and therefore instability). Vickers says, "I find it ridiculous to try to reduce the second to the first as so many people try to do" (p. 21).

Erich Jantsch (1980a) shaped formative ideas about how the universe self-organizes itself. He stressed that the primary orientation of self-organizing structures is process, and that solid components and structures are of secondary concern because "the interplay of processes may lead to the open evolution of structures. Emphasis is then on becoming-and even being appears in dynamic systems as an aspect of becoming" (p. 6).

Social systems are processes without the deterministic equilibrium and solidity of technological structures. They are concerned "primarily with renewing and evolving themselves, which are essentially learning and creating processes" (Banathy, p. 162).

> When a system, in its self-organization, reaches beyond the boundaries of its identity, it becomes creative. In self-organization, evolution is the result of self-transcendence. At each threshold of self-transcendence a new dimension of freedom is called into play for the shaping of the future (pp. 183-184).

As individuals and societies, we embody the progress and meaning of life. "We are not the helpless subjects of evolution-we are evolution" (p. 8).

MULTIPLE PERSPECTIVES AND ETHICS

Decisions about what should be are constantly facing designers. Such decisions should be authentic, informed, and viable. They are authentic "to the extent that they are made by all the people who constitute the designing community, namely by all those who are affected by the future system" (p. 172). They are informed "to the extent that they are based on all attainable information and knowledge that is relevant (1) to the overall context and content of the systems we design and its systemic environment and (2) to each and every choice and decision we are to make in the course of the design" (p. 172). They are viable "to the extent that they are made in consideration of multiple perspectives and a variety of points of view that are relevant to the inquiry" (p. 172). Decisions are authentic, informed, and viable to the extent that they employ all necessary perspectives.

Linstone and Mitroff (1994) propose three classes of perspectives that designers should utilize: the technical, the organizational, and the personal or individual. Each of these perspectives provides insights that are not available in the other perspectives. The characteristics of the technical perspective (T) are:

1. problems are simplified by abstraction and isolation from the real world, assuming that this approach permits solution of problems;
2. data and models are the focus of the inquiry;
3. logic, rationality, and objectivity are presupposed;
4. order, structure, and quantification are pursued;
5. experimentation and analysis aim to attain predictive validity;
6. validation of hypotheses and experimental replicability are expected (p. 174).

The T perspective is the very successful perspective employed in science and engineering. In social situations, the T perspective works well with the tame and well-structured problems that are easily handled with operations research, systems analysis, and econometrics. It does not work as well in complicated situations such as urban planning, criminal justice, and health care. Moreover, any attempt to separate the technical part of the problem for analysis leads to the formulation of peripheral and irrelevant problems.

The organizational perspective (O) focuses on process rather than product; it attends to actions and solutions instead of problem analysis. For it, "the critical questions are: Does something need to be done? If so, What? Who needs to do it? How?" (Linstone and Mitroff, p. 98). Within any organization there will be a number of O perspectives that need to be taken into account.

The personal perspective (P) is that of the unique individual. It sweeps in aspects that the T and O perspectives lack. P perspectives open onto the vista of personal motivations, moral feelings, mental and spiritual values. The uniqueness of individuals is an important consideration for complex systems.

These perspectives are integrated in thoughtful systems design principally by informal, consensus-building methods. During the whole process of successful design, "cross-cueing and feedback assure that important information is not overlooked. Bringing differences among the perspectives to the surface facilitates constructive resolution and integration" (Linstone and Mitroff, p. 117).

Banathy suggests a fourth perspective that is not overt in the T, O, and P perspectives. It is the sociocultural perspective (C) that is particularly important in dealing with public and non-profit social systems. The C perspective sweeps in the viewpoints of all the stakeholders in a design situation-even the victims of the design.

All four perspectives should be involved in the work of design. Particularly when the stakeholders are the designers, they should be empowered to:

1. use the T perspective, to apply the strategies and methods of the intellectual technology of systems design;
2. learn approaches and methods of how to "give voice" individually and collectively to the O and P perspectives; and
3. explore the C perspective and use the findings of this exploration as an input to design (Banathy, p. 177).

Also, if systems inquiry is to function at a high ethical level, it must recognize that "no single perspective can attain ethical systemic behavior. Ethical social systems inquiry implies balance among personal and moral action, rational and technical inquiry, and just organizational behavior" (p. 179). Integrating these perspectives requires that:

1. Norms of morality of the individual are to be linked to the larger enfolding sociocultural system;
2. Connection has to be created between the technical solution and organizational/institutional justice; and

3. Connection should be made between the technical inquiry and personal ethics (p. 180).

In a design situation that is genuinely participatory, "a key task . . . is to enable the designing community to use the appropriate T perspective, to understand and respect the P perspectives of its members, integrate those perspectives, and forge and O perspective that is acceptable to all" (p. 180).

A group at an International Systems Institute gathering approached the concept of ethics, as it applies to group decision-making with the following triggering question: "How should we deal with the 'blind spot' that each of us has as we see the world from our own moral perspective?" (p. 180). The group involved in the conversation answered the question with a metaphor: "I can see the other person (and vice versa) but cannot see myself seeing the other person, unless I hold a mirror before me. Ethics provides such a mirror" (p. 181). People discover their blind spots in conversations that focus on values when they listen to one another.

Morality concerns the norms, rules, and guidelines that individuals and groups employ to guide their behavior. "Ethics employs principles to solve conflicts between individuals/groups having different moral perspectives. Thus, ethics can be viewed as second-order morality" (p. 181). In the honoring of diverse moral viewpoints, a design group brings more complexity into its discourse. "It also empowers the conversation with the capacity to deal with increased complexity" (p. 181).

There are three parts of systems design in which ethics are a concern: the design methodology, the behavior of the designing group, and the product of the design. The design methodology should incorporate the multidimensional elements that we have been discussing. The designing group should be frank, earnest, and on its good behavior. It should work with the realization that (1) good designs are the result of good designing; (2) "no one has the right to design social systems for someone else" (p. 180); (3) "a design decision is ethical if the stakeholders (the designing community) give their informed approval" (p. 180); and (4) "participants are not coerced" (p. 180). Finally, "the product of the system design should include explicit ethical standards as guidelines of system behavior" (p. 181).

There are multiple values that inform the morality of individuals and groups and multiple perspectives for contemplating the rightness and wrongness of designs. The ethics advocated by systems theorists include: self-realization, social responsibility, ecological responsibility, evolutionary responsibility, honoring multilevel values, and caring for the whole system. All of these perspectives are to be honored in the design situation. A brief listing of succinct statements of those perspectives follows:

- The proper end of all individual experience is the evolutionary and harmonious development of the self (both as a person and as a part of wide collectives) . . . The appropriate function of social institutions is to create an environment that fosters that process (Markely and Harman, 1982, p. 115).

- A more human world will be a more socially responsible world and this responsibility will have costs and benefits, limitations as well as enlargements in terms of self-actualization (Vickers, 1982, p. 226).
- This [ecological] ethic views man as an integral part of the natural world. It implies the movement toward a balance between the economic, social, and ecological systems . . . It is a concern for the coordinated and balanced well-being of social groups (social ecology) and cultures (cultural ecology; Banathy, p. 183).
- Ethical behavior enhances evolution. Ethics emerges with evolution and it takes on a regulatory function. This regulatory function in the human world consists of rules of behavior but also of morality as a distinct human experience. As an integral part of evolution, ethics is experienced directly by way of the dynamics of self-organization and creative process (Jantsch in Banathy, pp. 183-184).
- [The ethics] that dominates the Western world is . . . an individual ethic in the disguise of a socially committing behavioral code. It is not a multi-level ethic in the true sense. Morality, in contrast, is the direct experience of ethics inherent in the dynamics of evolution. The higher number of levels and the intensity at which we live, the higher the number of levels and intensity at which our morality becomes effective (Jantsch, 1980a, p. 265).
- What is so remarkable is how similar the world's major moral systems are in considering "good" to be the achievement of the kind of harmony within consciousness and between people that we have called negentropy, and which in turn leads to higher levels of complexity (Csikszentmihalyi, 1993, P. 160).
- The problem of systems improvement is the problem of the ethics of the whole system (Churchman, 1968, p. 4).
- The moral law with respect to future generations is: We should undertake to design our societies and their environments so that people of the future will be able to design their lives in ways that express their own humanity (paraphrase of Churchman, 1982, p. 21).
- Ethics is an eternal conversation. The reason that ethical relativism ("different strokes for different people") is so bad is that it stops conversation (Churchman, 1982, p. 57).
- The realist is a down-to-earth, practical person who tries to solve the practical, hard problems of everyday life in a practical, coherent fashion. The realist goes to management development programs and expects to find out what to do next Monday to become a better realist. The idealist tries to understand the saga in terms of human ideals and their meaning in the very long run, and he sees that there is a constant struggle towards an ideal society. He tries, as best he can, to explain what that ideal might be (Churchman, 1982, p. 133).
- The new, the hopeful, and the creative appear to be more ephemeral than what was tried and worked. The realist, who deals with the here and now belittles the "impractical" idealist who invests energy in the stuff of a "blue-sky" world. Without the realist we could not survive. But without the idealist we could not evolve (p. 185).

THE PRIME DIRECTIVE

Design enables collective action. It is of crucial importance in our information age because of our constant need to create new systems and variations of old systems in response to turbulent change. In this age, "we design buildings, clothing, laws, processes, packaged food, power plants, all kinds of organizations, curricula, cars, and weapons-the list is endless. Take away designs and we strip the world of most of its enabling mechanism" (p. 35). Some designs are high quality; some are low quality; and some are destructive. "We are at the mercy of past and present designs, crafted for us by the experts" (p. 35). What can we do?

> We have a choice. We can continue to be uninformed design illiterates. We can accept design decisions made by the experts, and continue to live in ugly and crowded block-houses, drive dangerous and polluting cars, use shoddy goods . . . and suffer outdated educational designs. Or we can become design literate (p.36).

Taking control of our future and shaping our individual and collective destiny are the natural ideals of a free people. To do that, we have to develop literacy and competence in design. Therefore, "the building of a design culture is an inescapable necessity" (p. 35). When people are competent to make their own decisions, then they can personify a real democracy. "Participatory democracy comes to life when we individually and collectively develop a design culture that empowers us to create, govern, and constantly reinvent our systems" (p. 37).

CHAPTER 10

WARFIELD

ABSTRACT

Warfield's concern in *A Science of Generic Design* (1994) is the management and design of large-scale systems. In his preface, he says,

> For better or worse, our society has accepted the idea of large and complex systems. If we are going to have them, it behooves us to learn how to manage them. An excellent route to doing so is to learn how to design them (p. xxiii).

Large, complex systems present numerous difficulties to that overwhelm designers' abilities to design ad hoc. They call for a generic science of design that integrates the most important elements of universal knowledge with the knowledge that designers have thus far amassed.

Warfield lays a solid base for generic design science and articulates its ramifications in the straightforward manner of cybernetic circuitry. He explores first the bases of science in the pragmaticist manner of C. S. Peirce. In so doing, he incorporates elements that are usually omitted from the philosophy of science and integrates them into an embodied theory of science. In a peripheral way, he indicates how his design of science is more adequate than conventional conceptions.

He defines the roles of formal object language and natural metalanguage. He develops relational descriptions of qualitative reality and incorporates them into a non-numerical mathematics of dimensions. He integrates this qualitative mathematics with universal priors to create the science of generic design.

THE FOUNDATIONS

Peircean science incorporates the knower into the definition of knowledge. It deals with the triad: object, concept, knower (not with just the dyad of object and concept). This science is embodied. It deals overtly with the blind spot of conventional Cartesian science: the embodiment of the knower. It does not counterfactually assume a disembodied or cosmic mind that can objectively declare about truth and falsity.

In this tradition, Warfield is up front in dealing with epistemological questions and the communication linkages between foundations, theory, application, and management. He talks openly about the knower's involvement in creating realities, and the psychosocial nature of human participants at every stage of the design process.

In the science of generic design, questions of epistemology, methodology, and communication are critically important. If these ideas are thoroughly developed, they can lay to rest a major misunderstanding about social science. A thorough application of these sciences will prove that science does not have to be quantified. To the contrary, as Warfield shows, science requires precision but not necessarily quantification. One can use relational definitions in the formal-object language of a science with natural language as its metalanguage to create a mathematical space of scientific design.

Warfield identifies four Universal Priors of science: the human being, language, reasoning through relationships, and means of archival representation. In developing these concepts and weaving them together, he gives the essential context that fills in the lacunae of abstract theory. He joins these Priors with Peirce's Cosmic Partition: Library, Phaneron, and Residue. In this way he provides a sound foundation for a design science that has to deal with the many dimensions of social systems.

The Library section of the Cosmic Partition consists of all the information contained in human media. The Phaneron is the totality of ideas that exist in the minds of people. The Residue is the remainder of the universe with the Library and Phaneron removed from it. The process of science is to transform the Residue into the Phaneron and the Library; and to constantly recycle and revise its findings. Design science deals with the use and transformation of information from Residue and Phaneron into various media in the theoretical, methodological, technological, and managerial spectrum. Accordingly, it consists of observing, inferring, implementing, and archiving.

Design traditionally consists of three dimensions: the situation, knowledge about the design target, and the processes of design. Today in the design of large-scale and sociocultural systems, three further dimensions must be taken into account: standards of behavior, the history of critical incidents, and insight into human limitations. The proper priority structure of these dimensions is: standards of behavior, first, followed by history of critical incidents, insight into human limitations, knowledge of the design target, processes of design, and the situation.

If this priority structure is followed, the disasters caused by ill-formed design will be largely avoided.

THE HUMAN BEING

The first of the Universal Priors is the human being. Because human participation is omnipresent in knowledge and design, the nature of that participation should be explicit at every stage: in its foundation, postulates, theory, selection criteria, methodology, roles, environments, applications, and evaluation. To do this, relevant information about human beings must be garnered from the social and behavioral sciences and from philosophical writings.

Research has shown that people approach group behavior with both task-oriented and emotional behaviors. In unorganized group activities, the behavior within groups is aptly called "forming, storming, norming, and performing" (p. 38), all of which amount to social overhead costs that add little to the achievement of the task at hand. Often groups achieve a bogus consensus called "groupthink," in which the members of a group make decisions without access to substantive information. In general, groups have problems of communication, evaluation, control, decision-making, tension reduction, and reintegration. As Warfield puts it: "the unbelievable is normal, the bizarre is the social norm, and what would reasonably be expected in a society that adhered to reasonable behavior is abnormal whenever large systems are involved" (p. 41).

The origins of this bizarre group behavior can be found in our Virtual Worlds (VWs). A VW is largely inaccessible to its possessor. It constitutes "an unshakeable cognitive burden" (Peirce) that an individual brings into every life experience. "In the VW of every individual there may lie components of value to any Design Situation that may involve that person in some way; and likewise . . . every individual contains in that same VW components that may be harmful to that Design Situation" (p. 41). Within a design group it is important to have members who represent a full spectrum of knowledge, interest, and temperament. It is also necessary to somehow make the design situation palatable to all participants. The science of design should create tolerable conditions for creating choosing designs.

> Every human being who desires political and religious freedom has an interest in the creation of a science of design that offers the possibility of replacing what might be called evil designs with a widely understood and supported design process that is less antagonistic to open reasoning and open choice; and which takes into account views of the many as expressed through representation, even though it may be imperfect (p. 42).

A factor of major importance in dealing with complex situations is our weak human capability to form deep logic patterns. This weak capability is a salient aspect of our bounded rationality, our inability to comprehend a situation in all its interrelationships. The key aspect of this weakness is our general inability to hold more than approximately seven items in our short-term memory. In other words, we "cannot reason simultaneously about the interactions of more than a small, limited number of factors" (p. 43). For that reason systems need to be modularized in ways that are compatible with this limitation. In practice, each module should contain at most three interacting elements because these three will have four combinations and the sum of elements and combinations will be seven.

LANGUAGE

The second of the universal priors is language. There are our natural everyday languages. There are also specialty languages that remove ambiguity and make synergistic operation possible for people and machines. A specially designed language for a science is called its "object language." It requires a metalanguage to talk about it. It has been shown by Godel, however, that object languages cannot provide their own metalanguages. Therefore, an object language needs to be combined with a natural language that serves as its metalanguage. In other words, good science and design require meticulous object languages that are explained in and usable by natural language.

In addition to the language of oral and written prose, there are two other basic language types that are necessary for science and design: mathematics and graphics. Prose is sequential in its delivery and its reception; it is necessary for putting ideas in order, but deficient in its presentation of gestalts. Mathematics produces gestalts by condensing prose into powerful and unambiguous symbols.

Graphics present a landscape kind of description "that inherently incorporates the capability of the eye to see the contained structural relationships" (p. 47). Because of our bounded rationality, graphics are necessary to help our brains comprehend organization. For the purposes of design, a combination of structural graphics and prose is most useful, both for ordinary communication and for producing a precise object language.

A good hybrid language system expresses organizational relationships in precise prose and graphic representations. It works synergistically with ordinary language. Such a hybrid language must be an integral part of any computer interface that makes generic, large-scale design feasible. It respects our bounded rationality, makes deep/long logic both possible and visible, and enables referential transparency in the design process. The object language produced for the Science of Generic Design is called GRAILS, (a Graphically Integrated Language System).

REASONING THROUGH RELATIONSHIPS

The third of the priors is reasoning through relationships. To develop design science, Warfield extends analytic philosophy. He examines the following analytic topics: the types of inference and their respective roles in reasoning; the types of logic patterns; the propagation of relationships; and some structural metrics.

According to Peirce, there are four kinds of inference: perceptual judgment, abduction, induction, and deduction. Perceptual judgment is an internal process that interprets sensations and turns them into perceptions; it provides the bulk of an individual's Virtual World. Abduction is Peirce's word for the process in which intuitions achieve form in theories, conjectures, hypotheses, and explanation. It is a synthetic process that supplies new ideas to the Virtual World. Induction acts upon abductions to test them on the basis of observation and experiment. Deduction uses abductions as postulates for a system of necessary inference through formal reasoning. By using inferences, we move ourselves toward either doubt and disbelief or inclination and belief.

Mathematical logic deals with the patterns of relations that exist among explicitly defined generic concepts. It is an object language that functions with the utmost generality (a structure without content). It has blanks that can be filled by any appropriate content. It organizes knowledge "as the organic integration of contentless structure with structureless content" (p. 54).

Logic helps us overcome the limitations of short-term memory. It requires specific well-formed operations for its success. When it is properly used, it generates structures of deep and long logic that make transparent the relations that exist among parts of systems.

Logic and mathematics are two outstanding exemplars of transparent structure. Mathematics has at hand the universally accepted precision of numbers. Logic, however, often deals with qualities that cannot be numerically quantified. For that reason, logic must rigorously rely on semantic and logical precision.

Semantic precision requires definitions. These can be achieved in four principal ways: by naming, by extension, by intension, and by relationship. *Naming* is the simple operation that we are all familiar with; it is the most capricious form of definition. *Extension* defines a class name in terms of exemplars within that class; as is done, for example, in the prototype theory of Eleanor Rosch. *Intension* is the classical mode of definition which "defines the Concept by citing that set of Attributes that is perceived to be integral to the Concept" (p. 56).

To define a concept or class by *relationship*, one identifies how that concept is related to other factors be it operationally, by inclusion or exclusion, by relations of priority, and so forth. Then one gives a comprehensive context or map of the relationships that define it. The process has several stages.

In order to create accurate relationship maps, one must define both the types of relationship and their extent. First of all, the terms "relationship" and "relation" require definition.

The term "relationship" is used to represent an interpretive concept from everyday language that expresses how two or more entities are related. The term "relation," on the other hand, is reserved for a particular formal concept from mathematics (p. 59).

The association among these concepts requires an attention that it seldom receives. It cannot be assumed that logical relations easily model relationships. In the design of complex systems, every association, its logical basis, and the assignments that are implicit in it need overt consideration.

There are six major categories of interpretive relationships that need to be considered: definitive, comparative, influence, temporal, spatial, and mathematical. A *definitive* relationship is one in which A is a constituent or component of B; it is expressed by terms such as: includes, implies, covers, and is in the same category as. A *comparative* relationship is one that assesses A and B on the basis of some commonality such as greater than, preferred to, of higher priority, and so on. An *influence* relationship describes how B is affected by A. In other words it describes how A causes, aggravates, enhances, supports, or interrelates with B. A *temporal* relationship indicates that A occurs before B, that it must precede, requires more time than, overlaps in time with, and so on. A *spatial* relationship indicates relative position: east of, above, crosses in a plane, and so on. A *mathematical* relationship expresses "in mathematical symbolism, either a logical condition involving A and B, or a condition of calculation involving A and B, such as A is the square of B" (p. 60). A mathematical relationship can be expressed in such terms as: is a function of, affects the likelihood of, is computable from, and is disjoint with.

Unlike the relationships between them, which are few in number, the number of elements that may interact in a system may be innumerable. In this situation, it makes sense to concentrate on the relationships with which our intuition can grapple, rather than on the elements that are mind-boggling. The relationships can be starkly identified according to their properties. The elements can only be limited according to judgments of their saliency that are made by unaided or enhanced intuition.

Among relationships, the ones that are most important for deep and long logic are the ones that propagate. The relation "implies," for example, is transitive: if A implies B, and B implies C, then A implies C. Propagating relationships have a direction. Transitivity also holds for relationships like precede, is heavier than, and is west of. Not all relationships are transitive, for example, A loves B and B loves C does not necessarily imply that A loves C.

In complex systems, there tend to be long chains of propagating relationships that are difficult to handle with an unaided mind. For that reason, the "relationships" of natural language that operate through intuitions, need to be associated with the "relations" of a formal mathematical language. This association, especially when a computer handles the formal mathematics, can free creative workers from the constraints of narrow object languages. By programming this association, participants can assign deep/long logic and its documentation to machines.

Long logic refers to a number of four or more similar propagating relationships that follow each other in a rather predictable sequence. Deep logic refers to a number of four or more complicated interrelationships that concur to create a variety of results. In the GRAILS language these logical sequences are portrayed as walks in graphic space.

> Normal problems involve local logic or occasionally intermediate logic, but complex problems involve deep logic. Since deep logic is generally absent from representations, or if present is often masked by being embedded in thicket-like prose, a consequence often is underconceptualization and underdocumentation, as well as poor communication (p. 102).

MEANS OF ARCHIVAL REPRODUCTION

The fourth universal prior, means of archival representation, refers to the ways that new knowledge is recorded, how it is stored, how purely logical manipulations are performed on it, and how it is retrieved. In today's designing context, these functions should be highly automated in order to remove this kind of cognitive burden from decision-makers and designers. With that burden removed, the involved stakeholders can concentrate on substantive matters.

VALIDATION

Warfield presents a Law of Validation and its First Corollary. The Law states:

> The validity of any science depends upon substantial agreement within the scientific community of meaning at its highest grade, that is, meaning attained through Definition by Relationship (p. 96).

The corollary states:

> The validity of any science depends upon the capacity of the scientific community to construct definitions through Definition by Relationship for the full complex of relevant concepts involved in the science (p. 96).

This law and corollary establish merely necessary conditions for validation. Additional laws are required to establish sufficient conditions.

It has become customary in the social sciences to require quantification of any research that claims to be scientific. Some special attributes of number and counting give this quantitative bias surface respectability. Numbers, because of their universally understood order, can be used to count animals, vegetables, and minerals. They lend total agreement and validity to discourse when it comes to clearly distinguished units. Still, in the human sciences many important qualities are not clearly distinguishable. The cool clarity associated with numbers cannot be transferred to those qualities by attaching quantifications to them. While quantification can add to the validation of science, it does not confer validity.

"Validity in science is attained through Referential Transparency, and the latter implies deep logic" (p. 98). If a science has Referential Transparency, one can trace the logic of that science in both directions through its structure. To attain Referential Transparency, one specifies the Foundation, Theory, Methodology, and Applications of a science. One also explicates the circular relations between these stages by identifying:

- the Postulates that intervene between Foundation and Theory
- the Selection Criteria that are used to create Methodology from Theory
- the Roles and Environment in which the Methodology is Applied
- the evaluation feedback that Applications supply toward improving the design.

The Methodology and Applications part of the science is the Arena in which "action research" is conducted. The Foundation and Theory part of a science are its Corpus. Constant circulation between the Corpus and Arena creates a vital science; lack of circulation limits a science's effectiveness and validity. It is true, of course, that action research can discover feasible solutions within certain environments. But action research is "insufficient from the point of view of creating science. In the absence of any articulated steering Corpus, action research lends itself much better to problems of modest scale than to problems that are large in scale" (p. 115).

The physical sciences deal successfully in the Arena areas of methodology and application without much reference to their foundations and their theory. Warfield examines why this is so. He first constructs a GRAILS representation of the deep logic of high-utility physical science. In it, he includes foundational issues such as the nature of the human being, of literacy, language, integers, arithmetic, and representational systems. He shows how "the measurement system buffers the applications from the Universal Priors" (p. 98). The standardization that is available in physical science relieves scientists from a constant need to refer to foundations.

The social sciences lack such standardization. Until they find pervasive and universally recognized standards of measurement,

> They must rely on disciplined management of the Universal Priors for validity. Among the implications are that the language has to carry the burden that is automatically enforced by physical standards. In view of the vagaries of language, it seems very clear that creative representations with very careful delineation of relationships are absolutely necessary for validity (p. 100).

This creative and careful use of language requires definitions by relationship and graphical/prose representations.

Even this necessary attention to the Universal Priors is not sufficient for a well-formed science. If it is to gain control of its quality, a science needs additional steering mechanisms. For that reason, a model of the science is necessary.

Such a model will help overcome numerous other defects presently at work in society wherever it is related in some way to science. Use of this Model will not only make easier the integration of the Universal Priors in the Generic Design Science, but will also promote Referential Transparency in the Science itself (p. 102).

A DOMAIN OF SCIENCE MODEL

Warfield develops a general Domain of Science Model that describes the essential elements and relationships that are necessary for a science in the Peircean mode. This model explicates the foundation, theory, methodology, and applications facets of a science and pays special attention to the linkages between them. Warfield believes it is possible to integrate "parts of the recognized scientific disciplines into newer integrative sciences" (p. 114) on the basis of this model. He offers tentative classifications of the sciences in general and the conceptual sciences in particular, but these classifications are offered only in passing. His main interest in this volume is Generic Design Science.

Warfield remains true to the Peircean ideal of the embodiment of the knowers and users of his science as can be seen in the following quotation.

To provide adequate Referential Transparency in a science of generic design, at least three communication interfaces are critical. One is the interface between the producers of basic science and the users of methodology in problem-solving and design. A second is the interface between the technical people engaged in design and the managerial and administrative people who provide oversight and steering. A third is the interface between the managerial/administrative people and the general public (p. 126).

MANAGING COMPLEXITY

The principal use of a Science of Generic Design is its contribution to the management of complexity, especially the cognitive complexity that is introduced into situations where people try to function together in groups. "Cognitive complexity is the dilemma presented to the human mind when it engages with conceptualizations that are beyond its unaided powers" (p. 152). When cognitive complexity is joined to the complexity of difficult and urgent situations, it "will either be managed, or it will overwhelm the individual and the society" (p. 154).

Linkage escalation is a major cause of complexity in cooperative endeavors. It arises when a manager realizes that he or she is enmeshed in a complicated problem that prevents its clear perception and takes the next reasonable step. He or she seeks out "others who will be part of a (formal or informal) problem-solving team" (p. 135). Thereupon, a double escalation begins.

Linked to the original perception of the problem are new components of it
introduced by those who join the problem-solving team. This is the first type of
linkage escalation. A second type occurs by virtue of linking to the enlarged set of
problem perceptions all of the process-related difficulties that go along with the
attempt to get a group of people to work together to solve a problem (p. 136).

Dual roles in such situations are particularly troublesome. If managers assume the
dual roles of contributor and leader, they present an apparent conflict of interest.
Even if a group becomes somewhat functional in this first degree of escalated
complexity, it will run into further escalations because of:

the values, policies, practices, identities, or self-interests of the organization(s)
represented by the members of the group . . . If contributions toward a possible
resolution or solution or even dissolution of the problem appear to be incompatible
with organizational factors, e.g., with the value system of the organization(s),
progress may be halted. Now a solution must be seen as one that solves the
escalated problem (p. 136).

Escalations such as these frustrate the normal feedback loop in which actions are
taken, consequences are observed, matches or mismatches are observed, and actions
are taken (again). A single feedback loop such as this is adequate for short-term,
simple problems, but not for complex ones. To handle problems of escalated
complexity, a double loop methodology is required. This double loop encompasses
the single loop to study the underlying variables of the original problem and create a
solution adequate to the more complex situation. In the case of escalating
complexity, the revised problem is: how to work on the existing problem under the
constraints imposed by an organizational culture.

Double-loop learning is essential to the design process in complicated
situations. Unfortunately, organizations frequently do not know how to do it.
Warfield quotes Argyris (1982) who originated the double-loop concept as saying:

People who want to double-loop learn (1) do not know how to do so, (2) do not
know that they do not know, (3) do not know that they have designs in their heads
to ensure (1) and (2) above, and (4) do not know that they cannot learn to do so by
using their present skills (p. 137).

Even if the previously mentioned linkage escalations are surmounted, others turn
up, perhaps as an inability to communicate the solution to its implementers, perhaps
because of insufficient staff, perhaps because a new training program has to be
implemented.

In addition to problems of linkage escalation, one must manage complexity
in other areas. One must reduce the cognitive burden on individuals, eliminate
distractions in the design situation, and provide enhancement for constructive
behavior. Often individuals are thrust into positions where they are expected to
provide opinions and decisions for which they are not cognitively prepared. Good
design must provide preparation and assistance for them. The design situation can

detract from group performance; its detractions should be eliminated or minimized. There are possible enhancers for individual and group performance; these should be provided in a good design.

Warfield synthesizes the features relevant to effective systems design. He lists potential enhancements. He organizes numerous detractors that authority, science, and the working environment might impose on the design situation. Then he designs processes that incorporate the necessary elements of systems design, maximize its enhancements, and minimize its detractors. His overall design encompasses designed processes, designed environments, role specialization, leadership criteria, quality control, and attention to the cultural environment.

The designed processes are the most interesting. Three heuristic keys direct them. The first key is to concentrate on the element that is common to all design processes, that is, on operations carried out with ideas. The second key is to use set theory with its Theory of Relations as an underlying mathematical language of design. The third key is to use applicable research on human behavior and a vision of desired outcomes to specify the needed roles and working environments.

By concentrating on operations that can be carried on by ideas, one can limit one's field of concentration to five operations: to the generating, clarifying, structuring, interpreting, and amending of ideas. "The limited number of 'idea actions' means that the variety of processes needed can also be quite limited" (p. 145). Ideas can be generated by discussion based on triggering questions. They can be structured by applying knowledge available in the Theory of Relations. Ideas can be clarified through normal facilitated discourse. Information can be interpreted and amended by unaided and enhanced intuition. All of these conceptual processes are made concrete in the Science of Generic Design and in its applications.

Designed processes limit problems of escalating complexity by specifying separate professional roles in the discussion. They limit cognitive burden by sequencing the processes of design and providing computer assistance. They eliminate the poorly designed processes that lead group participants and observers to "think that group behavior itself is pathological" (p. 146). Experience with well-designed group processes is encouraging and not discouraging.

PRESENTATION OF THE SCIENCE

In the second part of this book, Warfield presents the foundations of the science of generic design, its theory, methodology, roles, and environments. In the third and fourth parts he presents its applications, documentation, and amplifications. For our purposes, the foundations and theory of the science are salient because they respond to the triggering question: what are the standards for the valid application of systems theory? The methodology is discussed in connection with the CogniScope™ technology, which is used in Chapter 19. The remainder of Warfield's book, and its companion volume, *A Handbook of Interactive*

Management (Warfield and Cardenas, 1994), are invaluable for the practice of the science, but are not germane to our investigation.

FOUNDATIONS OF THE SCIENCE

Everything that has gone before in this chapter lays foundations for the Science of Generic Design. This section focuses on conceptualizing those foundations. It filters, strains, congeals, and otherwise synthesizes the earlier expressed foundations into a highly organized set of expressions. It organizes those foundations according to the following criteria:

1. It must be organized to provide Referential Transparency
2. It must overtly incorporate the Universal Priors
3. Its organization must be disciplined by a model of a science
4. The Representation that it presents must be organized formally to (a) reflect sensitivity to the Universal Priors (b) incorporate self-documentation capabilities, and (c) provide for machine operations that facilitate the application of human Reasoning through Relationships, without imposing additional cognitive burden on already-overburdened human minds (p. 160).

The foundations of the science are spelled out in detail in Warfield's volume. They will not be repeated here. They will appear, however, under one guise or another in the chapter on the "Practice and Ethics of Design."

DIMENSIONS

The primary functions of theory are to explain key concepts and anticipate their consequences for methodology. The theory of the science of generic design is built upon the postulates that have just been mentioned. It develops the mathematics of dimensionality and relations that structure and give precision to the science. It prescribes laws of good design and principles for conducting design. Finally, it provides screening criteria for selecting design methodologies.

The mathematics of dimensionality (mathematical space) and relation integrates the science of generic design. "Mathematical space and relation represent two ideas that are absolutely fundamental to systems thinking" (p. 170). These concepts provide a non-quantified mathematics that is appropriate for the social sciences and design.

Mathematical space is built upon the concept of dimensions. Dimensions are familiar to us as the three dimensions of space and the fourth dimension of time. Mathematicians and scientists have expanded that concept so that it includes scales of dynamic relationship in n-space (a space that contains measurements in any number of dimensions). This space of n-dimensions is a mathematical space.

Within this space, physicists can plot the spatial and temporal behavior of multiple processes.

The social sciences generally have not used mathematical space as anything more than a metaphor. They have been blocked by the conventional requirement that the entities in that space be quantifiable. Because the social sciences deal with realities that are seldom quantifiable in their important characteristics, they have been shut out from using the obvious organizing power of the concept of space. Deprived of this power, the social sciences have lacked a necessary means for integrating their research.

Still the idea of mathematical space has a generality beyond numeration. It can be defined as follows:

> Designate the set of all elements by the symbol **E** and the set of all relations by the symbol **R**, all with reference to an imagined space that contains them. We may then assert formally that a space is given by

$$S = \{E, R, Z\}$$

> where **Z** consists of all those components of the space, if any, not included in the sets **E** and **R**" (pp. 172-173).

According to this definition, space is the universal relation; it consists of all relations, that is, "all n-tuples . . . of the form $(x_1, x_2, x_3 \ldots x_n)$ in which the x's range over the set of all possible assignments" (p. 173). Within this universal relation of space, specific relations are combinations of its elements. The important conclusion for social science is that the mathematics of sets, relations, and spaces does not require quantifiable elements. This mathematics can work with logical relations and gradations.

Warfield and Christakis (1987) demonstrated by example that the concept of dimension does not require quantification as a necessary condition of its use. They constructed a mathematical space based upon relations. In their test case, they used definitive, influence, and temporal relations. Within this space, they defined a dimension as "a set of elements of the same type, residing in the same place" (Warfield and Christakis, p. 130). In this way, they were able to expand the idea of dimensionality to cover non-quantifiable realities. They believe that their expanded definition of dimensionality dominates older definitions. It does this not in the sense that older interpretations are violated. Instead, "the new interpretation extends substantially the sense of older interpretations, thereby opening up the use of the term in a much broader class of situations" (Warfield and Christakis, p. 127).

The whole process of creating these dimensions is computerized; so detailed understanding of its techniques can be left to the experts. To use the technology one simply responds to generic questions on the level of their content, and allows the computer and its technician to process that content, organize it, display it, and document the whole process. What is important for our theoretic purposes is that a dimension is a technical term in the object language of social science. As such it is defined in terms of relationships: a definitive one, "is in the same category as"; one of influence, "is interdependent with"; and a temporal one, "must precede."   In turn, these relationships are expressed by mathematical relations that make their computerization possible.

LAWS OF GENERIC DESIGN

There are six laws of Generic Design, two of which are external to design science and four that apply explicitly to design. The external laws are the Law of Limits and the Law of Gradation. The internal laws are the Law of Success and Failure for Generic Design, the Law of Requisite Variety, the Law of Requisite Parsimony, and the Law of Requisite Saliency.

"The *Law of Limits* asserts that to any activity in the universe there exists a corresponding set of Limits upon that activity, which determines the feasible extent of the activity" (p. 177). This law requires a designer to discover "(a) what the Limits may be upon design in general and how these Limits may relate to any particular Design Situation and (b) those additional Limits that are at work in a particular Design Situation" (p. 177).

"The *Law of Gradation* asserts that any conceptual body of knowledge can be graded in stages, such that there is one simplest stage, one most comprehensive stage (reflecting the total state of relevant knowledge), and intermediate stages whose content lies between the two extremes" (p. 178). This law instructs a designer to fit the design process to the design Target and design situation. It points out that the whole panoply of generic design science does not have to be applied to every design system.

"The *Law of Success and Failure* for Generic Design asserts that there are seven critical factors in the Success Bundle for the Design Process. It further asserts that inadequacy in any one of these factors may cause failure" (p. 180). In the design process, if all seven of these critical factors are attended to, the designer has some assurance of success. If one critical factor is ignored, the designer opens the door to failure. "The seven factors are: leadership, financial support, component availability, designer participation, documentation support, and design processes that converge to informed agreement" (p. 180).

"The *Law of Requisite Variety* indicates the need for a match between the dimensionality of the Design Situation and the Target of the Design Process" (p. 180). This law, discovered by Ashby (1958), says that the complexity of the Design must match the complexity of its Target. If the Design is too complex or not complex enough, the designer will be frustrated.

"The *Law of Requisite Parsimony* indicates a need for controlling the rate of presenting information for processing to the human mind, in order to avoid its overload during the Design Process" (p. 181). This law recognizes the embodied and therefore bounded rationality of the Actors in a design process. Warfield expresses this law as follows:

Every individual's short-term brain activity lends itself to dealing simultaneously with approximately seven items (a number that is reached with three basic items and four of their joint interactions). Attempts to go beyond this scope of reasoning are met with physiological and psychological Limits that preclude sound reasoning (p. 182).

The *Law of Requisite Saliency* addresses the inevitable wide variability in the ways task force members assess relative importance in complex situations. It insists that "the Design Process must incorporate specific provisions for human learning that offer the strong possibility of diminishing significantly the variability in perception of relative saliency of design options" (p. 182). The Law states:

> The situational factors that require consideration in developing a design Target and introducing it in a Design Situation are seldom of equal saliency. Instead there is an underlying logic awaiting discovery in each Design Situation that will reveal the relative saliency of these factors (p. 182).

SCREENING CRITERIA FOR METHODOLOGY

The methodologies in the Science of Generic Design meet certain criteria. In general, they take into account the human and the logical aspects of the Design Situation. The specific criteria for screening methodologies are the following:

1. Provide constructive group capabilities for generating and structuring ideas, for designing alternatives, and for doing tradeoff analyses.
2. Possess explicit dual (anthropological and logical) design basis.
3. Provide in the anthropological basis for full role definition, enhancement of facilitator (Pilotos) credibility, and means for group maintenance.
4. Couple to a strong, sound, historical basis.
5. Show openness.
6. Enhance transferability of the product of its use by providing on-line documentation.
7. Promote group design efficiency.
8. In promoting group maintenance, do not demand infeasible behavior from participants; promote full participation; and provide opportunity for focused group dialog in structuring, designing alternatives, and doing tradeoff analysis.
9. Offer special properties (such as some unique benefit when compared to competing methodologies), such as anticipating future automation to increase utility, and being transferable from source organization to client organization without a major training requirement.
10. Do not require those who are content-knowledgeable to dilute their group effort by requiring them to sit at a computer terminal; delegate computer-terminal operation to an individual who is expected to be very capable in that role, but not to contribute to the substantive discussion of the issue (p. 187).

EVALUATION CRITERIA FOR THEORY

The two eminent tests for a theory's validity are its consequences and its internal consistency.

> Putting the two test concepts together, we arrive at the double requirement that (a) the Science be logically consistent internally (which provides the necessary internal Referential Transparency) and that (b) its faithful use provides success in Applications. These two criteria are tempered by the eternal requirement upon science that findings from Applications be used as appropriate to produce amendment in the relevant science (p. 189).

*   *   *

The great thing about the Science of Generic Design is that most of its complexity is computerized in its accompanying methodology of Interactive Management (IM; cf. Warfield and Cardenas, 1994). This methodology is further formalized in the CogniScope™, which is employed in generating synthesis in Chapter 18.

PART THREE

ADVANCES IN THE AREAS
OF SOCIAL, COGNITIVE,
AND EVOLUTIONARY THEORY

ADVANCES

The previous five chapters presented ways that sociology and design science have adapted to critiques from critical theory, how they have increased their theoretical sophistication and improved their ability to handle complex situations efficiently and democratically. The following seven chapters present advances in social, cognitive, and evolutionary theory that have developed

Luhmann presents a complete and integrated theory of social systems in four chapters that deal with meaning, the subject, communication, system, environment, structure, time, contradiction, conflict and self-reference. Kampis examines the ways that we model reality in terms of material and formal implications. He elaborates the theory of component-systems in this context. Goertzel develops a mathematical model of a component-system (the cognitive equation) and extends it in psychological and social directions. Lakoff demonstrates the metaphorical nature of our abstract concepts. The General Evolutionary Research Group explores how our social cognitive maps direct our cultural evolution.

CHAPTER 11

LUHMANN (1)
MEANING, SUBJECT,
AND COMMUNICATION

ABSTRACT

In *Social Systems* (1995), Niklas Luhmann weaves a complex theory of social systems from several areas of inquiry. He uses phenomenology to develop a theory of meaning with its related concepts. He uses the lessons of critical theory to distance himself from a philosophy of consciousness that would consider the human (transcendental) subject as the basis of meaning. He uses the concept of autopoiesis to ground his conception of psychic and social systems as self-referential systems. He uses various sources to frame his theory of communication and language. He extends self-reference to epistemology and explicitly recognizes the circularity of cognition. He uses differentiation theory to explain the development of social subsystems.

Some of these themes are more accessible than others. Three of the more accessible ones are considered in this first of four chapters: the phenomenology of meaning, the idea of the subject, and the process of communication. The discussion of meaning involves the concepts of complexity, contingency, information, selection, and negation. The discussion of "subject" gives a brief history of the concept, describes how seminal subjectivity is active in every selection, and identifies the intersubjective origin of "person." The discussion of communication considers the paradox of double contingency in the interaction of ego and alter, and it explains how meaning provides redundancy and difference to society.

THE PHENOMENOLOGY OF MEANING

Luhmann does *not* base his concept of meaning on the content of a statement, some particular fact or matter in the world, or some "meaning of life." Instead he focuses on how human experience is ordered. "A direct and presuppositionless approach to the problem of meaning is then best sought in a phenomenological description of what actually presents itself in meaningful experience" (1990, p. 25).

He begins with the fundamental insight of Husserl that every experience is an experience of something; that knowledge is *intentional*; that is, every experience intends (points to) something outside of the experiencer.

The basic experience of intentionality contains the following interrelated elements:

- transcendence
- a recognition of actuality and potentiality
- an overabundance of possibilities (complexity)
- a need to reduce this complexity that requires choices about what to notice and what to do (selection)
- the contingency of this selection
- the negation of possibilities that are not selected.

Underlying this experience is the assumption that a self-referential subject experiences the difference between its actuality and potentiality; makes contingent selections; and retains its negated choices in memory.

Transcendence is built into our experience because we are always attending to things that are beyond our boundaries. Not only that, we attend to outside things with an awareness that they are outside us. This outside/inside awareness is accompanied by actual/*potential awareness*. We experience what we are at the moment (our actuality) and anticipate what we might become (our potentialities) should we choose one or another of our present options.

To continue living, we have to process the *overabundance of possibilities* that confront us by attending to one and negating the others. By that *selection*, we integrate our actuality with the transcendence of one of our possibilities. We have no guarantees in our selections; we never know if we are actualizing the *right* possibility. "In practice, then, complexity means the necessity of choosing; contingency, the necessity of accepting risks" (1990, p. 26).

The *negation* involved in selection has "functional primacy . . . in mean-ing-constituting experience" (1990, p. 27). When we negate the possibilities that we do not select, we do not eliminate them entirely; we merely bracket them; we keep them as the world-background of our experience. Negation makes this bracketing possible because it has a reflexive quality; it can be applied to itself. Therefore, with every selection/negation we maintain the assurance that our negations will not obliterate everything, and that the world will still be there when we need it.

> Negation is a reflexive (and a necessarily reflexive) process of experience. It can be applied to itself, and this possibility of the negation of negation is indispensable in any experience that can negate at all. This, however, means that all negation remains irredeemably provisional and does not permanently block our access to what has been negated. Only time, not negation, eliminates possibilities definitively (1990, p. 28).

Luhmann sums up how the processing of meaning enables living in the following statement:

> What is special about the meaningful or meaning-based processing of experience is that it makes possible both the reduction and the preservation of complexity; i.e., it provides a form of selection that prevents the world from shrinking down to just one particular content of consciousness with each act of determining experience (1990, p. 27).

Provisionally, then, at this point in our discussion, meaning is a process that simplifies experience and maintains the complexity of the world. It simplifies experience by selecting the objects of experience and action. It maintains the complexity of the world through the reflexivity of negation that brackets the non-attended-to world out of our immediate experience, but maintains it as background reality. Because of this reflexivity of negation, meaning emboldens us to face the risks of contingency that would otherwise immobilize us in irreversible life and death decisions.

THE SUBJECT

Nature of the Subject

"Subject" in philosophical discussions is closely identified with the notion of "consciousness." Both notions are tied into the history of Western epistemology. At the close of the Middle Ages, the traditional authorities that justified the truth of theoretical and ethical discourse (the magisterium of the Church and Aristotle) were in recession.

Alternative strategies for guaranteeing truths were being offered. Nominalism, as proposed by William Occam, put forth the radical idea that truth was contingent upon the concepts that we happen to apply to reality-it was too radical for its day. The authority of the literal world of Scripture was put forward by various Protestant religions, but this solution did not appeal to philosophers and scientists.

Descartes filled the need for an acceptable authority with his *Discourse on Method*, in which he proposed the *subject* as the basis of certainty with his formulation, "Cogito, ergo sum." This ego-subject was conceived as a disembodied consciousness that could view the world objectively and derive mathematical truths through careful methodical thinking. This idea of the subject relied heavily on the theological concept of the soul. In the ensuing history of Western philosophy, the idea of the subject was refined by Locke and Kant among others, but it eventually broke down as a guarantor of truth. It is now in general disrepute.

The Demise of the Subject

Friedrich Nietzsche, Maurice Merleau-Ponty, Michel Foucault, and other modern philosophers have demolished the idea of a transcendental subject of consciousness. Nietzsche dramatized the emptiness of the whole Platonic conception of transcendental knowledge (for example, "God is dead"; *The Gay Science*, 1977, pp. 202-203). Merleau-Ponty showed the primacy of the lived body in the creation of consciousness (1964, pp. 4-8). Foucault documented how the abstract idea of the subject was coupled with increasing bureaucratic specifications in ways that facilitate the manipulation of "subjects" in non-sovereign power relations (cf. e.g., 1973, pp. 250, 308, 386-387). He said, "We have to promote new forms of subjectivity through refusal of this kind of individuality which has been imposed on us for several centuries" (quoted in Rabinow, 1984, p. 22). Foucault also showed the independence of cultural artifacts from their authors. Modern psychology and philosophy generally recognize that social reality and individual personality are, at least partially, constructed and not pre-given.

The question of the subject is not, obviously, the question, "Do we experience ourselves as subjects and persons?" The question is one of priority and posteriority: "Are we the natural-born arbiters of social reality; or are we, as persons, the productions of social reality?" Priority is discredited in post-modern philosophy and social psychology. Posteriority is in ascendancy.

The Formation of Subjectivity

For Luhmann, the semantics of transcendental "subjects" is replaced by the semantics of "psychic systems." A psychic system is an autopoietic system based upon meaning. In this, it is similar to social systems, which are also meaning-based autopoietic systems. The word "autopoietic" refers to the fact that psychic and social systems create the very elements and processes that constitute them as systems.

To understand the derivation of individuality and subjectivity in Luhmann's thought requires a grasp of how he describes autopoiesis and the system-environment interface.

AUTOPOIETIC SYSTEMS

Autopoietic systems straddle the system-environment interface. They define themselves in oppositional reference to their environment. They constantly recreate themselves in their elements and processes while they (1) maintain their separateness from their environment and (2) interact with it constantly. They constantly lose elements and renew themselves by creating new ones. They exist only by creating themselves. If they stop creating themselves, they die.

At the same time, they gather energy, nutrients, and the building blocks of their elements from their environments. They "use the difference between system and environment within themselves for orientation" (1995, p. 9). They are closed systems that maintain their closure at all costs, but they establish that closure by the rules of selection they use in utilizing aspects of their environment.

Luhmann took the concept of autopoiesis from the biological theory of Humberto Maturana and Francisco Varela. He is interested in showing that living systems are not the only autopoietic systems. He shows how psychic systems ("subjects") and social systems (interactions, societies, and society) are also autopoietic.

Psychic and social autopoietic systems reproduce processes of meaning. They make selections from the information that faces them. The reflexivity of negation preserves the non-selected elements of complexity as the enduring horizon of their experience. Their self-referentiality provides them with a way of self-determining their selections so that they are active creators of their meaning and are not merely altered by outside influences. In their self-referentiality, they process new meaning on the basis of old meanings. They are constantly circulating meanings.

ASPECTS OF MEANING

The Static Description

Meaning is an evolutionary achievement that is co-created by psychic and social systems. Luhmann puts it this way:

> Psychic and social systems have evolved together. At any time the one kind of system is the necessary environment of the other . . . Persons cannot emerge and continue to exist without social systems, nor can social systems without persons. This co-evolution has led to a common achievement . . . We call this evolutionary achievement "meaning" (1995, p. 59).

Meaning reduces the risk involved in our experiences and actions by providing redundancy, that is, the assurance that lots of people and lots of our own ideas redundantly indicate the same thing. Through redundancy, our meanings gather psychological and social conviction, and they simplify the process of making new selections. In this manner, meaning does not supply any particular answer or course of action; it supplies only the assurance of redundancy for something or another. Meaning, then, has a reach that is universal but without content.

An autopoietic system is constantly referring back and forth across the system-environment interface as it generates meaning from information. Its closure enables it to selectively appropriate (meaning) elements from the total expanse of complexity that faces it. Without closure, it would have no continuity. With closure, it can generate meanings secured by redundancy and negatable negation from anything in its complex internal and external environment.

As it generates meanings, an autopoietic system recognizes its own relatively simple and redundant meaning as something it can change. It neglects the overwhelming and variable complexity of its environment except when that complexity intrudes on its meaning as information (in Bateson's sense). In this sense, it values its "self" over its environment.

A system uses its autopoiesis to develop meaning from the complexity that faces it. On the basis of its autopoiesis, it separates complexity into inner ("self") and outer (environment). It also organizes external complexity insofar as it simplifies relevant environmental aspects in representations and incorporates them into its overall meaning. By means of these autopoietic processes, meaning enables our openness to the world. Autopoietic meaning and openness to the world are corollaries of each other. Luhmann puts it this way: "As an evolutionary universal, meaning . . . corresponds to the hypothesis of the *closure of self-referential system formations.* The closure of the self-referential order is synonymous here with the *infinite openness of the world* (1995, p. 62).

One consequence of this universality of meaning is that "meaninglessness" is impossible in experience because a decision to call something "meaningless" is itself a selection that creates meaning. "Meaning is an unnegatable category, a category devoid of difference [whose negation as] meaninglessness is . . . possible only in the domain of signs and resides in a confusion of signs (1995, p. 62).

The Instability Description

"Meaning is basally unstable, restless, and with a built-in compulsion to self-alteration" (1995, p. 65). Moment-to-moment, our meaning orientation has to change to deal with new information. "The instability of meaning resides in [this] untenability of [our] core of actuality" (1995, p. 65). Still we keep our balance because our actions open new possibilities within new horizons.

> [Our] ability to restabilize is provided by the fact that everything actual has meaning only within a horizon of possibilities indicated along with [it]. And to have meaning means that one of the possibilities that could be connected up can and must be selected as the next actuality, as soon as what is actual at the moment

> has faded away, transpired, and given up its actuality out of its own instability . . .
> Meaning is the unity of actualization and virtualization, of re-actualization and re-
> virtualization, as a self-propelling process (1995, p. 65).

In this recurrent losing of balance, we incorporate new information, and re-balance as we focus on new horizons. We demonstrate an auto-agility that is a marvel of autopoiesis. Our balancing act incorporates an intricate pattern of necessity and contingency. Our actuality is contingent and transient. It could have been something else should we have made different selections in the past, and it does not remain the same because we face moment-to-moment selections. Only the necessity of making a selection, not the content of the selection, is mandated. The choice we make is contingent.

Our choice negates the rest of our options by bracketing them and thereby puts us at risk and in a state of imbalance. We deal with this risk by being extremely sensitive to the new situation that confronts us with some of the old bracketed world and with new possibilities; that is, we adjust to a new horizon. Our actuality is unstable and faced with necessary selection; our potentiality is contingent and maintains stability through its gradual change. We pick our way through rough country by making steps that may advance us on our way and may not. We alertly check our relation to our new horizon with each step as we select our next movement.

DOUBLE CONTINGENCY

Statement of the Problem

What happens when two psychic or social systems interact? Luhmann examines this situation by reconsidering Parson's problem of double contingency. He states this paradox as follows: "Action cannot take place if alter makes his action dependent on how ego acts, and ego wants to connect his action to alter's" (1995, p. 103). Double contingency involves two closed self-referential circles with no asymmetry and therefore no communication or common action. This situation "concerns a basic condition of possibility for social action as such. No action can occur without first solving this problem of double contingency" (1995, p. 103).

Luhmann does not follow Parson's solution to this problem that involved value consensus. Instead, he concentrates on contingency and the self-generativity of chance using Von Foerster's "order from noise principle." Luhmann states his program for solving this problem as follows:

> No preordained value consensus is needed; the problem of double contingency (i.e.,
> empty, closed indeterminable self-reference) draws in chance straightaway, creates
> sensitivity to chance, and when no value consensus exists, one can thereby invent

it. The system emerges *etsi no daretur Deus* [even if God doesn't exist] (1995, p. 105)

Parable of the Black Boxes

Luhmann reframes the problem of double contingency in terms of "black boxes" in order to make it independent of existing personalities and applicable to social systems as well as psychic systems.

> Two black boxes, by whatever accident, come to have dealings with one another. Each determines its own behavior by complex self-referential operations within its own boundaries. What can be seen of each is therefore necessarily a reduction. Each assumes the same about the other. Therefore, however many efforts they exert and however much time they spend (they themselves are always faster!), the black boxes remain opaque to one another (p. 109).

How can these black boxes become white for one another?

Both black boxes exist in an unstable actuality that is balanced by continual contingent selections. When they encounter one another, they are forced to integrate each other into their meaning systems. Because they lack insight into each other, they devise stratagems. Ego attempts to determine alter's behavior by assuming what it will be. Then ego sensitively observes alter's behavior to see what happens. Ego soon discovers that alter's behavior cannot be calculated by ego's expectations. In the mutuality of this discovery, ego and alter grant freedom to one another and to themselves. In addition, ego comes to realize that it is assuming how alter will act *and* that alter knows that ego is assuming.

Building Social Structures

A situation of mutual recognition is established. Ego and alter concentrate on each other's input/output behavior, and learn to see themselves as functioning as the other does. They try to influence each other and learn from what happens. In this process, they create a social system. Luhmann says:

> In this way, an emergent order can arise that *is conditioned* by the complexity of the systems that make it possible *but that does not depend on this complexity's being calculated or controlled.* We call this emergent order a social system (p. 110).

A social system, then, does *not* rely on ego and alter seeing through each other and definitely prognosticating each other. A social system is *not* built upon such basal certainties; it *is* built upon social uncertainties and the mechanisms that ego and alter put in place to regulate their dealings with those uncertainties. In Luhmann's words:

> The social system is a system because there is no basal certainty about states and no prediction of behavior to be built thereon. Only the uncertainties *that result from this* are controlled, and they are controlled only with reference to the participants' *own* behavior. System formation constrains (= structures) the possibilities of *safeguarding one's own behavior* in any such situation . . . The absorption of uncertainty runs its course by stabilizing expectations not by stabilizing behavior (p. 110).

In the ongoing dynamic set up by the condition of double contingency, ego and alter build up structural values by projecting and testing their expectations. They structure their experiences according to these structural values and thereby create an emergent social reality. These realities develop as meaning formulations that have a degree of transparency for reciprocal observation and communication.

Some of the basic concepts, so formed, are those of person, intelligence, memory, and learning. These realities cannot be observed within black boxes. They are projected substrates for understanding mutually observed behaviors, substrates reinforced by redundant projections. "Person" indicates that ego cannot observe how acquaintance with alter provides security about what alter will probably do. In other words, alter acts autonomously but with a logic similar to ego's. "Intelligence" indicates that ego cannot observe how alter chooses one and not another solution to a problem. "Memory" indicates that "one cannot observe how one complex, actual state of a system passes over into the next, so that one must fall back instead on selected past inputs" (p. 111). "Learning" indicates that ego cannot observe how information triggers far-reaching consequences in alter. These examples show that:

> It would be futile to seek a psychic or even organic substrate for such things as person, intelligence, memory, or learning. All this concerns observer stratagems for interpreting what cannot be observed and transferring it to the emergent level of contact between systems (p. 111).

Reality, person, intelligence, and learning are emergent concepts of social and psychological reality. They are fabrications based upon a concession that "it could always have been otherwise" (p. 111). They are not, however, fictions. They are elaborations of the selective connectivity of meaning. They provide transparency to black boxes despite their opaque complexity.

Freedom and Contingency

Because ego and alter cannot calculate each other's behavior, they concede freedom to one another and limit themselves to knowledge that helps them handle contingency. With this stratagem they reduce the complexity of the selections that they face. Because they do not see into their companion black box, they are "bound to the experience of action," what their companion does, and they are "steered by the concession of freedom" (p. 112).

The experience of contingency relates to environments outside the black boxes and therefore beyond the bounds of intra-system relations. Ego and alter both experience themselves as having environments. Therefore, they relate their experience to those environments as well as to actions. "The generalized result of constant operation under the condition of double contingency is finally the social dimension of all meaning, namely, that one can ask for any meaning how it is experienced and processed by others" (p. 113). In this way, the improbability of the social order (how could a society be exactly such and so) is explained as being normal.

The problem of double contingency solves itself when ego tells alter, "I will do what you want if you do what I want" (p. 117). This proffered agreement is built on the freedom of both parties: "I do not allow myself to be determined by you, if you do not allow yourself to be determined by me" (p. 117). This agreement is the extremely unstable core structure around which social systems crystallize. The core structure and social system built upon it will collapse if they are not constantly renewed.

In the experience of double contingency, ego sees alter as an alter ego, that is, as an ego with a different perspective. Ego also realizes that alter sees him as an alter ego. Both ego and alter experience this situation as unstable and unacceptable. "In *this* experience the perspectives converge, and that makes it possible to suppose an interest in negating this negativity, an interest in determination" (p. 122).

This interest in determination leads ego and alter to tackle the problem of communication. It is the factually effective catalytic agent that brings it about.

COMMUNICATION

In their mutual recognition of alter egos, ego and alter reach the understanding that is basic to communication. To overcome their impasse of double contingency, they externalize their expectations for each other with the action of utterance. The intended content of that utterance and alter's reaction to it constitute information for both ego and alter. In this way, the weaving of understanding, utterance, and information creates a social communicative reality.

Understanding, utterance, and information constitute the essence of communication for Luhmann. Each of these elements is a meaning selection event in the ongoing communications that constitute society. The underlying understanding of mutual recognition, for example, is a contingent selection to treat alter as alter ego. An utterance, such as a smile, is an expectation selectively put forward to express tentative friendliness and to test alter's friendliness. Alter's return smile constitutes information that is either selected or rejected by ego as a return offer of tentative friendliness.

The Transmission Metaphor

The common metaphor for communication in our society is that of the pipeline wherein communication is the transmission of information from one person to another. Some of the implications of this metaphor are:

- Information is a thing that can be sent.
- Utterance is a simple act of sending.
- Reception of information is a simple process.
- There is a simple identity of what is sent and what is received.
- Communication is a two-part process of sending and receiving.

Luhmann considers all of these implications of the transmission metaphor to be misleading. First of all, using "thing" as a metaphor for information obscures the fact that information is a selection that reduces complexity. Information as it is experienced is a contingent and partial event. Even if we call an information-event a "thing," it is neither substantial nor permanent nor well defined.

Second, utterance is not an act of mere sending. Utterance involves selection of some meaning to share and some code to share it; it involves, therefore a selection of a selection. Utterance is the proposal of a selection.

Third, the completion of communication requires more than a simple reception; it requires understanding. The nature of this understanding needs to be probed in the nature of the communication itself and not merely semantically.

Fourth, the identity of what is sent and what is received is not a simple matter. Any achieved identity is guaranteed not by the content of the information, but by the whole context and the communication process itself.

Fifth, communication is a three-part process and not a two-part process. It involves three selections: information, utterance, and understanding. As has already been described, information is a selection that reduces complexity in our lived experience. When faced with the perplexing problem of double contingency, we choose from among the information selections that we have already amassed and do something with them (act them out; utter them in some way). Utterance, therefore, is the second selection. Understanding (the third selection) underlies the whole process as the pre-existing recognition of double contingency and is the outcome of every successful communication.

The Evolutionary Importance of Differences

Each of these selections is made on the basis of difference *and* unity. The overall unity is the system-environment in which systems define themselves self-referentially as different from their environment. Information is the selection from difference that enhances a system's unity. Utterance is the selection from a system's store of information and coding. Understanding is a selective action that distinguishes between information and utterance.

The difference *between information and utterance* is crucial for the emergence of communication. In recognizing this difference, ego and alter realize that their utterances are contingent, that they must be conditioned upon commonly accepted meanings. On the basis of this realization, ego and alter can build stores of shared meanings. Luhmann says,

> Communication emerges only if this last difference [between information and utterance] is observed, expected, understood, and used as the basis for connecting with further behaviors. Thus understanding normally includes more or less extensive misunderstandings; but these are always . . . misunderstandings that can be controlled and corrected (p. 141).

The link *between utterance and understanding* is coding. Luhmann does not spend a lot of time on the origins of coding. In fact, he seems to make a slight bow to thinkers like Habermas when he says, in passing, "Because it is the operative unification of information and utterance, coding must be treated by ego and alter in the same way. This requires adequate standardization" (p. 142). He also recognizes the value of John Austin's typology of speech acts into their locutionary, illocutionary, and perlocutionary aspects. He holds, however, that these distinctions are "marginal in comparison to the question of under what conditions their unity emerges" (p. 142).

In communicative practice an elemental event of communication is the smallest proposition or demand that can be negated by phrases such as: "not this but that, not this way, not now, and so forth . . . Reaction is what terminates communication, and only then can one tell what has emerged as a unit" (p. 154). Most communicative events exist as parts of processes in which they reciprocally condition each other. These processes presuppose an internal order. "But how can communication in general become a process?" (p. 155).

Three Improbabilities of Communication

Luhmann identifies three major improbabilities of communication: (1) that ego and alter would understand one another at the zero-point of communication; (2) that alter's communication would be able to reach ego, its addressee; and (3) that alter would have success in getting ego to do what alter wanted. In addition, he considers the discouragement that must have accompanied ego and alter's facing up to these improbabilities. The difficulties involved with the first improbability have already been addressed in discussing the paradox of double contingency. The second improbability arises when expanding relationships outstrip a society's existing ability to communicate over expanses of space and time. The third improbability is that of successfully getting alter to do what you want even after your message has been received. These three improbabilities are interrelated like hydraulic plumbing. Repressing differences in one area distributes pressure in another area. "Once one problem is solved, the solution of others is even less probable" (p. 159). For example,

> As soon as alphabetized writing made it possible to carry communication beyond the temporally and spatially limited circle of those who are present at any particular time, one could no longer rely on the force of oral presentation; one needed to argue more strictly about the thing itself (p. 160).

In this example, the new medium of alphabetic writing generated the need for a new medium of argumentation, "philosophy."

Luhmann defines "media" as those evolutionary achievements that overcome improbabilities of communication "and serve in a functionally adequate way to transform what is improbable into what is probable" (Luhmann, 1995, p. 160). There are three generic media that correspond to the previously mentioned communicative improbabilities. "The medium that increases the understandability of communication beyond the sphere of perception is *language*" (p. 160). Media that expand communication over space and time are *"media of dissemination,* namely, writing, printing, and electronic broadcasting" (p. 161). The media that developed when society became too complex for barter and persuasion are the *"symbolically generalized communication media,* which are functionally adequate to . . . [a] particular problem" (p. 161).

Symbolically generalized media, like "truth, love, property/money, power/law; and also, in rudimentary form, religious belief, art, and today, standardized 'basic values'" (p. 161), help condition the selection of communication so that it can achieve acceptance. Each symbolically generalized medium, in its own peculiar way, generates consensus and creates social systems in our contemporary society.

> Language, media of dissemination, and symbolically generalized communication media are thus evolutionary achievements that interdependently ground the processing of information and increase what can be produced by social communication. This is how society produces and reproduces itself as a social system (p. 162).

CONCLUSION

Some of the major points that Luhmann makes concerning meaning, the subject, and communication are the following:

- Meaning results from selections
- Selections require a provisional determination of the future, which is enabled by the background of negatable negations that are involved in every selection
- The notion of "subject" develops from the difference between system and environment inherent in every selection
- The notions of "person," "intelligence," and "freedom" derive from solving the problem of double contingency , which arises when two people confront each other and are unable to predict each other's behavior
- Communication emerges as an externalization of expectation in the form of codes that are sensitively monitored and modulated. This activity results in more or less coordinated behavior
- The extension of communication beyond face-to-face interaction creates additional improbabilities that are overcome by media of communication
- To transmit communication over a distance, media of dissemination have been generated: writing, printing, electronic broadcasting, and so on.
- To get people to do what we want, we have developed symbolically generalized media, like truth, love, property/money, power/law, religious belief, art, and basic values.

CHAPTER 12

LUHMANN (2)
SYSTEMS AND ENVIRONMENTS

ABSTRACT

The previous chapter dealt with Luhmann's ideas of meaning, subject, and communication. This chapter details how autopoietic systems use differences to create themselves, develop their structures, and establish their identities.

   Organic, psychic, and social autopoietic systems use differences to create their identities and boundaries. They create their meanings by differentiating themselves from their environment, information from noise, and useful actions from useless ones. In so doing, the entities become meaning-processing systems.

   As they create boundaries and selectively process their inputs and outputs, systems create themselves as centers of their worlds. They come to relate to each other through the processes of structural coupling, which is also called interpenetration. In interhuman relationships, the process of binding, which creates these structural couplings, is aided by binary schematisms in the forms of morality and socialization.

IDENTITY WITHIN A DIFFERENCE

Autopoietic systems exist in a state of flux; they constantly reproduce their elements and structures. They orient themselves by the difference between themselves and their environments. Using this difference, they create their self-reference and their identity. The environment of a system is its negative correlate: its "everything else" (p. 181). It is a welter of complexity that the system needs to simplify and the underlying reality specific to the system that is assumed in all its cognitive operations.

189

An autopoietic system creates the differences between itself and its environment by the selections it makes. When a system makes selections, it differentiates its internal complexity from the greater complexity of its environment. It places greater relevance on internal events/processes and endeavors to arrange them. It considers external events/processes to be rather irrelevant, and largely ignores them. In this way, "the system acquires freedom and the autonomy of self-regulation by indifference to its environment" (p. 183). By leaving environmental events/processes to chance, the system simplifies its own complexity while enlarging its ability to respond *ad hoc* to greater environmental complexities. Thus, the difference between an autopoietic system and its environment is not a metaphysical either/or. It is the result of selections made by a system as it reduces its complexity.

A system can work out its relationship to differences in three areas: operative, structural formation, and reflection. On the basis of relative complexity, the system *operates* (endures) by keeping constant those things that maintain it and enable its reproduction. Using differences, to generalize and re-specify events, it builds the internal *structure* that steers its internal processes without "point for point correspondences with this relevant environment" (p. 184). It creates its identity in *reflection* by identifying itself as what it can control and what continues to recur, and identifying its environment as everything else. On all three levels, the difference of relative complexity provides the standard of autopoietic identity. "Identity in contrast to everything else is nothing more than the determination and localization of difference in relative degree of complexity" (p. 184).

EXTERNAL DIFFERENTIATION

In their ongoing selections of complexity, autopoietic systems create distinctions in their environments and co-create the objects they perceive. They gradually build a world of objects as these objects undergo a continual flux of contradiction, reorientation, and redundancy in a constantly reorganizing environment.
This understanding of the origin of differences radically de-ontologizes objects. Under this interpretation,

> [There is] no unambiguous localization of any sort of "items" within the world nor any unambiguous classifying relation between them. Everything that happens belongs to a *system* (or to many systems) and *always at the same time* to the environment of other systems (p. 177).

Luhmann understands systems as existing in the world. Accordingly, he reserves the term "system" for existing realities. He uses the terms "systems references" and "systems models" for the conceptual structures that attempt to describe existing systems.

INTERNAL DIFFERENTIATION

Autopoietic reproduction guides the internal differentiation of a system. These systems reproduce not only their internal elements, but also their ability to reproduce. Luhmann puts it this way:

> The connection between reproduction and differentiation becomes comprehensible if one views reproduction, not as the identical or almost-identical replication of the same (e.g., as replacing supplies), but as a constantly new constitution of events that can be connected. Reproduction always implies reproducing the possibility of reproduction (p. 189).

For social systems, this autopoietic reproduction continually re-enacts the condition of double contingency. Every momentary event of a social system presents it with a challenge to adapt its structure to fit new complexity, *and* the possibility to generate within itself a new subsystem. In this process, a system evolves complexity and includes system/environment differences within itself.

Luhmann illustrates this situation by describing the situation in which a man might offer a light to a woman smoker.

> At a party one sees a woman reach for a cigarette, and (if she dawdles suggestively), one may offer her a light from one's own cigarette lighter. [In this situation,] settled differentiations stabilize the possibilities for reproduction by constraining conditions on the comprehensibility of communication and the suitability of behavioral modes. But the meaning surpluses that must be produced alongside provide ever further chance for innovative system formation; in other words, they provide the chance to include new differences and new constraints and thus to increase the ability to constrain the initial situation via differentiation. Only thus can system complexity increase (p. 189).

Subsystems like this social one make improbable states become probable. They make possible the increase and normalization of organized complexity for the system. In our example, the interaction that provides the light may lead to a date and may lead to many otherwise improbable eventualities (like the man and woman going to a parent-teacher conference with their child).

Such improbable internal self-differentiations build subsystems upon an already-pacified environment. They presuppose a social situation that is already rather regulated. The woman's dawdling with the cigarette is understood as an invitation to conversation. The conversation is understood as the basis for further mutual self-revelation. And so on.

The internal differentiation of a system *increases* its complexity but also makes possible new forms for *reducing* complexity. Because of differentiation, each subsystem orients itself only to its own systems/environment difference and its part of the system's overall complexity. It assumes that other reproductive requirements of the system are fulfilled by other subsystems. At the same time, each subsystem reconstructs the whole system. Because of the circular dependence of parts and wholes, each shift within a system re-articulates the self-reference of the whole system.

BOUNDARIES

Social systems constitute themselves by communicating meanings. Through meanings, they establish their boundaries. In Luhmann's words, "The difference between system and environment is mediated exclusively by *meaning-constituted boundaries*" (p. 194). How do such boundaries come about?

A system re-forms its boundaries every time that it reproduces itself in an elemental meaning event. Each such event "makes a relation and with it a boundary decision" (p. 195). This is so because every communicative event employs the system-environment difference to convey meaning.

Every communication is based upon an expectation that it will be heard. It "stakes a claim. At the very least, it demands time and attention" (p. 195). The customs of language and other generalized media of communication enhance its probability of acceptance. If it is accepted and if it contains new themes, it extends the system's boundaries, and enmeshes its speaker in the ongoing communicative action.

Everyone entering a communicative situation faces a question and a decision.

> What is acceptable as communication to whom? One must decide this and thereby orient himself socially before one actively participates in communication, and one must distinguish himself by communicative action if one wants to avoid unacceptable communication. Thus drawing boundaries is finally a process of (tacitly anticipatory, whether covert or open) negotiation. It occurs as the system simplifies itself by tolerating or not tolerating communicative action (p. 196).

If social boundaries are processes of negotiating inclusion and exclusion, what are the criteria that guide the decision-making? One measure of acceptability is provided by thematic expectations: Is this theme acceptable in this conversation? Another is by time concerns: for example, are there deadlines? A third is by social exclusion by, for example, the need to belong to a formal organization. With these criteria, "one can regulate what is considered action in the system and which actions are to be attributed to the environment" (pp. 196-197).

Meaning-constituted boundaries are self-generated boundaries that are regulated within the system itself. Systems are guided in their processing of new themes "by what has already occurred, by what is possible in a situation, and by general structures of expectation" (p. 197). Such structures of expectation help one to "foresee in detail how and about what one should communicate in the supermarket, on the football field, at lunch, at home . . ." (p. 197).

COLLECTIVE ACTION

Any complex social system faces "the problem of *specifying* environmental contacts" (Luhmann, 1995, p. 198). It requires some mechanism that focuses its *capacity for collective action.* Organizing for collective action decisively improves a system's relationship to its environment. Societal organization restricts the activities of its members in order to enable actions that are unattainable by individual effort. Such organization is triggered, of course, by individual action, but it binds the whole system. This binding comes about only by the designation of symbols that are based on a rationale.

The rationale can be an *ad hoc* consensus or a ritualized appeal to divine forces. It can also be generalized to many decisions by the creation of hierarchies. As societies progress, the actions of hierarchies often require justification because agreements that can be given can also be withheld. As a result, expertise, argumentation, the majority principle, or regulated procedures (like law) become very important for the ongoing reproduction of the societal system.

Given proper conditions, like sufficient space, means of communication, objects that it can handle, and a willingness to change, a system can use its expectational structure to try to change its state or get some other result. By orienting itself to conditions of before and after, a system gains greater accuracy in its selections. The system learns to program its actions "by providing either conditions that trigger action or goals that action should aim for or both" (p. 204). Through this process, the system creates actions that are quickly determinable as correct; it makes its environment manageable.

THE WORLD WITH MANY CENTERS

> With meaning-constituted boundaries, systems begin to have a *world.* They experience themselves, their environment, and everything that functions in it as an element, as a selection within a horizon that includes all possibilities and indicates further ones, that indicates an end and a beyond, that is both necessarily and arbitrarily bounded from anywhere within it (p. 207).

Systems correlate meaning and world. They position every meaning that they create within a world of previous meanings. They assume the existence of that world as background for the specification of every meaning.

> The *world* becomes an ultimate concept that describes the *unity of the difference between system and environment* . . . an ultimate concept, one free from further differences . . . [It is a] unit of closure subsequent to a difference. It is the world after the fall from grace (p. 208).

In this world, the pride of centrality is removed from the subject. The center of the world is created by each system/environment difference that constitutes the world.

Every difference becomes the center of the world, and precisely that makes the world necessary: for every system/environment difference, the world integrates all the system/environment differences that a system finds in itself and its environment. In this sense the world has multiple centers-but only so that every difference can fit the others into its own system and environment (p. 208).

Centrality is meaningless in the state of undifferentiated innocence. It creates itself in opposition to an environment by creating that difference.

That difference, and every difference, excludes *and* presupposes an underlying excluded middle that is the world. In this, the logic of the unity between differences goes beyond classical logic. It is related to dialectics. It recognizes that every difference is imposed upon "a substratum where it does not apply" (p. 209). As Adorno puts it, "objects do not go into their concepts without leaving a remainder" (Adorno, 1992, p. 5).

PSYCHIC AND SOCIAL MEANING PROCESSORS

Luhmann distinguishes several types of autopoietic systems. Living things are autopoietic because they constantly reproduce their cells, their organic structures, and their ability to reproduce. Meaning systems are autopoietic because they constantly reproduce and modify their meaning units and structures of understanding. Meaning systems are divided into psychic meaning systems which process meaning as consciousness, and social meaning systems which process meaning as communication. Social meaning systems are further subdivided into interactions, and society (as a whole). These relationships can be presented in outline form as follows:

                    Autopoietic Systems
        Living
        Meaning
                Psychic
                Social
                    Interactions
                    Society.

As we have already observed, traditional thinking considers human beings (the subjects of consciousness) to be the only subjects that process meaning. In our postmodern world, however, one observes that communication media have increasingly separated meanings from their authors. If one recognizes this fact as an evolutionary accomplishment, then one is led to conclude that communications constitute meaningful relationships that are relatively independent of individual human subjects. One then recognizes activities, like the interpretation of a text, the influence of myth, the language of romantic love, and the culture of a society, as meaningful communications between social systems that are not reducible to individual attribution.

Since social systems process meaning in relative independence from psychic meaning systems, the stage is set for communication between them. In this condition, society is not considered as an overarching system that includes psychic systems. On the contrary, societies and individuals are environments for each other. Society is the environment for individual psychic systems *and* human beings are the environment for social systems.

The implications of this different perspective are profound. No longer are human beings considered as organs of the leviathan of society. Instead, they are considered to be autopoietic systems that evolve in conjunction with an autopoietic society. The task of human beings and society, therefore, is to work out an acceptable *modus vivendi*. Both human beings and society are complex works in progress that evolve through mutual interpenetration.

INTERPENETRATION

Systems that are environments for each other can couple their structures in such a way that they increase their independence from environmental determination by increasing their dependence on each other. By this process of "structural coupling" (Maturana and Varela), two or more systems form a more complex integrated whole while retaining (even increasing) their individual capabilities. In meaning (psychic and social) systems, structural coupling is described as interpenetration.

When a system makes its own complexity, which involves indeterminacy, contingency, and the pressure to select, available to another system that can use that complexity for its own system-building, then it is said to have "penetrated" the second system. From that point on, the behavior of the penetrating system is co-determined by the receiving system. When the Internet presents itself to me, for example, what happens on the net is co-determined by the use I make of it.

Interpenetration exists when penetration occurs reciprocally, "when both systems enable each other by introducing their own already-constituted complexity into each other . . . This means that greater degrees of freedom are possible in spite (better: because!) of increased dependencies" (p. 213).

Interpenetrating systems present complexity to each other and thereby supply each other "with adequate disorder" (p. 214). As individuals try to communicate, they create the noise that generates social systems. "Social systems come into being on the basis of the noise that psychic systems create in their attempts to communicate" (p. 214). This "noise," or complexity, creates the pressure to select a meaning without specifying what meaning. As a result, it presents risk, freedom, and negatability to a system.

In this noisy situation, a system tentatively advances an expectation and then sensitively attunes itself to psychic and social feedback. If its expectation is denied, it learns to negate that expectation. If its expectation is affirmed, it is reassured and, in the course of time, it may become redundantly reassured. With redundancy, two or more systems set the stage for further ventures into noise.

When systems interpenetrate, they do share elements, but they share those elements from individual perspectives. As Luhmann says,

> Interpenetrating systems converge in individual elements-that is, they use the same ones-*but they give each of them a different selectivity and connectivity*, different pasts and futures. Because temporalized elements (events) are involved, the convergence is possible only in the present. The elements signify different things in the participating systems, although they are identical as events: they select among different possibilities and lead to different consequences (p. 215).

In this way, autopoietic systems share meaning elements while maintaining their own autonomy.

Interpenetration connects the autopoieses of organic life, consciousness, and communication. These three forms of autopoiesis condition each other. "Autopoiesis qua life and qua consciousness is a presupposition for forming social systems" (p. 218). Conversely, "the social system, based on life and consciousness, makes the autopoiesis of these conditions possible in that it enables them to renew themselves constantly in a closed nexus of reproduction" (p. 219). In all of their interpenetrations, autopoietic systems use the complexity of their contributing systems, but they do so on the basis of their being autonomous. On both sides of structural coupling, there is

> the *difference* between and *interlocking* of *autopoiesis* and *structure* (the one continuously reproducing, the other discontinuously changing) for relationships of interpenetration between *organic/psychic* and *social* systems to come about (p. 220).

BINDING

How does structural coupling become binding for systems? How do the demands, for example, of memory and information storage bind neurophysiological processes? More to the point, how do social systems bind psychic possibilities?

The evolutionary explanation of binding does not evoke supernorms or appeals to requirements for order. Bindings come about because of the temporal sequence of selections in which every selective event excludes some possibilities and opens others. In the course of time, purely factual responsibilities are redundantly reinforced and become obligations. In brief, "bindings come about by selection, specifically, by selections that eliminate (more or less securely) other possibilities" (p. 223).

INTERHUMAN INTERPENETRATION

The interpenetration and binding between human beings is special for us: we call it *intimacy*. This kind of interpenetration is historically conditioned. It "comes into being when more and more domains of personal experience and bodily behavior become accessible and relevant to another human being and vice versa" (p. 224). When alter is perceived as situating himself or herself in the world in an autonomous way, then ego finds alter interesting and wants his or her meaning.

> [This meaning] lies in interpenetration itself, not in performances but in the other's complexity, which is acquired via intimacy as a feature of one's own life. It lies in a new kind of emergent reality that, as the semantics of love has been saying since the seventeenth century, is at odds with the conventional world and creates its own (pp. 224-225).

The interesting person acquires special significance as part of one's world. He or she is seen as an unfathomable, unique relationship with the world that provides richness to one's life that is otherwise impossible. To attain security in this relationship, one attributes the desirable qualities that one experiences to alter's ego. In this way ego's "*I*" is located in alter's world, and vice versa. The increasing demands of such intimacy result in the heightened sensitivity between ego and alter that is called empathy. Such heightened sensitivity requires a great deal of psychic and social sophistication. "The genesis and reproduction of intimacy presupposes a very refined acquaintance with situations and milieus, thus a great deal of culture, because adequately nuanced observation and attribution is possible only on such a basis" (p. 226).

BINARY SCHEMATISM

In social and interhuman interpenetration, systems are faced with information processing tasks that they cannot solve. A system can never transfer the complexity of another system into itself. It has to somehow simplify the other system's complexity. To do this, a meaning-processing system uses schematisms.

   At an early stage of socialization, a social system sets up norms that separate activities into conforming and deviant. It provides itself with a guarantee of order by building a structure of norms. With these norms, it also provides human beings with a certain security and sense of direction. Humans within that order have few autonomous choices, but they can choose to live "on the sunny side or on the shady side" (p. 230).

   As personal formation becomes more individualized, however, the "excluded third" that is ignored in binary schematisms is reactivated. In this situation, the norm schema is still necessary for ordering techniques, but it is deprived of its compelling power as an affirmation of ultimate meaning.

Individualized persons can now launch protest movements that treat prevailing norms as unreasonable. How, in this situation, can a system constitute itself using the complexity of another system?

Systems continue to use binary schematisms, but they use them within a movable horizon that shifts with every observation and exploration. They orient themselves to the depths of the other system and sound out the capacity for consensus, perhaps in the form of yeas and nays. In principle, this sounding can always be carried further; but, in practice, it must be broken off at some time. "Thus a binary schematism is built into the horizontal structure of all meaningful experience: to continue on or to break off" (p. 232).

On the basis of the break-off binary schematism, systems can adopt parallel schemas using the same differences. In so doing, they create a structured openness in which they can share their elements. With agreed difference schemas, they can process information that they receive from each other.

Contingency is built into this whole process. Break-offs focus on one agreed difference while temporarily negating other possible differences. Further difference schemas, and the mutually shared elements that they create, are based on the contingency of their own and previous break-offs. This contingency underscores the autonomy of the interpenetrating systems. It emphasizes their difference and autonomy; it does not fuse their beings; it coordinates their operative reproduction.

Thus, for example, different schools of Western thought settled on using the logic of Aristotle from the thirteenth century until the present time. In so doing, they remained idealists, empiricists, and pragmatists, but they built an imposing edifice of thought and technology with incredibly refined difference structures on that logical basis.

In interhuman terms, one may schematize that another's behavior will be friendly or unfriendly, useful or harmful, and so on. In this way, one controls the situation and is prepared to deal with either contingency. Still, one binds oneself to recognizing the contingency of the other's behavior and his or her autonomy. The other is considered as an alter ego who has true or false opinions, who acts correctly or incorrectly, and who, moreover, has a history of such opinions and activities; in short, the other is seen as a "subject."

Seen from this perspective, "subject" is not a transcendental entity. It is a historical result of a series of human interpenetrations. "Thus the subject is 'subject' . . . only as the biographically unique constellation of designations and realizations that binary schematisms have held open. It owes its possibility to this feature, not to itself" (p. 234)

MORALITY

Morality uses binary schematisms to reduce complexity in social and interhuman interpenetration. It provides rules for awarding esteem and registering disdain. If a person acts in socially acceptable ways, morality awards him or her with a generalized positive recognition and evaluation. Luhmann defines the morality of a social system as "the totality of the conditions for deciding the bestowal of esteem or disdain within the system" (p. 236). In general,

> Morality is a *symbolic generalization* that reduces the full reflexive *complexity* of double contingent ego/alter relations to expressions of esteem and by this generalization open[s] up (1) room for the free play of *conditionings* and (2) the possibility of reconstructing complexity through the *binary schematism* esteem/disdain (p. 236).

According to this concept of morality, there is a convergence of social and interpersonal interpenetration around the difference esteem/disdain. This convergence is easy to maintain in relatively simple societies. In modern societies, however, the demands of complexity have broken this convergence. On the one hand, the decline of religion places strains upon the social order to uphold morality. On the other, "the semantic codes for intimate relations and for public policy drifted apart" (p. 237). The novels of the eighteenth century lampooned this divergence of private and public morality with the theme of ridiculousness. The conclusion to be drawn from this divergence is that "special developments in private social sensibility and in public sociality could no longer be unified in a single canon of aristocratic morality" (p. 237).

Thus, moral bindings have been loosened. In the modern world, bindings are more strictly specified concerning performances, fashions, and similar things that no longer concern the whole person. Bindings come about by membership in religious, social, and environmental movements, for example, or leisure and business groupings. In their cumulative effect, such bindings have more effect than the moral schematisms of morality. They are less universal than moral schematisms. They are based upon a weakened and temporary capacity for commitment, but one that is very sensitive to interpersonal and social nuances.

SOCIALIZATION

Socialization is the process in which social circumstances mold the minds and bodies of human beings. Luhmann bases his systemic definition of socialization on five principles that he has previously developed. They are:

1. that problems of causality are secondary to problems of self-reference;
2. that all information processing "takes off" not from identities (e.g., grounds) but from differences;
3. that communication (as constituting and reproducing autopoiesis) is distinct from action (as the constituted element of social systems);
4. that human beings are the environment of social systems; and
5. That the relationship of human beings to social system is one of interpenetration (Luhmann, 1995, p. 240).

On the basis of these principles, Luhmann *defines* socialization as "*the process that, by interpenetration, forms the psychic system and the bodily behavior of human beings that it controls*" (p. 241). In human-social system interpenetration, human beings make the complexity of their society their own (by binary schematisms, and so on).. At the same time, they reduce the complexity of their environment by selecting social mores that increase their meaning base and social effectiveness. Their success in communication has a redundant effect. It enables them to move on to more challenging meanings as they go on their autopoietic way, accumulating redundancies and reproducing their societies in their own meaning systems.

For societies, socialization is also a process of autopoiesis; it presupposes basal self-reference and deviant reproduction. Societies reproduce themselves through ongoing communication; they evolve through deviant successful attempts of coping with their environments (human beings and other social systems).

Human bodies are not mere supports for capabilities or mere instruments for social use. They too are autopoietic systems. They interpenetrate with psychic and social systems and produce their own difference schemata. These schemata embody the self-environment difference and all the differences that they enact in staying alive. In this way they produce internal (non-symbolic) models of their physical, psychic and social environments. "[They have] a meaning that allows complexity in social systems to appear as available: one immediately sees, takes into account, and anticipates that one can behave in one way or another" (p. 251).

CONCLUSION

Systems create and maintain themselves by imposing circularity on their self-reproductions. They gain the ability to do this in their recognition and selection of differences in their worlds. In their structural couplings, psychic and social systems enable collective action by generating schematisms (differentiations), which begin in the binary form (us/them, good/bad, true/false) and later differentiate into smaller schematisms regarding role-performance, esthetic preference, and so on. These later differentiations undermine the original binary schematizations by introducing the previously "excluded middle" into moral and cognitive expectations. Thus, progress evolves by generating binary schematisms and then progressively undermining them.

CHAPTER 13

LUHMANN (3)
STRUCTURE AND TIME

ABSTRACT

The previous two chapters discussed Luhmann's concepts of meaning, subject, communication, system, and environment. This chapter describes how he explains the growth of structures through the process of expectation-driven reproduction.

The question of how social structures can change over time involves two parts: how can part of the structures change while the structures themselves remain stable? The classical answer to this question held that structures were "things" that had both essential and peripheral "qualities" and that the essential qualities were stable while the peripheral ones were transitory. The theory of autopoietic systems rejects this kind of explanation.

Autopoietic systems reproduce themselves constantly. Their existence beyond the span of a momentary event depends upon their processes of reproduction. For them, the question is not, "How do they change?" The crucial question is, "How do they remain the same?"

Social systems endure in time because of their structure of self-reproduction. They constantly reproduce themselves in a manner that embodies constraint and openness. For this purpose, they use a strategy of expectations that includes anticipating expectations, mutual anticipation, ambiguity, generality, and several symbolic abbreviations such as: person, role, program, value, norm, and cognition.

STRUCTURE

Structure has been a fundamental concept of social theory (Levi-Strauss, 1958, pp. 303-51; Parsons, 1977, pp. 100-117). "The scientific analysis of systems, texts, language games, and so on is attributed a reference to reality and this reference to reality is guaranteed by the concept of structure" (Luhmann, 1995, p. 280). Structuralism (for example, Levi-Strauss) has referred to social systems as stable entities. Structural functionalism (for example, Parsons) has regarded social systems as open systems that perform functions on inputs and yield outputs. In this conception, societies are stable things that manipulate things external to them; or they are some "whole" equipped with static and variable "qualities" in which the static qualities are the essence of the whole.

Within the context of societal self-reproduction, structure is subsidiary and consequent to the way self-referential systems maintain themselves in time. In other words, societal structure is both chicken and egg, but structure is secondary to the process of reproduction.

A social system is not a "thing." It is a totally temporized relationship of interconnected elements that reproduce themselves and their relationships. The elements are action/events of minimal duration whose connections are also momentary. Given this minimal duration, the problem of having a different constitution from moment to moment is not significant. More significant is the problem of enduring as a stable system.

In this temporal context, structure is information that directs action. Information is the content and measure of a system's ordered complexity or pattern. There is little or no information, coherence, or pattern in unstructured and unconnected complexity. There is increasing information in interconnected complexity (structure).

REPRODUCTION

The idea of action is based upon the fundamental and radical notion of an *event*. An event is a temporal atom, "an indivisible, all-or-nothing happening . . . Its representation on a spatio-temporal model would be merely a point" (Floyd Allport, quoted by Luhmann, p. 287). An action is such a momentary all-or-nothing happening. As such, it passes away as do all events in time, yet it "brings about a total change in past, present, and future . . . It gives up the quality of being present to the next event and becomes a past for it (i.e., for its future)" (p. 287). By passing away, an action provides maximum freedom vis-à-vis time to its successors and to the system that they constitute.

A social system, catalyzed by internal or external change, acts to reproduce itself using its embodied informational pattern, its structure. It survives only by constantly reproducing. It is constantly dissolving and reconstituting itself. A social system uses its very dissolution to create greater organized complexity.

"Action systems use time to force their continuing self-dissolution and thereby guarantee the selectivity of all self-renewal; and they use this selectivity to enable self-renewal in an environment that makes continuously varying demands" (p. 290). Through their flexibly patterned structure and their action/event elements, social systems are equipped to perform structural changes upon themselves without lapsing into either rigidity or chaos.

The "structure" [In Maturana and Varela, the "organization"] maintains the continued existence of an autopoietic system by constraining the course of its actions. Structure defines how a system's elements relate over time. It does *not* directly maintain the relationships between its system's elements, because those elements come and go from within the system. It *does* maintain the relationship between a system's relations and the elements of those relations. It exists on a level of order different from the order of elements, the level of active, patterned information.

Structure provides constraints on the selection from complexity that a system makes in its ongoing renewals. It limits the possibilities that a system can select. At the same time, it opens the system to contingent selection from among those possibilities. Its standard of selection is in the form of a dichotomy: with successful selections, the system lives (and possibly evolves); with unsuccessful selections, it dies. Structure, therefore, is the constraint that procures a system's self-reproduction.

Self-reproduction replaces concrete elements in the system with other concrete elements. It provides invariance in time for a system by constraining selections to those that have survival potential. It supplies (contingently) a system's momentary concrete elements while preserving it invariant on the level of structure. In this sense, self-reproduction *is* the structure of an autopoietic system. In the ongoing activity of autopoietic systems, it provides a sense of self.

The structure of self-reproduction is the process of before/after difference, the unity of that difference. In its contingent provision of exclusion and a search for connection, this structure gives a system its relationship to time. It allows a system to determine itself internally through a combination of self-identity and self-diversity.

EXPECTATIONS

Expectations are crucial functions of social systems. They constrain a system's search for possible new connections and thereby shape its structure of self-reproduction. "The structures of social systems consist in expectations . . . they are structures of expectation, and . . . there are no other structural possibilities for social systems, because social systems temporize their elements as action events" (p. 293). Expectations direct decision-making. They create redundancy for autopoietic systems including systems of scientific inquiry. Through their creation of habitual patterns in systems, they provide assurance that a habitual procedure or assumption can be trusted.

Expectations provide a link to the future for social systems. They tentatively predict what a system's future environment and its future adjustment to that environment will be. Expectations concerning natural phenomena are relatively unproblematic because nature's behavior is more or less predictable. Expectations in social situations involve the problem of double contingency. They are much more complicated.

The problem of double contingency is exemplified by the meeting of two [as yet unsocialized] people. They need to anticipate each other's behavior in order to make life-important selections. They discover, however, that they cannot predict each other's behavior. In this process, they conclude that the behaviors of both ego and alter are contingent; that is, their behaviors can go in several ways (acceptance, denial, non-comprehension, and so on).

Because they lack insight into each other, they devise stratagems. Ego attempts to determine alter's behavior by assuming what it will be. Then ego sensitively observes alter's behavior to see what happens. Ego soon discovers that alter's behavior cannot be prognosticated. In the mutuality of this discovery, ego and alter grant freedom to one another and to themselves. In addition, ego comes to realize that it is assuming how alter will act *and* that alter knows that ego is assuming (cf. 1995, pp. 109-110).

In social situations, expectations have to become reflexive. Ego has to become aware not only that he or she anticipates that alter will act in a certain way, but that ego is anticipated (by alter) to be anticipating. In this way, expectations form a social field of more than one person. "Ego must be able to anticipate what alter anticipates of him to make his own anticipations and behavior agree with alter's anticipation" (p. 303).

Progress Into Insecurity

In this working out of a *modus vivendi*, ego and alter separate out a subdomain of social activity: events that can be expected. Through the risky sharing and mutual adjustment of expectations, they create a structure of shared expectancy. By amplifying their insecurities, ego and alter create order (that is, a shared differentiation) between them. High levels of insecurity require the creation of more expectations about the behavior of others. These expectations result in "symbolic abbreviations," which eventually become the names: "person," "role," "program," and "value," in a differentiated social world.

As ego tries to secure mutual anticipation of a particular course of action, he or she faces difficulties and perils. Ego finds it difficult to maintain expectational security. When ego makes an appointment, for example, and alter makes him or her wait, ego soon learns to shape expectations to the peculiarities of alter's behavior. In making this simple adjustment, ego acts on a prognosis of the situation that is made on a higher level of reflexivity. Ego has anticipated that his or her anticipation of expectations is contingent upon the behavior and expectations of alter.

This kind of reflexivity makes "corrections (and even a struggle for corrections) possible on the level of expectation itself" (p. 305). In offering one's expectations, ego provides a provisional structure for future relationships, a structure that can be revised up until the time that a mutual activity is performed. Thus, ego's expectations of expectations provide it with a chance of reversibility. In this way, ego and alter dovetail their expectations and generate sociocultural evolution.

This progress into insecurity is made efficient by the use of "symbolic abbreviations representing highly complex expectational situations" (p. 306). These abbreviations stipulate matters of propriety, tact, values, obligations, custom, and normality: whole categories of do's and don'ts. These categories exist on a "meta-level of expectations that are expected and serve there as a surrogates for tedious investigations, enumerations, and publications of the actual expectations implied in any given situation" (p. 306). They provide the social horizon within which reversible expectations can be offered.

Strategies for Reducing Risk

The risk that brings about this differentiation of social structures of expectation would be ominous indeed if there were no strategies and techniques for lessening it. These strategies ensure that expectations have a reasonable chance of fulfillment, and provide mechanisms for dealing with disappointed expectations. They build upon a crucial quality of expectations: their reversibility.

One strategy for minimizing the risk of disappointment is to make expectations ambiguous. Ambiguous expectations are unlike specific expectations that carry a high risk of failure. Ambiguous expectations can often be achieved. "Therefore, making expectations ambiguous is a strategy for creating relative security and for protecting them [systems] from environmentally conditioned disturbances" (p. 308). By using this strategy, a system takes some control of its situation. It decides what aspects of the environment connect with what aspects of its ambiguously formulated expectations. In the course of time, it builds up a web of interconnections, a structure of expectations and decisions. In this way, "it no longer depends on point-for-point agreement with its environment" (p. 309).

Risk and security are intertwined in the creation of social structures. Systems are caught in environments that put them at risk. They try to minimize the risk through strategies like tentative and ambiguous expectations. They create structures of mutual expectation that increase their internal sense of security, but simultaneously open them to new and unanticipated insecurities. "All evolution seems to rest on massing and amplifying insecurities" (pp. 309-310). Systems can come to a point, upon sufficient reflection that they voluntarily accept insecurity and actively seek to enhance it. "Evolution is an ever-new incorporation of insecurities into securities and of securities into insecurities without an ultimate guarantee that this will always succeed on every level of complexity" (p. 310).

REIFIED EXPECTATIONS

In addition to forming mutually agreed upon expectations, systems need to have expectations that are relatively fixed over time. This is especially true because autopoietic systems are completely temporized. Psychic and social systems give their expectations (which, in themselves, are momentary action/events) endurance by attaching their expectations to something that is not an event: a name. They attach their expectations to names; then they factually order those names, establishing connections and distinctions. By identifying different expectations with the same name, they turn the becoming of nature into the being of language. The categories that they utilize create selective "identities" that are more or less useful.

The general category that has been most used in the history of Western philosophy and science is that of "thing." The "thing schema" was hardly challenged until late in the modern period. In this schema, "the distinction between *res corporales* and *res incorporales* functioned as the guiding difference . . . The world itself was viewed as a *universitas rerum* and, in its coming to be and passing away, as Nature" (Luhmann, 1995, p.64). By classifying expectations as genus and species, for example, we thought that we grasped the essence of things such as human beings. We now know that we were mistaken.

> With the increasing complexity of the societal system, with the increasing analytical capacity of function systems, with the increasing instability and need for change, conceptualizations based on the thing, and especially on that special thing, the "human being," no longer suffice. This is linked to the collapse of the system of stratification, after which one can expect *all* behavior from *every* human being (p. 314).

CLASSES OF EXPECTATIONS

There are other perspectives, beyond the thing perspective, for classifying behavioral expectations. The ideas of "persons, roles, programs, and values . . . are perspectives for factual identification of expectational nexuses. Expectations, which are bundled together in such identities, can be more or less standardized depending on how one handles possible disappointments" (p.315).

For Luhmann, a "person" is not a "thing" called a "human being." Nor is it the living autopoietic system, the human body, even though it has multiple intimate connections with that body by interpenetration. A person is not a psychic autopoietic system either, because psychic systems embody multiple personalities.

A person is an encapsulation of psychic expectations. "A person is constituted for the sake of ordering behavioral expectations that can be fulfilled by her and her alone" (p. 315). Persons, both ego and alter, arise and are continually reproduced by selections that resolve complexity, especially the problem of double contingency. Being a person means drawing a lot of expectations to oneself, with which to face others and to direct one's attitudes and behavior.

We create our persons and several subordinate personas by mimicking the behavior of our parents and other role models, but we still have our own distinct ways of believing and behaving because we are distinct corporeal psychic systems. We use our different *personas* when we are fitting into different bundles of expectations: one persona as a parent, one on the job, one at a bar or cocktail party.

At some stages of history, a "role" and the person who fulfilled it were practically indistinguishable-a tribal shaman, for example, or an absolute monarch. A role is differentiated from the person who performs it when civilization becomes more complex. A role, with respect to any individual who might perform it, is both more general and more specific. It is more general because it can be filled by more than one person. It is more specific because it defines a set of expectations that is much narrower that the set of expectations that constitute the person occupying the role. Once the difference between persons and roles is established, people "can identify themselves as persons and orient themselves to roles" (p. 317).

A further step in the progression of expectational identifications that begins with person is the emergence of "program." "A program is a complex of conditions for the correctness (and thus the social acceptability) of behavior" (p.317). By the means of programs, the behaviors of more than one person are regulated and made expected. People are coerced to "get with the program." Programs are bundles of expectations that go into correctly performing activities like "the reconstruction of an automobile engine, the preparation of a department store for an 'end of the season' sale, the planning and performance of an opera . . ." (p. 317).

"Values" are the next degree of abstraction and the highest attainable level of establishing expectations. As understood by Luhmann, values do not establish the correctness of specific actions. "Values are general, individually symbolized perspectives which allow one to prefer certain states or events" (p. 317). They serve in the communication process "as a kind of probe with which one can test whether more concrete expectations are also at work, if not generally, then at least in the concrete situation one faces" (p. 318). They are generic indicators of the kinds of expectations that one might espouse.

These four levels of abstraction (person, role, program, and value) provide a graduated scale for assessing the expectations that are put on human behavior. They enlarge the range of expectations that we can safely have in the social arena by allowing us to function together without requiring conformity of one another. They broaden "the mere opposition of actual behavior and normative, morally charged rules for correct behavior with which earlier societies could manage" (p. 318).

Role and program have become highly differentiated in our society. They provide us with ways to deal with the complex behavioral expectancies of our societies; they make possible the formal organizations that are so much a part of our modern civilization.

As role and program deal with the requirements of complexity in our lives, person and value become peripheral to the mechanics of society and become "private." Therefore, our society is permissive concerning the ways that individuals present their persons, and it is tolerant of value differences because neither personality nor value bear directly on how commerce and society are regulated.

In addition to their enabling function for social differentiation, these four constraints on behavioral expectations increase the range of choice for members of society. In modern society, freedom and constraint increase together. As types of constraints grow, we have more choices about how to respond to them.

NORMATIVE AND COGNITIVE EXPECTATIONS

Autopoietic systems progress by increasing their levels of acceptable insecurity. In this process, they establish a confidence within themselves for dealing with environmental disturbances. They develop cognitive and normative methods for dealing with disappointment: what Luhmann calls, modalizations.

Modalizations

Modalization refers to the way one responds to disappointment. It is insurance, in the form of risk-assessment and contingency-planning, that prepares ground rules for dealing with possible disappointment of expectations. It enables one "to anticipate how one will behave if one is disappointed. It gives the expectation additional stability" (p. 320). In particular, modalizations enable us to anticipate more expectations in our search for cognitive and normative structure.

With modalization, one handles what is insecure, disappointment, as if it were secure. One confronts a possible disappointment with the question: In this case, should I "give up the expectation, or change it, or not?" (p. 320). In other words, one can be disposed to learning or not learning. In Luhmann's lexicon, "Expectations that are willing to learn are stylized as *cognitions*" (p. 320); while "expectations not disposed toward learning [are] *norms*" (p. 321).

We use norms and cognitions together in daily life. We skillfully mix them in social situations to deal with our reactions to disappointments. In this way, we maintain our readiness for future learning and behavior in complex situations where we cannot blindly trust in an assumed course of action. When surprise strikes, we manage to muddle through. We grow through the experience and develop new sensibilities and skills. As Luhmann says,

> Despite everything, the difference [between expectation and experience] works itself out. Once admitted, it recruits chance, forms sensibilities, strengthens the capacity for making distinctions, and forces ever more decisions. The difference becomes a point of reference for further form building, symbolization, and information processing, and thereby strengthens expectations that are experienced as insecure. Above all, normative, counterfactual expectation can be consolidated by entitling the one holding the expectation to continue to retain it despite disappointment and to reassert it. Knowledge of disappointment does not then decide the fate of the expectation (p. 322).

The semantics of cognitions and norms become reified in our use of the words, "is" and "ought." By using these words unreflectively, we come to believe that norms and cognitions are things; we do not appreciate them as our socially-created buffers for absorbing and dealing with disappointment. Norms and cognitions are styles of expectation. With norms, we attend to the difference between conformative and deviant behavior. With cognitions, we attend to the difference between knowledge and ignorance. Normative expectations lead us to construct security nets of regulations and laws. Cognitive expectations lead to the methods of science.

Through cognitions and norms, we establish some control over our insecurities. We use them as contingency buffers. In the security that they provide, we make contingency plans and harbor highly improbable expectations, things like hypotheses and positive laws.

The Origins of Social Structures

The derivation of norms and cognitions as generic ways to envelop and process greater insecurity creates a new neighborhood of perspectives on the origins of social structures. In this neighborhood,

> Structural formation is not preformed in a principle, an *arche*, nor does it occur according to objective historical laws that establish how state A is transformed into state B. Instead, the decisive point seems to be the translation of problems in system formation into differences. If a decisive point is reached-and we believe that socially double contingency and temporal expectations that can be disappointed constitute such a point-order emerges out of chance events in the course of time (p. 324).

In other words, difference leading to tension drives social systems far-from-equilibrium where they spontaneously create new structures of adaptation and growth.

The structures that crystallize in this manner allow us to form expectations that can be tested. They provide us with alternate ways to deal with frustration: retaining an expectation (by maintaining norms) or giving it up (by generating new cognitions). They elaborate our options by providing "justifications, opportunities for consensus, allowances for exceptions, and so forth" (p. 325).

Anticipatory structures that develop over time are sensitive to disturbances. They adjust to and incorporate new strata of meaning, more abstract semantics, deviant procedures, and flexibility. Over time, norms incorporate cognitive elements, and cognitive systems take on a cognitive character. Norms abetted by casuistry become resilient. Cognitive systems come to encompass so much of a society's belief system that they become normative. To give them up would cost the society too much, especially if there is not an alternative system capable of taking their place.

Traditional social thought postulates norms as the foundational structure of social life. Examples of this kind of thought can be found in the theological traditions, the natural law tradition developed by the Stoics, modern judgments made on the basis of conformity or deviance from Nature, and Parson's normative derivation of social structures. Luhmann does not postulate norms; he derives them from society's need to attain security for its expectations: they come into demand and are generated as counterfactually necessary.

Norms are pervasive aspects of social life. They bind societies together. They assume different forms in every society. The norms of one society may conflict with the norms of another (an instance of difference that may create a higher order of norm). But in a generalized sense, every social order has them. Norms, as generalized structures of expectation, have a certain independence from the actual existing laws and customs, just as those laws and customs have a certain generalized independence from the events to which they refer.

Generalizations, of course, lie behind meaning and all sorts of sophisticated expectations. Generalized expectations leave their anticipations more or less indeterminate, leaving open details like the shape of an occurrence, its time, its actors, who perceives it, and how. In their generality, expectations can often be fulfilled. "Insecurity is incorporated and absorbed through temporal, factual, and social generalizations. Expectations remain valid, nevertheless and satisfy their requirements" (p. 327).

STRUCTURAL CHANGE

Autopoietic systems undergo real structural change because they are structures of events. They are temporized systems that constitute their elements as events and are compelled, as a condition of survival, to change (through self-reproduction). Such systems have momentary static existence in their elements; that is, they remain the same for only the duration of an event, the temporal atom. Such systems endure through their autopoietic structures of self-reproduction. They change to the extent that their temporized complexity varies.

The structure of an autopoietic system provides not only for change, but for structural change, because structure keeps "what can be continued (and therefore changed) relatively constant" (p. 345). Such a structure possesses a kind of memory, which provides a basis of reversibility. While events are momentary and irreversible, structures have a history that enables them to reverse relationships. Thus, while structures cannot learn on the level of actions (which are events), they can learn on the level if expectations. They can "dissolve what has been established, and can adapt to external or internal changes" (p. 345).

For self-referential systems the classical dichotomy between static and dynamic systems is voided. In the ultimate analysis, the only thing static about such systems is the momentary atom of action; everything else is temporized. There are no constraints upon change imposed by static qualities resistive of change in such systems; and therefore, the classic theories of dynamic change are irrelevant. The constraints upon change are themselves changeable; they are the autopoietic structures of reproduction.

In self-referential systems, stability and variation are structural relations among events. The questions for them are; will those relations be reproduced or not? With similar elements or novel ones?

If a system is unsuccessful in its attempts to self-reproduce, it exits from history. If it tries to reproduce on exactly the same template every moment in spite of external and internal changes, it is likely to perish. It must reproduce in situ, at a moment in space and time. Its chances of continuing are enhanced if it masters techniques of expectation. Because of shifting conditions and the immediacy of constant change, it makes constant adjustments. "Structural changes constantly occur without being announced, wanted, or answered to . . . Often one becomes aware of structures and they become communicable only when they must be changed" (p. 348).

CONCLUSION

The autopoietic notion of structure overturns traditional social thinking by temporizing social life. It explains structural change. It provides functional (almost operational) definitions of hoary philosophical notions like action, event, expectation, thing, person, role, program, value, norm, and cognition. It orients social life to its future in terms of managing insecurity. It accounts for social evolution as a process of differentiation in which differences are accentuated and creative structures ensue.

CHAPTER 14

LUHMANN (4)
CONTRADICTION
AND SELF-REFERENCE

ABSTRACT

Contradictions are central to the existence of autopoietic systems because they survive as the unities of contradiction that they have mastered. The contradictions that arise from our observation of nature are the source of logic. Communicated contradictions, or conflicts, provide the flexible solutions that enable societal evolution. When they are successfully balanced and embodied, contradictions provide richness, strength, and flexibility to societies.

Societies and interactions are both autopoietic systems. They cannot be reduced one to the other, but neither can exist without the other. Interactions have generated societies. Societies make more sophisticated interactions possible. Experimental interactions advance the development of societies, which enable more advanced interactions . . .

In their reproduction, societies find their boundaries in basal self-reference; they proceed to reflexivity as communication about communication; and then to reflection when they recognize their separations from their environments. With rationality, societies move beyond reflection to the prudent exploitation of the system/environment interface. With epistemology, they attend to the circularity involved when the process that creates them is the same process that they use to discover how they were created.

CONTRADICTION

Autopoietic Experience With Contradiction

By their very nature, autopoietic systems structure their self-reproduction. They require structure to make the selections that maintain their reproduction in the face of novel circumstances. With structure in their reproduction, they prosper. Without structure they perish.

Selection requires a structure that is a unity of contradiction. It requires a system to decide for one alternative and against the others and, in the process, to hold all alternatives simultaneously as possible expectations. A system cannot develop self-reference without somehow defining itself against something. The meaning it creates in its selection is defined by the meanings it rejects. As Luhmann says, "Every meaning includes its own negation as a possibility" (p.362).

An autopoietic system incorporates these contradictions-the negations of selections made-into itself in the subjunctive mode, where they exist as provisionally excluded possibilities. It uses this accepted meaning-accepted/meaning-rejected differentiation as additional structure in its future selections. In its ongoing self-reproduction, it experiences contradictions as demands for selection, and uses past differentiations (unities of contradiction) as pointers in novel situations. It does not necessarily recognize that it is dealing with contradictions. It can function without any analysis of disturbing factors in its environment. It merely discriminates things as not belonging.

Systems use these provisional contradictions in their memories as the "antigens/antibodies" of their immune systems. By keeping these opposing elements on the edge of instability, systems keep alert for new strangeness and prepared for flexible reaction. In this way, they are able to respond to novel situations in flexible and tentative ways that enhance their survival prospects. "Contradictions serve to reproduce the system by reproducing necessary instabilities that can, but need not, set the mechanisms of the immune system into action" (p. 370).

By incorporating contradictions, a system knits together a delicately articulated range of responses, from tightly conservative to undisciplined. It structures its expectations in such a way that it can use the reversible quality of negations to cope with both successful and unsuccessful expectations. It is able to increase its ordered complexity because it can be adventuresome and stay alive.

Autopoiesis And Observation

Contradictions are basic to the constitution of meaning (semantics). They enter into every actualization of meaning by constituting the environment (the not-x) for every x. This presence of contradictions in autopoiesis is mirrored in the observations that we make of autopoietic systems. Autopoietic contradictions in biology are evolutionary ancestors to the differentiations that populate the ordered complexity of our phenomenal world.

It might seem that language, especially as formal logic, excludes contradictions instead of embracing them. Formal logic is, after all, built upon the principle of non-contradiction. Upon further reflection, however, logic is revealed not as a way of ignoring contradictions but as a method of managing them. It deals with the unity of contradictions, their quality of capturing all of reality in a single juxtaposition. Logic is an elaborate system designed for ordering our contradictions and multiplying their number.

There is a mirroring relation between the principle of non-contradiction and autopoiesis. The exclusion of a third alternative (a middle), which is mandated by non-contradiction, reflects the process in which an organism compartmentalizes itself to exclude elements that are outside its boundaries. Just as in life, this exclusion is not rigid; it is temporized. Moment-to-moment, the logical system of thought forms new categories (differentiations, contradictions) that fit natural and social situations more closely. Living logic modifies old contradictions constantly in manner that is both conservative and progressive by making old differentiations both redundant and obsolete.

Do Contradictions Exist In Nature?

Sociologists often speak of social conflicts and contradictions. Scientists, on the other hand, argue that contradictions do not exist. If contradictions were to actually exist, goes the argument, then no knowledge of the world would be possible. Accordingly, any claim that contradictions exist in reality is considered to be a logical mistake. And so, a right understanding of social reality should reveal that social contradictions are merely misunderstandings.

On the basis of the preceding discussion, the question of the existence of contradictions can be replaced by the question: Do autopoietic structures exist independently of the observations that we make concerning them. The answer to the latter question is a qualified "yes." Such structures seem to be a part of the real world independent of our observations; even though their individuality and generic identities cannot be captured in our descriptions. They have a reality that can be only approximated in our expressions.

If autopoietic systems exist (in this sense) in nature, then the contradictions that are inherent in autopoietic meaning-structure exist in the same sense. Since social systems are existing autopoietic systems, then contradictions exist in the same sense.

But, are not those contradictions parts of a meaning-system? Are we not imposing those contradictions upon an unstructured complexity that has no differentiation without our meaning-impositions. "Yes" and "no." It is human hubris to presume that only we make use of meaning. Any autopoietic system is structured, is based upon information, and has memory. An ameba has a meaning structure that maintains its self-reproduction, builds up differentiations, and enables adaptation. Therefore, "yes," contradictions are part of a meaning-structure, but the meaning-structure is not of mere human construction. Therefore, "no," the world that faces humans is not an unstructured complexity; it is structured by billions of years of organic evolution and millions of years of primate social structure.

To be precise, contradictions are semantic realities, as are all meaning structures. They result from the existence of autopoietic systems and the meanings that are constituted by their selections. They exist because autopoietic systems exist and make selections. They are the differentiations that structure the reality of autopoietic systems. Meanings and contradictions exist beyond the bounds of self-enclosed human comprehension and have a fractal relationship to the meanings that structure our reality.

Stop Or Go?

As we have noted, a contradiction *in autopoiesis* signals the need to make a decision, "Don't just stand there, do something!" *In observations*, on the other hand, contradictions signal a mistake; they are a stop sign. Autopoiesis works on a parallel network in which a stop sign on any road is not much of a hindrance, but the same is not true for observations. Observations are expressed (for the most part) in linear language; therefore, they must heed the stop sign of contradiction if they are to keep their coherence. In scientific endeavors, the appearance of a contradiction signals the need to stop, to rethink, to invent new explanations. In this way, contradictions provide impetus for further scientific exploration.

> Contradictions have an entirely different function depending on whether one is dealing with autopoietic operations or observations. In the context of autopoietic operations (which must always carry on if observation is to be possible at all), contradictions shape a specific form, which [demands a reaction] . . . The situation presents itself altogether differently to an observer. For him, and only for him contradiction means undecidability [because contradictions render his obser-vations suspect] (p. 360).

A contradiction takes us out of the mode of business as usual. When it persists, we face an imperative to act, to make a creative selection and to face the novel situations that emerge from that selection. A contradiction serves a double function. It blocks observation and triggers action. "Contradiction is a semantic form that coordinates autopoiesis and observation . . . Switching off operations that connect with observation means simultaneously switching on operations that precisely then are still possible" (p. 360).

Luhmann uses the parable of Buridan's ass to illustrate this point. The ass is faced with two equally appealing bales of hay. Therefore, according to a paradigm that ordains that a subject must always choose the more desirable of two alternatives, the ass will starve because of its inability to choose. In the parable, the ass faces a logical problem of contradiction that forces it to stop. The ass, of course, even if it realizes that it cannot make a logical decision, will eat some hay. Contradictions stop observation; they trigger action.

The Need For Conflict

Communication is the life of social systems. It is the connective expectation of social activity. It requires a unity among information, utterance, and understanding (with or without acceptance). Information is novel meaning; utterance is the activity that contains that meaning; and understanding is a coupling of semantic expectations. Any of these requirements can be manipulated or frustrated. Information can be false or irrelevant; utterance can be insincere; understanding can be absent.

Conflicts are a particular form of communication: communicated contradictions. Like all contradictions, they combine "semantic features under the perspective of incompatibility" (p. 385). Conflicts enable social systems to go on in situations of contradictory expectations by continuing communication in an oppositional mode. They signal that the social system can stop, and suggest that some action needs to be taken in response.

As conflicts go on, they generate creative adaptations, which enable social systems to endure. Conflicts-endured generate a society's immune system. They enter its social memory (history) to maintain alertness and instability among opposing expressions. They strengthen a social system's ability to reproduce under difficult circumstances.

If a social system is to increase the interconnection of its elements and achieve greater ordered complexity, it needs to incorporate a multitude of conflicts. That is, it must have a history of maintaining communication in spite of conflicts. Through such a history, a system sharpens and balances both sides of a conflict; and, by the spontaneous generativity of unstable connectivity, it works out creative modes of existence. Also, if both the antigens and antibodies generated by the conflict are sharp, then the system is prepared to deal expeditiously with new and unexpected complexity.

> Complex systems require a high degree of instability to enable on-going reaction to themselves and their environment, and they must continually reproduce this instability-for example, in the form of prices that constantly change, laws that can be questioned and changed, or marriages that can lead to divorce (p. 367).

Social systems need to incorporate conflicts and the more, the better. They find a rich source of conflicts in logical disputes, but utilize other sources as well. They establish time differentials, for example, to increase synthesized conflicts. They also restrict the means for resolving conflict and limit the use of physical force to political authorities. By this means, they prevent the premature suppression of conflict and create an arena in which there is adequate freedom for conflicting behavior. In so doing, they lower the threshold of dissent and "enable autopoietic reproduction despite the absence of agreement. Here the extreme mobility of 'no'- which is logically as potent as 'yes'-is exploited and domesticated at once" (p.397).

This exploitation and domestication of conflict can be explained within three basic principles of General Systems Theory. Luhmann formulates them as follows: "(1) the loosening of internal bindings, (2) the specification of contributions that are enlisted for interpenetration, and (3) the creation of effects by randomly beginning and then self-amplifying effect cumulation" (p. 398). The minor novelty in his presentation of the general theory is his second step, the specification of contributions. This step seems to refer to the role of specific conflicts, regulators, and critical thresholds of instability.

Its embodied conflicts provide the differentiation that enables a society to cope with its complexity. Luhmann formulates this in terms of a problem and its solution.

> The problem [is] to reproduce enough social systems, and enough different kinds of them, to correspond to the complexity of a specific developmental stage of society. Normally this occurs according to prescription, that is, on the basis of structures of expectation. The immune system secures autopoiesis when this normal way is blocked (pp. 403-404).

A society's immune system "does not reside in a merely negative copy of structures or in a 'critical' consciousness of what is at hand. It resides in its own peculiar forms of continuing communication-in forms that, for example, vary so much through struggle and victory that normalizations are again possible" (p. 403).

SOCIETY AND INTERACTION

Luhmann refuses to ground social systems in aggregations of individuals. Unlike Maturana and Varela, he does not consider social systems as third-order autopoietic unities (cells being first-order, and organisms being second-order; 1987, p. 78; 180-181). Unlike Habermas, he does not ground society in the rules of communicative action (1989, p. 137; 141-144). He also rejects George Herbert Mead's theory of "symbolic interactionism," in which, "society, as distinct from interaction, exists as individuals (or as individuals in interaction). But the individuals are constituted only in the interaction, and thus are psychically internalized social artifacts" (Luhmann, 1995a, p. 405). [Habermas agrees with this position (cf. 1989, 11-17).] Luhmann feels that any socio-psychological explanation is unsuitable "for comprehending the highly complex problems of the societal system, which cannot be ascribed to individuals or to their interaction" (p. 405). He excludes "the system reference of psychic systems from the analysis of social systems and understand[s] the difference between society and interaction as that between two kinds of social system" (pp. 405-406).

Society and interactions have a problematic relationship with each other. They cannot be reduced one to the other; and they cannot exist without each other. "Societal systems are not interaction systems and cannot be conceived simply as the sum of the interaction systems that occur . . . although interaction systems always presuppose society and could not begin or end without it, they are not societal systems" (p. 406).

In their temporal aspect, interactions are episodes that require societal communication before they begin, and they rely on the expectation that societal communication will continue afterwards. The ongoing presence of society makes interactions possible and gives them a certain freedom of expression, because the failure of an interaction is not likely to crash the whole societal system.

In return, interactions "achieve structures that cannot be made congruent with society and yet equip it with complexity by building in differences. Thus interaction brings about society by being relieved of the pressure of having to be society" (p. 407). "Society acquire[s] complexity and interaction acquire[s] its qualified improbability. And only through it [this complex relationship] is the evolution of improbable complexity possible" (p. 407).

Boundaries

Society is a concept that expresses the "unity of the totality of what is social . . . It is the all-encompassing social system that includes everything that is social and therefore does not admit a social environment" (p. 408). Society finds its unity in its self-referential closure. Its every activity (communication) leads to further communicative activity. It includes all social activity.

Society's only environments are physical, organic, and psychic systems with which it cannot communicate. It has

> no environmental contacts on the level of its own functioning. Just as an organism does not live outside its own skin . . . so a society cannot communicate with its environment. It is completely and without exception a closed system. This distinguishes it from all other social systems, in particular from interaction systems, which include communicative relations with their environment, welcome what is new, utter decisions, and so on (pp.409-410).

Interactions find their closure differently. They "include everything that can be treated as present and are able, if need be, to decide who, among those who happen to be present, is to be treated as present and who not" (p. 412). Interactions are ad hoc; they make up their boundaries and adapt them as they find fit. They determine their content, participants, and time on the basis of the collective perception of their members. They determine their closure not only in outright communication, but also in the perceptions that underlie communication.

Perception and Communication

Perception is a psychic phenomenon that becomes social in the context of double contingency when ego and alter perceive each other as perceiving each other. This reflexive perceiving lies at the basis of articulated communication and it has specific advantages. Perception achieves:

1. great complexity in absorbing information with limited analytical precision—thus a far-reaching but only 'approximate' mode of intelligibility, which can never be communicated;
2. an approximate simultaneity and rapidity in information processing, whereas communication depends on a sequential mode of information processing;
3. slight accountability and capacity for being negated, thus great security about the commonality of an item of information (however diffuse) that one possesses;
4. a capacity for modalizing communication through parallel processes of weakening, strengthening, and contrary utterance on a level of (intended or unintended) 'indirect' communication where the high risk of explicit action can be avoided (p. 413).

Perception influences interactions in utterly pervasive ways.   It forces communication to go on, because interacting participants interpret every behavior as a communication.   If one examines one's fingernails or looks at a watch, one communicates. "In practice, one *cannot not* communicate in an interaction system; one must withdraw if one wants to avoid communication" (p. 413).   This pervasiveness of perception renders interactions susceptible to disturbance, either constructive or destructive. It creates openings for new insights and new activity. It also allows disruption of communication by intrusion and sabotage.

In interactions, perceptions and communications contribute to each other. There is "a double process of perception and communication in which burdens and problems lie partially in one and partially in the other process and are constantly redistributed" (p. 414). This sharing and reinforcing of ideas, feelings, and actions is constrained by a pervasive insistence upon communication in interaction systems. This sharing, reinforcing, and constraining intensify the tumult that generates the novel contributions, which are of use to a larger social system.

This tumultuous kind of interaction forces the formation of structures that separate communication from perception in factual, temporal, and social domains. In interactions, people are forced to put their perceptions into words. "The relevant events must be placed in a sequence; they must be structured by factual themes; and not all those who are present to one another may speak at the same time, but only, as a rule, one after another" (p. 415).   In this creating of themes, schemata, and sequencing, the people in a conversation generate structures that order discourse. In forming these structures they either adopt existing societal mores or they create new ones.

Society/Interaction Relations

Society and interaction lend each other mutual support. "Society guarantees the meaningfully self-referential closure of communicative events, thus the capacity to begin, end, and form connections of the communications in each interaction" (p. 416). Society provides context and continuity for interactions. On the other hand, interactions anchor society in corporal and psychic reality.

> Interaction systems activate the hydraulics of interpenetration. The push and pull of presence works on those who are present to one another and induces them to subject their freedom to constraints. Therefore society is not possible without interaction nor interaction without society, but the two types of system do not merge. Instead, they are indispensable for each other in their difference (pp. 416-417).

In the temporal dimension, episodes of interaction make no sense if they are not a continuation and reproduction of society. Why would one drive in traffic every morning if driving were not part of larger processes, such as, going to work, earning a living, and raising a family? Conversely, societies would never evolve without interactions. How would a society survive without an economy in which people worked? How would it progress unless people shared good ideas in interactions?

In the social dimension, society provides roles for people beyond their roles as participants in any particular interaction. Insofar as anyone is committed to a role outside the interaction, he or she must be excused from total participation in the interaction. Therefore, "the difference between society and interaction transforms commitment into freedom" (p. 419). If you are asked to work overtime, for example, but you are also obliged to pick up your daughter at day care, you are doubly obliged, but you also have the freedom to work out different courses of action.

In the fact dimension, interactions pick up themes from their societies, for example, from newspapers, TV, scientific journals, and gossip. They innovate those themes and create new ones, like jokes on late night TV and scientific position papers. Some of these new themes go beyond gossip and contribute to the differentiation of society. Such a new theme was child-abuse, which emerged from a timeless taboo only in the past 20 years, and has resulted in community awareness, court cases, therapies, and so forth.

Historically, societies develop from interactions. In the course of time, they differentiate from those interactions. "Modern society separates its system formation from possibilities for interaction" (p. 425).

In this differentiation, society largely ignores interactions and renders them free, unconstrained, and trivial. "Less and less can one count on solving societally relevant problems by interaction; for example, by using people's physical presence to gain a consensus or to prevent uncontrollable activities" (p. 426). There is no metanarrative (myth, moral code, or dialectical theory) that is a shared operative description of the social world. As a result, people find themselves without shared motivations, without responsibility, and without respect for the way society functions.

SELF-REFERENCE AND RATIONALITY

Luhmann distinguishes four grades of self-reference: basal self-reference, reflexivity, reflection, and rationality. *Basal self-reference* is the minimal grade needed for autopoiesis, in which elements constrain each other into recurrent consistent reproduction; here the "self" functions only as an element. With *reflexivity*, autopoietic systems work within the temporal dimension. They use the before/after differences that are generated as they apply self-reference to the process of self-reference. Here the "self" is a process that refers back on itself, like the generation of mold in an abandoned cup of coffee or like communication about communication. *Reflection* arises from the recognition of the system/environment difference; here the system is the self. It occurs "in all forms of self-presentation that assume the environment does not immediately accept the system in the way it would like itself to be understood" (p. 444). *Rationality* integrates a recognition of the unity of the system/environment difference into the system. In brief, basal self-reference indicates self as an element of a relation; reflexivity indicates self as a process; reflection indicates self as a system; and rationality reflects upon the unity that overarches the system/environment difference.

Basal Self-Reference

Basal self-reference requires ongoing activity and constraints upon that activity. The combination of activity and constraint leads to consistent activity. The activity adheres to type by staying within its parameters of ongoing self-reproduction, and by acting on its ever-present imperative to select amid complexity. It uses remembrance of past action to guide future action. This memory and foresight can be as basic as the polymeric activity of a compartmented hypercycle (cf. Eigen, 1992, pp. 42-44).

Basal self-reference brings irreversibility into a system because it is a relation between action/events that have no permanence and can never be repeated. It links those events in a process of self-renewal, and thereby creates a new, autopoietic reality, which has a permanence that is altogether lacking in action/events. Within this new process/structure with its temporal endurance, changes occur during reproduction; changes that are impossible within the atomistic constitution of action/events (because of their merely momentary existence).

Basal self-reference brings about autopoietic process/structures that have a before and after: an after that can be different from a before not only temporally but also structurally. As they evolve, these autopoietic structures increase their ordered complexity by differentiating out organizing functions. In the direction of causality, they develop technological functions that self-organize processes in terms of before and after (what Luhmann calls reflexivity). In the direction of philosophic rationality, they develop functions that self-observe and self-describe themselves in terms of system/environment (what Luhmann calls reflection).

Reflexivity

Reflexivity is self-reference upon the before and after of ongoing self-reproduction. It applies self-reference to the process of self-reference. A reflexive system discovers the idea of causality as it observes how an event comes about thanks to the selectivity of earlier and later actions. It gains the ability to intensify its selectivity by controlling the non-occurrence of certain of its activities.

In the social arena, reflexivity is communication about communication. It is exemplified in processes that guide and control themselves. In romantic love, for example, "that one communicates about love and how (bodily behavior is very much a part) is also a proof of love, and there is no possible proof outside of this self-reference (p. 453). Also, money is a reflective use of the relations of exchange; with money, we exchange the possibility of exchange with one another. The power of the government is likewise reflexive because it "concentrates precisely and exactly on directing other's use of power" (p. 453). With such media of communication as romantic interaction, money, and bureaucratic power at their disposal, societies join a sensitive perturbability with a formidable capacity to recuperate.

Reflection

Reflection is a third mode of self-reference, in which an autopoietic system attends to its difference from its environment. It introduces the difference between system and environment into the system. It is present in germ as a part of every autopoietic system. "A system that can reproduce itself must be able to observe and describe itself" (p. 457).

Reflection develops in the psychic realm, as ego and alter deal with double contingency, when ego realizes her separateness and finds herself mirrored in alter. It makes its appearance as self-observation and self-description. Reflection builds upon the difference in complexity between system and environment and expresses that difference in its purest, most abstract form. "Identity in contrast to everything else is nothing more than the determination and localization of difference in relative degrees of complexity" (p. 184).

Reflection in social interactions, that is, reflection upon interaction, occurs when an interacting group "must mark individual actions as binding the system" (p. 455). Reflection occurs when a collective obligation is acknowledged due to the collective or representative action of the group. Social reflection also occurs in groups "when they break off contact between those who are present to each other and arrange for them to meet again, that is, if they must maintain their identity over latent phases" (p. 455).

Every social system has at its disposal, "a rudimentary procedure for self-observation [because] every communication declares, whether consciously or not, thematically or not, that it belongs to a system." (p. 456). Social self-observation is always in the form of communication. Self-observation becomes thematic when it takes the form of self-description; that is, when "it produces semantic artifacts to which further communication can refer and with which the system's unity is indicated" (p. 456). Self-observation becomes reflection when it produces "a semantics that can represent the relationship between system and environment within the system" (p. 457). The degree of refinement required for reflection is demonstrated in the following passage: "Neither Greek/barbarian nor *corpus Christi/corpus diaboli* suffices; one must take into consideration that pagans are not pagans for themselves" (p. 457).

Theories Of Reflection

Theories of reflection work out comparisons and relationships between reflections. Epistemologies and theories of science, for example, try to "explain how identity in the difference between knowledge and object is possible at all-whether as the conditioning of transcendental consciousness, as a dialectical process, or as a pragmatics open to confirmation" (p. 458). Similar theories exist for political, educational, legal, and economic systems and even in the domain of intimate relations.

Universally accepted thematizations for the whole, however, seem to have gone out of date. The traditional thematization in terms of "Old Europe's cosmically hierarchical consciousness of order, which was oriented to the primacy of politics and/or religion" (p. 458), is obsolete. More recent formulations in terms of dialectics, evolution, modernity, or fear of catastrophe are only relatively convincing and do not completely escape the charge of "ideology." It seems that we no longer have myths or metanarratives that bind society as a whole.

The Circularity Of Reflection

No matter what degree of refinement we achieve through "intensified self-observation self-description, reflection, or theories of reflection" (p. 459), we never reach objectivity. We do not achieve "a privileged access to knowledge" (p. 459). In all our elaborations of self-observation, we remain bound to *self*-observation. We never completely escape a circle.

Still self-reference is not a tautology or "a complete duplication of whatever functions at the self at any time" (p. 460). An autopoietic system carries out its reproduction in such a way that it reintroduces its unity back within itself as "accompanying" self-reference. It conceptualizes itself (reduces its own complexity); and then juxtaposes the semantics that it generates in this way to the semantics of its every selection. Its self-conceptualization acts as a cord of reference in all of its operations. In this way, "the system operates always, but not only, in contact with itself. It functions as an open and a closed system at once" (p. 460).

Asymmetry

Autopoietic systems use a variety of methods to break out of the tautological circle. Those methods can be grouped under the heading of *asymmetrization*, in which "a system, to make its operations possible, chooses points of reference that are no longer put into question" (p. 466). By positing indubitable (at least for the moment) facts, the system treats parts of its own structure as being in its exterior. In so doing, it breaks the interconnectedness of such "facts" with the rest of itself. It needs such a procedure in order to cope with the complexity of its world.

There are a variety of methods for creating asymmetrization: "externalization, finalization, ideologization, hiercarchization, punctuation, and so on . . . [which] clarify the form in which additional meaning is recruited and the tautology of pure self-reference is interrupted" (p. 466). In the time dimension, one can externalize the past and strive to overcome it, and one can posit goals in the insecurity of the future. In the fact dimension, one can posit a distinction between environmental variables that can and cannot be controlled. In the social dimension, the acceptance of hierarchies of either ancient or modern provenance, or the recognition of individual decision-making, give foci and environment to action. All forms of asymmetrization point back to the position of the primacy of action.

Rationality

Luhmann deals with *rationality* as rational action, as it develops from autopoiesis. He does not, of course, accept the hierarchical justifications of traditional metaphysics and morality. Nor does he try to build an alternate structure that might provide norms for judging the suitability of action, as does Habermas, who finds a foundation for rationality in the claims of communicative action and who describes how rationality builds the lifeworld. Instead, he works with the concept that has traditionally been called "prudence," what one takes into account in deciding whether to do something.

Planning is a normal way of taking things into account. As systems planning, it "fixes specific future aspects of a system and tries to actualize them" (p. 469). The specific aspects of planning that concern Luhmann are: "whether a *social* system can plan *itself*, and which problems one must reckon with if this is attempted" (p. 469). According to Luhmann, all such "planning is notoriously inadequate. It does not achieve its goals, or at least not to the extent that it would like, and it triggers side-effects it did not foresee" (p. 469).

This ineffectiveness can be explained by the fact that: any planning that is done in a system is observed by the system. As a result, when a system plans, "it produces implementation and resistance at once" (p. 470). In particular, planning always "introduces a simplified version of a system's complexity into the system" (p. 470), which can be challenged by those adversely affected by the planning on the basis of its bias. The attempts to create a mutually acceptable simplified version of the system involve consideration of societal and political relationships (from the conservative side) and consideration of public opinion, parliamentary discussion, and binding decisions (on the liberal side). Because of this self-observation and its ensuing conflicts, planning and consensus-formation are two sides of the same coin. Effective planning is geared to both management of complexity and creation of consensus.

Planning produces a system's future-oriented description of itself. In the past, planning could perhaps assume a fixed foundation in a universally accepted description of itself. Today, it must ground itself on a variable consensus. In other words, the way the system represents itself to itself is experienced as contingent. If a system's relationship to its past is so variable, then its relationship to its future is even more uncertain.

Many social systems today are *hypercomplex*: they are oriented to their own complexity. They try to deal with the fact that their planning produces a likelihood of unforeseen reactions. In their planning, they try to plan their futures by factoring in those unforeseen effects. In such systems, because "the difference between planning and observing planning cannot be eliminated . . . there can be no point of equilibrium in the system for this difference or for the tensions it creates" (p. 471). Any spokeswoman in such a system confronts the system as her alter in a condition of double contingency. Luhmann asks, "Does this mean that rationality is no longer possible? Or does it only mean that one must think differently about rationality than heretofore?" (p. 472). In response, he builds a different conception of rationality.

Rationality builds upon reflection, but it is not identical with it. With reflection, a system constrains complexity by attending to the difference between itself and its environment. In certain European formulations, which recognize the inadequacy of the traditional bases of rationalization, this difference between self and environment is elevated to a kind of rationality in terms of "a natural self-esteem, a self-grounding reason, or a will to power" (p. 472). Luhmann considers these formulations to be "anthropological packaging."

Rationality is based upon the difference between a meaning system that is trying to behave responsibly and its environment. Its environment is not some encompassing system; it is "a world horizon that corresponds to the system's internal horizon" (p. 474). As a result, a system's rationality is not to be explained in terms of some superordinate system. Also, consensus upon a preordained norm of rational action or rational values is precluded.

There is a way out of this impasse. "One can imagine a kind of convergence if planner and observer both use the system/environment difference as a schema for acquiring information" (p. 474). If both parties to a conflict project their designs in terms of their environmental impacts upon the society, then creative adaptations can emerge. Theoretically, this approach can be characterized as using the unity of the difference between system and environment.

Practically, using the unity of this difference means that "a system must control its effects on the environment by checking their repercussions upon itself if it wants to behave rationally" (p. 475). Many actions may be prudently asymmetricized by considering them as being absorbed by the environment without repercussions upon the system. Other actions, however, produce such repercussions. By encompassing both possibilities a system works upon the most demanding level of rationality. Action upon this level demonstrates that, reflection on the unity of difference need not annul the advantages of difference. The "self" at this height of rationality transcends the system-environment difference.

EPISTEMOLOGY

Functional Analysis

Self-referential systems have created us as observers in the course of evolution. If we acknowledge this fact, we can recognize the self-referential nature of our knowledge and make consistent use of our thinking skills. We can realize that any theory of knowledge, any epistemology, that aims for universality must apply to itself the standards that it applies to its field of study.

Traditional epistemology has posed as a lawgiver that tells us what we must do if we are to create true and valid statements. It proposes its laws from a viewpoint of the assumed unity of truth. The epistemology of self-referential systems requires a view of truth that requires statements to be in harmony with the theory that explains how self-referential systems come to be. It requires adherence to both self-identity and self-divergence. It finds its justification in the context of the basic schemata of evolution such as General Systems Theory and autopoiesis.

Luhmann's name for this brand of epistemology is functional analysis. The functions of this epistemology are indicated by the overall schemata of evolutionary and autopoietic change. The analysis locates particular aspects of a science (such as sociology) within those schemata. Thus, meaning, difference communication, differentiation, and so on are analyzed within the context of autopoietic reproduction. Functional analysis uses the traditional analytic methods of science within the *nomos* of evolution.

From the standpoint of functional analysis we can grasp the paradox of the incongruent perspective in which we overburden the reality that is observed with a complexity that it does not experience. In functional analysis, we do "not simply trace how these systems experience themselves and their environment, [we cover them over] with a procedure of reproducing and increasing [their] complexity that is impossible for [them]" (p. 56). We do not do justice to the ongoing concrete self-experience of the systems with our conceptual abstractions.

Every observation entails this kind of overburdening of the reality that is observed, because an observation is a simplification of the object system that presents itself to the observer. Self-reference imposed upon living systems is necessarily an imposition and only one thematization among many that can be imposed on them. Because observations are necessary, the overburdening is necessary. The adequacy of these thematizations is the domain of epistemology.

The decision to characterize a system as self-referential entails epistemological claims. "In the domain of the societal system, what we have called functional analysis is a principle of scientific system observation and not *eo ipso* a principle of self-organization for societal relationships that reproduce themselves every day" (p. 300). Nevertheless, "action and observation do not necessarily exclude each other" (p. 300). The observations we make about the self-reproductive action of systems, such as societies, point to a "morphogenetic principle of decisive significance . . . that steers the evolutionary selection of successful structures" (p. 300). In other words, our observations can identify fundamental principles that underlie self-reproduction.

Natural Epistemology

One does not need to establish epistemological foundations in transcendental theory. One can return to natural epistemologies that understand the accumulation of knowledge as a process of adaptive reiteration. When one uses a naturalized epistemology, one understands that knowledge is self-referential; one expects circularity.

> Precisely because it [naturalized epistemology] understands itself as a science of natural processes, it has already admitted this [self-reference], and precisely this distinguishes it, as *post*-transcendental, from *pre*-transcendental epistemologies, which appealed to common sense, associative habit, or the certainty of ideas as the basis of knowledge (p. 479).

One does not, however, accept an identity between the real and what knowledge indicates as real. First of all, one supplies reasons for knowledge. Then since these reasons "merely transform the circle into an infinite regress" (p. 479), one places one's hopes in approximating reality evermore closely.

> If one in turn justifies the reasons and keeps every step of this process open to critique and ready for revision, it becomes more improbable that such an edifice could have been constructed without reference to reality. The circularity is not eliminated. It is used, unfolded, de-tautologized (p. 479).

By using this self-reference of knowledge, one keeps the environmentally sensitive structure of one's knowledge intact.

> Traditional epistemologies consider circles of this sort as grounds for suspicion that statements are false, if not gratuitous. The opposite is true. They force themselves upon us. One cannot avoid them . . . But one can . . . build them into the theory of science, for they contain precise instructions for self-control (p. 483).

Autopoietic Knowledge

Within an autopoietic view, "knowledge is a nonhierarchical quality that emerges out of a recursive covering inside the system" (p. 483). Knowledge is not grounded in unified final elements because unity of any sort is produced autopoietically. Knowledge is not grounded either in direct observation of any object, process, or unity. Because it is a recursively-closed system, knowledge produces all the units that it uses. An observer has no alternative to the inferring of unities based on postulated differences that are both personal and contingent.

"How then can one guarantee that observation maintains contact with reality when it claims to be knowledge, even scientific knowledge?" (p. 484). First, one can move the site of knowledge claims from psychic systems to social systems. Social systems can be made independent of individual motives and reputations. The knowledge of social systems "can be subjected to its own conditionings, perhaps in the form of 'theories' and 'methods'" (p. 484). In modern science, such knowledge can be evaluated on its productivity, its ability to generate new knowledge.

Second, in answer to the concern about common ground, one could require a common logic for universalistic theories. One could force these theories "to test on themselves everything that they determine about their object" (p. 485). If one proceeds in this direction, every science, when it discovers that objects in its domain do not conform to a priori principles, would acknowledge that such principles do not hold for it either. One would simply acknowledge that one's knowledge grows from physical, biological, semantic, or historical processes and can be no more certain or necessary than those contingently generated processes. If science discovers autopoietic development, it must consider its knowledge also to be autopoietic. If self-reference is "built into a conditional and increasing nexus of closure and openness" (pp. 485-486), then science needs to ask: "Does science do it in the same way? And if not, why not? And anyway, how else?" (p. 486).

Science is a self-referential system that concerns itself with self-referential objects. For that reason, "Science's relationship to its object is, for its part, a relationship of double contingency" (p. 486). Science can investigate its object only if it sets in motion processes of communication and interpenetration with its object within the context of self-reference. This double contingency "forces a new level of reality" (p. 486).

As knowledge interacts with its objects, it produces (and is) "an emergent reality that cannot be reduced to features already present in the object or in the subject" (p. 486). Knowledge with its observations and differentiations is in neither the observer nor the object; it is emergent and it exists in the plane of semantics. "This insight bursts open epistemology's subject/object schema without disputing (indeed, while presupposing) the possibility of pregiven characteristics and projections onto the environment that are relative to the system" (p. 486).

Emergent realities spring spontaneously from situations of double contingency, and from situations of interconnected complexity generally. This conclusion follows from applying to epistemology the principles that apply to all self-referential system.

> [Knowledge] works auto-catalytically, that is, [it] reorganizes 'material' that already exists on an emergent level of reality. On this emergent level the world is viewed in a new way, although there remain specific uncertainties and therefore specific techniques for reducing uncertainties by interaction with the object, namely, by stimulating self-referential processing (p. 486).

CONCLUSION

Systems create and maintain themselves by imposing circularity on their self-reproductions. They gain the ability to do this by recognizing and selecting differences in their worlds. In their structural couplings, psychic and social systems enable collective action by generating schematisms of differentiations. These schematisms begin in the binary form (this-not that, us-them, good-bad, true-false). They later differentiate into smaller schematisms, regarding role performance, esthetic preference, and so on. In so doing, they undermine the binary schematisms and introduce the "excluded middle" into moral and cognitive expectations. Thus persons and societies evolve by generating binary schematisms and then by progressively undermining them.

Stress is built into every structure as a balance of forces. Stresses for an autopoietic system exist in the selections that structure its self-reproduction. These stresses are potentially negatable affirmations incorporated into the system by selection. They create a precarious internal balance that enables further tentative selections and structural couplings. In this way, stresses secure the system's autopoiesis, enable its survival in novel circumstances, and constitute its evolutionary advancement.

A stone-and-mortar metaphor for the precarious process of structural coupling, which is accomplished by balancing stresses, is the construction of a gothic dome. The medieval masons built from each corner of the gothic dome so that the stones would meet in the center in a four-sided keystone, which is the intersection of two congruent and perpendicular arches. The four arcs that compose these two arches can be considered as autopoietic systems in this metaphor.

Each arc was built on a solid foundation and was meticulously constructed by balancing stresses. Stones were laid upon stones and gently cantilevered out. These cantilevers were temporarily balanced by temporary supports until two opposing arcs were joined in a keystone. In addition, buttresses compensated the gravitational stress that would be transferred to the sides of the arcs when their temporary supports were removed. These buttresses reinforced the sides of the arcs so that the total structural coupling (the arch) was self-supporting. In further structural couplings, this arch in concert with other arches and structural variations (like flying buttresses) generated marvels like Westminster Abbey and the cathedral at Chartres.

In architectural marvels like Chartres, the stresses of gravity are balanced against each other through the solidity of stone and the patterns of gothic construction. In mechanical wonders like an airplane, gravity is compensated by the combination of forward thrust and the Bernoulli effect. In information (autopoietic) systems, all manners of opposing forces are contained and processed as unities of contradiction (selections that carry their negations in their backpack). Using these unities of contradiction, autopoietic structures structurally couple with information in their environment (differences that they can react to), including other autopoietic systems. In this way, they generate the world, our ability to know it, our ability to communicate, our language, and our epistemology.

CHAPTER 15

KAMPIS

ABSTRACT

In *Self-Modifying Systems*, George Kampis examines the ways that we model living reality. He finds our usual models, based upon dynamic, thermodynamic, and complexity theories, inadequate because they reduce material reality to formal (logical structures). Kampis claims that these theories cannot explain the creativity of living systems because of that very formality.

To explain the creativity of life, Kampis rehabilitates the notion of material causality. He proposes that living systems contain material implications that escape our formal description. In their material implications, living things do constantly recreate their components and processes; but they also create new living systems.

Adequate models of living systems, therefore, cannot be formal and first-order in terms of dynamic, thermodynamic, or complex equations. They must reflect the spontaneity (non-formality) of living systems at all levels of their material functioning. They have to deal with the emergence of order in our phenomenology and semantics.

Such models are not second- or meta-order in the formal/mathematical sense of having higher abstraction. They do not presuppose formal mappings between coarse and fine grain reality. To the contrary, they explicitly recognize that our mappings exist in a different phenomenal domain than that inhabited by the systems being described. Our models may be formal and accurate on a certain level of abstraction and probability, but the behavior that they describe is not formal and deterministic—it is material and historical.

The theory of component-systems provides a non-formal model that describes the material constraints and processes of self-modifying systems. It shows how systems go beyond self-ordering to self-organization and generate a class of systems based on meaning. It also opens avenues of perception into the origins of categories and epistemologies.

MODELING

In a very general sense, a model is a representation of something in a simpler form. A model is both a tool and a goal of inquiry. It is a tool of inquiry because it helps us envision how complicated processes work. It is a pragmatic goal of inquiry because its formulation (consider the double helix) enables further understanding and research. A model severs some aspects of reality from an amorphous background and, in that way, brings our phenomenal universe into being.

Models, in a narrower sense, represent reality by identifying processes that are seen as underlying our phenomenal experience. They are extremely useful; they introduce new predictability into our lives. Our solar mechanical models allow us to predict eclipses and tides. Our Boolean networks enable us to use computers.

Paradigms and Constructs

There are two sources of models: paradigms and constructs. Paradigms are (often unreflective) assumptions that are postulated upon reality. *Paradigmatic models* postulate that a certain order exists in reality; they are valuable to the extent that they can predict the behavior of a system. *Constructed models* are put together by following specific material steps. In this, they are composed in the way computer programs are created-step by step. Paradigmatic models are conceived abstractly at a remove from the material implications of real systems; they are deterministic. They are more amenable to formal mathematical treatment. Constructed models pay meticulous attention to material detail and aim at a causal, step-by-step explanation of phenomena. They are causal but not deterministic.

> The ideas of causality and determinism are closely related, but different. Causality says that the future events happen *because* of the present. Determinism says that the future events happen *as* they are determined by the present (p. 38).

The idea of deterministic causality is related to paradigmatic models; it says that a present configuration necessarily yields a specific future configuration. The idea of non-deterministic causality is related to constructed models; it says that a present configuration yields a subsequent one but can do so contingently. Kampis defends a theory of material causality that does not entail determinism. Mainstream theory holds that causality and determinism are inextricably linked.

Classical Dynamics

Kampis claims that formal (mathematical) efforts to model biological and cognitive systems are inadequate in their very essence. He argues that theories of dynamics, thermodynamics, and complexity cannot explain the spontaneous innovations of life and thought because they equate the formality of their reasoning with the material implications that are exhibited by their subjects. In their unavoidable course-graining of reality, these theories ignore numerous implications that influence the behavior of life and thought.

Formal theories model material sequences in terms of equations. In abstract simplicity, they define a specific value of x (position in space) as a function of time:

$$x = f(t).$$

In such an equation, every value of t yields specific values of x.

A system defined by deterministic equations exhibits a strangely static kind of dynamics. If one space-time state of a system is determined, then all the other states of the system are likewise determined. For example, if the scientists of NASA know the specific position of Mars in relation to Earth at one specific time, they can compute its position at any time in the future or in the past. Because of this state-determined quality, classical dynamic systems can be run both forward and backward. They are closed, circular systems; they do not permit innovation. They portray a basically static universe.

Classical dynamic theory postulates a descriptive frame of reference that excludes extraneous variables; it assumes that states can be well-defined; and it assumes that an observer can have simultaneous access to temporal information. All these assumptions are counterfactual. Each of them disables attempts to understand the processes of life and thought.

Because they postulate a descriptive frame of reference that excludes many material implications, classical dynamic models miss the evolutionary possibilities of natural systems. In their assumption of well-defined states, these models are locked into a world that consists of interacting, well-defined things (Democritean atoms). In their assumption of simultaneity of observation, such models posit a static universe. For all these reasons, classical dynamic theory denies the possibility of systemic change over time; it posits a world in which Zeno's paradoxes are insoluble.

More Complex Formal Dynamic Models

Quantum dynamic theory introduces probabilities into the $x = f(t)$ equation. In quantum dynamics, one cannot know both the position and momentum of subatomic particles at the same time (the Heisenberg indeterminacy principle). As a result, given x's momentum at time (t), one can only deduce a stochastic (statistical) probability of the location of x. The relation between quantum indeterminacy and quasi-classical (macro) indeterminacy is still a matter of conjecture.

Chaos theory introduces a radical indeterminacy into macro-level reality by spotlighting the influence of minute differences in initial conditions. The most famous formulation of this influence relates to the weather: If a butterfly flaps its wings in Hong Kong, it can create a thunderstorm in New York City.

The theory of far-from-equilibrium thermodynamics (Prigogine) demonstrates how classical dynamic systems give rise to indeterminacy when they lose their equilibrium. Catastrophe theory (Thom) also shows how determinate dynamic equations yield divergent and surprising results. The theory of synergetics (Haken) chronicles the increases of complexity and efficiency that occur when systems cooperate with each other. The principle of noise (Von Foerster) explains the importance of shifting intensities and conflicts in the creation of higher degrees of complexity. Theories of complexity (Kauffman, for example), find self-organization in stochastic patterns to be inherent in nature as modeled by Boolean networks.

These sophisticated theories of indeterminacy and self-organization explain many phenomena that are unexplainable in a classically understood universe. They fail to explain, however, the processes of life and thought, according to Kampis. In their reliance upon formal theories of dynamics, these theories equate material phenomena with the equations that describe them. They ignore and, if necessary, distort the non-conforming material influences that exist in the measured and unmeasured results of any dynamic experiment.

Of special importance in these theories is their attention to the multiple frames of reference in which we view reality. These levels of observation correspond to a hierarchy of description levels, for example, a particle, a molecule, or an animal.

The micro-processes of particles, molecules, and organisms are linked together in parallel, connectionist schemes. The processes of the molecular level, for example, are linked to other molecular processes materially and, to some extent, formally (in our theories). In addition to these "horizontal" linkages, the same molecular processes are joined "vertically" to particle processes and often to organic ones. The vertical connections are often many-to-one interactions that bunch many elements of one level to an element of another more coarse-grained level. These bewildering connections sometimes branch to create novel connections, and sometimes they vary their speeds of connections. They are seldom available to our inspection. They are usually ignored when they crop up. Nevertheless, they sometimes create new objects on higher levels of observation.

Some simple mathematical equations produce chaotic and elegant fractal geometric figures. The pixel generation achieved by the Mandelbrot equation, for example, is completely unpredictable on the macro level. Still every pixel is mathematically determined on the micro level. And, back on the macro level, these pixels eventually create awe-inspiring beauty. We can also generate pseudorandom numbers that appear to be unpredictable but are, in fact completely pre-determined. It would seem that changes in grain size are closely related to changes in phenomenological domains.

By focusing on both micro- and macro-levels, theories of thermodynamics and complexity explain how increased informational complexity arises from deterministic and stochastic microprocesses. These theories produce formal models that mimic the behaviors of natural systems.

While macrophenomena are generated by microprocesses, they are not created solely by them. Macrophenomena require observers and describers. The increase in descriptive complexity, in which non-deterministic new components appear and are named in an ever-increasing information set, is a human act. An increase of complexity for us always involves naming new phenomena. The more complex the item, the more complex is its description and the more numerous are the names involved in its composition.

In our descriptions, the meaning systems of living things are extremely complex. They are not reducible to formulae. They require exhaustive description and naming and are extremely hard to explain. Biology and cognitive science labor under the obligation of extended descriptions that try "to explain all the information that accumulated over very long times" (p. 338). To explain a living system, one needs to describe the contingent branchings of its history.

MATERIAL IMPLICATIONS

The theories of Prigogine, Thom, Haken, and others assume that formal representations of material reality are the sum total of the material implications of that reality. Because of this assumption, according to Kampis, they cannot explain the spontaneous creations of living things.

Kampis proposes that material systems have processes and implications that escape any observation or description. These material implications are unknown to us prior to the appearance of their creative effects. After their effects are created and observed, we can postulate the material implications that are necessary for their existence, thus turning material implications into formal ones. But, before their occurrence, we can neither predict effects nor discern the material implications behind them. Material implications are extraneous to our theories and opaque to our inspection. They become available to us usually only when we are searching for explanations of unexplained phenomena.

Material implications should not be mistaken for the "flies" of hidden variables that are debunked by Gell-Mann. The "flies" for Gell-Mann are unknown quantum effects that interfere with quasi-classical phenomena. The material implications of Kampis are real relations that exist in macrophenomena, but they are opaque to our observation. They are not hidden "variables" because "variables" are elements of a formal mathematical explanation. They are material connections that escape our formal theories.

Material causality is an idea that goes back to Aristotle. In Rosen (1985), then Kampis, it finds a new language. Material implications express the causality of the material systems themselves, not the determinacy of a formal system postulated to represent them. A material implication is not expressible in measurements; it is, basically, just a word for causes that we do not perceive. Still, the effects of material implications are everywhere in living, cognitive, and historical processes.

Material causes link events. Any science of material causality can yield no detailed analysis other than the recording of facts. The first task of such a science is the naming of newly discovered facts. The second task is to discern the sequence of events. Only subsequently can *ad hoc* explanations be offered.

Categories, Qualities, Forms

Material causes are not measurable. They are not even describable prior to their appearance in phenomena that beg for explanation. At their first instance of discovery, they are unavailable for treatment in the descriptive plane of mathematics. As material implications burst into our awareness, they can merely be described in terms of phenomenology and semantics. In other words, they require *naming*. For *us*, as observers, this must happen. For the system in question, however, the material implications have *organizational meaning*-nature does not have a problem with this.

The categories with which these material implications receive their conscious reality for us do not pre-exist. They are co-created by newly created observables and our activity in naming them. They are fuzzy sets shaped by metaphoric and metonymic mapping processes. They do not conform to the frames of classical logic because they operate with components and categories that are ever-fluctuating.

An object is created (in our conscious awareness) when we notice and describe It. Usually, it is described in blunt terms that do not do justice to its inherent material complexity.

> An object is an object because the typical mode of our interaction marks and separates it from the background . . . We do not fine-tune our interaction mode to suit it to the individuality of every object. Instead, we use uniform interaction frames. From the point of view of the objects, we tend to approach them with a steam hammer-violently and unspecifically (p. 449).

Because of this imprecision, we create crude objects that ignore many curious but seemingly useless facets of reality. New objects get names only when they enter our limited frame of observation; until that time, they do not exist for us. Because we are often oblivious of our role in creating objects and ignorant of their "useless" facets, we find paradoxes in our depictions of reality.

Atoms and Things

The *units* that underlie *formal* theories are grammatical units (nouns, verbs, adjectives). They have logical characteristics that befit their created formal nature. They are complete and permanent atoms. In their completeness, their interior factors completely determine their finite and knowable future if their external factors are kept constant. In their permanence or invariance, they live forever, never to be created, modified, or destroyed.

The formal approach postulates an isomorphism between the symbol manipulations of mathematics and the activity of material systems. It treats material realities as mathematical tokens. In so doing, it enables the advances of modern science and technology. But, in the process, it postulates an analogy that, upon close inspection, does not conform to our present conceptions of atomic reality and cannot account for the observed creativity of component-systems.

The approaches of formal theories, including various theories of complexity, are atomistic and reductionistic. They postulate units that do not resemble the components of human thinking and biological systems. They postulate, counterfactually, that units exist in the material universe that share an atomistic character with our symbolic definitions of them. Such units do not exist in our present understanding of physics. They certainly do not correspond to the constantly changing character of biological components and ideas.

Things become units for us as a result of our consistent interactions with them. We define them as units by the way they behave in our observations. If our observations are blunt and unsophisticated, we find units in everything we observe. If we use more discrimination, we find some things to be more consistent units than others.

A component-system, for example, has more consistency and unit-ness than the components that constitute it.

> Somewhat metaphorically, we can say that the components are bound together more closely by the invisible ties in the system than the building blocks are within the component. The components can be broken by any component-component interaction, but the system . . . defines the mode of the interactions by temporarily fixing a reading frame, and through that, it . . . defines the existence of the components (p. 269).

The actual information-carrying property of a component is determined by its interactions with other components in the system. The actual information that it carries can change within the meaning frame of the system as different material possibilities, implicit in the component, are used in the system's reproduction. In each case, the building blocks within the components appear and disappear, but the component-system keeps rolling along. The system is defined by the history of its interactions. "Only the complete history off the system defines it as a complete unity" (p. 443).

COMPONENT SYSTEMS

Mechanisms

Machines are often conceived somewhat negatively as mere mechanistic structures. "But machines can be seen in a dual way: as mechanistic systems or as *arrangements that serve a purpose*" (p. 274). In this second sense, the mechanism of an internal combustion engine offers a set of purposeful arrangements: "If the fuel is set to fire, the piston moves down in the cylinder and the wheel starts to roll" (p. 274). Such a statement does not offer a theoretical explanation of "thermocombustion and the physics of rolling . . . [It simply describes] a coupling between the two" (p. 274). It offers a descriptive set of mechanical implications. "A 'mechanism' in the sense used here is a particular sequence of individual events. That is, it is a little history" (p. 274). In this sense, "biochemical mechanisms are material implication structures. They do 'explain' things; but they do that in a curious language that refers to the things themselves and nothing else" (p. 272).

Component-systems, such as biochemical mechanisms, *differ* from mechanisms of our construction in the way they generate functions and boundaries. When we build a mechanism, we impose boundaries, rationality, and functions on something outside of ourselves; therefore, mechanisms of our construction are passive and unaware of their own construction. In a machine of our conscious construction, the purposes of the machine and the functions of its parts are prescribed by us. In other words, we give the machine its teleology.

Biochemical mechanisms, on the other hand, receive no direction, boundaries, rationality, or functional definition from outside sources. They form when the activity of one part of an ongoing process affects the activity of other parts of the process; such parts become components for one another.

In their supercyclic interaction, these components constitute a component-system with an inside and an outside. The inside corresponds to their supercyclic interaction, the outside corresponds to everything that is not (or not yet) part of that interaction. Component-systems have no externally prescribed teleology. They generate and evolve their functions as they go along, giving themselves new meaning input with the addition of every new component. They feed their own complexity back into their every re-creation while throwing away prior structures of complexity on their journey to higher complexity.

Component-systems do not endure as the kind of "things" imagined in Democritean and classical science. They are "genesis machines" that string together momentary presences. They constantly re-create themselves in different guises: exchanging molecules, losing components, creating new ones, incorporating alien structures, reorganizing structures, experimenting with new combinations. They accomplish the kinds of radical evolutionary change that baffle formal dynamics. They momentarily produce material implication structures that give way to further implication structures. They maintain their unity as *meaning* structures that maintain their self-production while searching for viable new material structures. Component-systems are self-modifying systems. They organize themselves from their bootstraps and autonomously alter their internal material dynamics.

Meaning Systems

The theory of component-systems explains how living things evolve in spite of refrigeration and/or component-rearrangement. The unit of living systems is a unit of meaning. "The causal, non-dynamical relationships of component-systems are of informational nature, and serve both as a model for information and as an explanation for its irreducibility to dynamics" (p. 439).

Among the multitude of distinctions about information, the one between information *about* a system and information *for* a system has particular importance. Component-systems do not know *about* themselves the way we know them. "The 'windows' through which the component-component interactions occur are different from the window through which we peep in" (p.443). They know only what is needful *for* maintaining and improving their processes. Their knowledge is not speculative; it is practical. In the human realm, information-for corresponds to the multitude of unconscious activities that automatically maintain our material existence. Information-about, on the other hand, is conscious and formal knowledge.

Information-for is the meaning-stuff of component-systems that do not necessarily see themselves as complex. Information-for does, however, make component-systems complex to our observation and description. Information-for is the pulsing pattern of a living organism. The organism does not observe this pattern nor describe it. When we observe and describe the organism, we produce names and information-about; we break up a seamless process into categories and discrete relationships.

The organism just lives. In its moment-to-moment reality (in its need to replicate), information-for is the presently actual need to act (to make a selection in Luhmann's terminology). Whether this momentary information-for also contains representations (proto-information-about) is a matter of disagreement between Varela and Bickerton. Kampis holds that this momentary information-for does not include an element of information-about because he restricts information-about to symbolic description.

The components of the system contribute information to the meaning pattern of the system. The coherence and endurance of its pattern of information is a component-system's essence. In living things, this pattern results from an alphabet of chemical interactions through, for example, sequences of DNA interactions. Communication within organisms uses molecular strings and the rule-based interchange of molecules.

Components in their interactions create the dynamic pattern of component-systems. They become functions by influencing each other's behavior. They activate and inhibit one another; they establish patterns of generation and decay. The pattern of their behavior sets the boundaries of their system while allowing additional dynamic interactions. This pattern is internal to the system and is expressed in alphabetic and grammatical chemical interactions.

FORMAL SYSTEMS

Non-component-systems do not develop their complexity internally. Such systems (Boolean networks, for example) are the result of numerous simple operations. They are seen as complex only in our information *about* them. For them, complexity arises only in the coarse-grained realm of observation and description-the realm of phenomenology and semantics.

Formal systems exist in the realm of syntax. They inductively posit relationships into material processes and deductively explicate the ensuing consequences. When formal systems are joined to constructed mechanical systems (digital computers and computing programs), they can generate some of the anomalies that are topics in complexity theory. In such conjunctions, within preset limits, bifurcations and self-ordering sometimes occur. Self-ordering of this sort occurs with the famous Mandelbrot set, and the Brusselator (Prigogine, 1980, p. 98). This kind of self-ordering is generated by formal rules and is an expression of deterministic chaos. The order that is generated is not complex in itself; it is only complex for us, at a coarse-grained level of observation.

Boolean webs, random grammars, and some of the schemes of connectionist artificial intelligence cannot be so easily classified because they do not offer much in the way of formal rules. For Boolean webs, the only requirement seems to be the metaphor of a random walk in topological space. For random grammars, the requirement is binary interactions. Some connectionist schemes will probably devise random grammars that function with fuzzy sets. Since these interactions are not very formal, are they somewhat material?

The learning effects observed in complexity theories are explained in terms of "attractors." These attractors can be modeled as goals of maximization on a topological landscape. Units on the landscape behave stochastically; they find some local maximum by randomly walking on the landscape. They are disturbed by noise that prevents their settling prematurely in a local maximum that is not as great as a better maximum that lies close by.

INSIDE/OUTSIDE

The learning of a component-system is not imposed from outside by the postulation of maxima and functions, even stochastic ones. The learning is generated internally through component augmentation/reorganization. From our point of view, we can say that this learning is catalyzed through couplings with its environment. We can also propose teleonomic interpretations to explain its behavior. From the organism's point of view, however, all these couplings are internal selections. The organism just maintains the interactions (component-functions) that are its life.

This simplicity of life is not easy for our busy minds to accept. We prefer to say that component-systems are momentarily recreating themselves as meaning systems. They experience numerous possibilities of mutual acceptance and rejection in their environments. They structurally couple with elements of those fluctuating environments including other component-systems. They maintain their internal meanings and survive in radical contingency. In saying this, we are not incorrect, but we inevitably adopt a phenomenological viewpoint that the organism does not have. We risk losing sight of the simplicity that this explanation breaks up and inevitably distorts.

EVOLUTIONARY EPISTEMOLOGY

"The basic thought of the evolutionary epistemology of Popper and Lorenz was that all our cognitive abilities are results of biological evolution and bear the marks of the process by which they emerged" (p. 445). By considering the origin of our categories, language, and so forth, we can understand why we believe in causality, the permanence of objects, and the rest. In this way, we can explain our convictions scientifically even when we cannot justify them philosophically.

By reflecting on the nature of component-systems, we can probe into an even more fundamental question of evolutionary epistemology: "How is it possible at all to have observers in the world?" (p. 446). How does it come about that there exist observers separate from what they observe where there previously existed no separation of parts?

The notion of information-*for* can explain the development of non-reflective distinctions in meaning-based systems. With its first functional interdependencies, a component-system establishes a separation (non-reflective distinction) between self and other. It builds non-reflective distinctions upon one another. In so doing, it increases its information-*for*, furthering its complexity and information processing ability. It builds its complex meaning structures by coupling with other complex meaning structures, developing methods of communication and mutual categorization. At the end of a long evolutionary process, human language and observation appear. They develop ultimately from the chemical language communication of interacting polymers.

THREE VISTAS ON CAUSALITY

Kampis demarcates divisions between natural systems on the basis of the causalities that those systems exhibit. Classical physical phenomena are both observable and deterministic. Formal sciences of complexity deal with the problematical linkages between micro- and macro-phenomena where the observed macro-phenomena are created by deterministic microprocesses but are unpredictable to the observer of the phenomenon. The science of component systems concerns itself with the radically indeterminate processes that are exhibited by living systems; processes that result from material, not formal causality.

Law

Classical dynamic theorists conceive change totally within the context of *law*. They presume that "every relevant property of a given object is observable" (p. 449) on a given observational level. They reduce dynamics to reversible formulae that make the whole natural system computable from any of its space/time coordinates. Regarding classical dynamics, Kampis offers a *determinism thesis:* "Every completely observable system is deterministic. Every deterministic process has an invariant transition scheme which expresses complete information" (p. 456). In other words, the processes of classical dynamics are observable because they are deterministic and vice-versa.

Form

Theorists of complexity and self-organization work in an arena where microprocesses and macroprocesses are both observable, but the linkages between micro and macro are not. They compute the effect of stochastic (and/or chaotic) dynamic changes over several levels of observation. On a macro-level, "only batches of (at least some of the) relevant properties are available" (p. 449). Therefore, there exists a probability that the relations interlinking the variables of a macro-level theory may "be different from the original relations that hold between the behavior-determining lower-level variables" (p. 450). As a result, non-trivial phenomena can occur that challenge macro-level expectations.

The formal sciences of emerging order (chaos, dissipative structures, catastrophe, and noise) deal with the linkages between micro and the macro in form-generating processes. They examine the challenges to expectations that occur because of randomness, bifurcations, catastrophes, and so forth. These challenges appear as new *forms* that cannot be explained by the information that scientists are able to gather. The emergence of these challenges "is always a consequence of a shift in the level of description" (p. 458). Regarding these form-generating processes Kampis offers his *form thesis:* "Every hierarchical system produces forms. Every form is characterized by choices among possible outcomes, corresponding to equivalence classes of the detailed behavior-determining information content" (p. 459).

Meaning

Truly creative processes, like those associated with life and thinking, go beyond the realm of the self-ordering processes that generate form. In these creative processes, the "information we can gain and represent . . . is not sufficient for calculating their future-not even the equivalence classes (a coarse-grained picture) of the future can be given" (p. 459). The information-for that is responsible for a creative manifestation exists and is responsible for the creation of the future. The information-about this creative manifestation is available only after the manifestation is recognized. It becomes available only through the recognition of its effects.

Component-systems are informational and creative. The relationship between creativity and information is made explicit in Kampis's *information thesis*: "Every creative system produces semiotic information. In turn, every genuinely information-laden system is characterized by creative processes" (p. 459).

Three Types of Natural Philosophy

The processes identified in the law, form, and information theses conform to three types of natural philosophy. These types of natural philosophy are related to different kinds of causality. In the following table, Kampis outlines some inter-relationships among self-modifying systems and relates them to other dynamical theories.

	LAW	FORM	MEANING
dominated by	matter	process	information
manifested as	substance	self-organization	system
exemplified by	Parmenides (Newton)	Whitehead (Prigogine)	Bergson
causal law	determinism	form-generation	creation
existence	being	becoming	re-creation

Three Types of Natural Philosophy,
Reprinted from George Kampis (1991) *Self-modifying Systems*, p.462,
With permission from Elsevier Science

In this table Newton is grouped with Parmenides because the classical dynamics fathered by Newton posits an unchanging (in the evolutionary context) world that corresponds with Parmenides's vision of the unchanging unity of truth and reality. Prigogine is cast as formulating a modified and popularized version of Whitehead's process philosophy. Bergson anticipated continual creation and re-creation through component-systems with his rather amorphous concept of the *elan vital.*

CONCLUSION

Kampis situates the theory of component-systems in the philosophical tradition. In the next chapter Goertzel forges new territory by constructing a mathematics of component-systems

CHAPTER 16

GOERTZEL

ABSTRACT

In *Chaotic Logic* (1994), Ben Goertzel puts together neural network theory, dynamical systems theory, and information theory so that "we can begin to understand significant aspects of the mind and brain" (p. 263). He also synthesizes insights from mathematics, biology, and physics so that "we can begin to understand biological evolution" (p. 264). His specific goal in this book is to see whether, "by combining current ideas regarding complex system dynamics with the pattern-theoretic psychology developed in my earlier books, it might not be possible to work out a dynamics of mind" (p.264).

To that end, he constructs a mathematical model of mind on the level of structure, not the level of details. He does this because "complex self-organizing systems, while unpredictable on the level of detail, are increasingly predictable on the level of structure" (p. 2). He shifts the "focus from numerical iterations to structure dynamics" (p. 2). More specifically, he shifts "up from the level of physical parameters, and take[s] a 'process perspective' in which the mind and brain are viewed as networks of interacting, inter-creating processes" (p. 2).

STATICS AND DYNAMICS

The strategy of reducing mind to neural nets has not, according to Goertzel, been successful. Goertzel utilizes the dynamics of neural nets but moves up to the level of patterns of neuronal activity. His strategy is to combine a *structural* explanation of mind with a *dynamic* explanation. He refers to the static component as the dual network and the dynamic component as the cognitive law of motion. The dual network consists of a hierarchical command component and a structured associative memory component. The components are related fractally with each other and

247

within themselves through a process of continuous compositionality. In other words, the structure of a memory is reflected in the structure of the command processes that brings it about and vice-versa. The fractal composition is plausible because it is a natural consequence of the chaotic interactions that generate mind.

The law of cognitive motion is based upon a theory of self-generating systems that closely resembles the theory of component-systems. The law is a "precise cognitive equation, hypothesized to govern the creative evolution of the network of mental processes" (p. 3). It applies the laws of self-organization to the processes of mentation and indicates attractors of the cognitive equation.

Using this cognitive equation and relying on the "basic article of faith underlying complex systems science . . . that there are certain large-scale patterns common to the behavior of different self-organizing systems" (p. 7), Goertzel lays a foundation of constructed definitions. He proposes a theory of logic, a theory of self-generating systems, and a cognitive equation of motion. He also develops the necessary mathematics as needed.

> There is at present no mathematical theory of direct use in exploring the properties of self-generating pattern dynamical systems or any other kind of non-trivial self-generating system. The tools for exploring these models simply do not exist; we must make them up as we go along (p. 155).

CHAOTIC DYNAMICAL SYSTEMS

A chaotic dynamical system "is one whose behavior is deterministic but appears random" (p. 21). Such systems are drawn by strange attractors that determine behavior in unfathomable ways. Goertzel's chaotic system works with patterns. It functions like systems working with elements. It requires, however, a proper definition of a "pattern chaotic entity," some special operations like "hyperfunctions," and an attenuated notion of determinism.

The behavior of a structurally chaotic entity is determined by strange attractors that generate patterns, such as those of fractal geometry, by drawing seemingly random activities to themselves. An *attractor* of a dynamical system "is a region of the space of possible systems states with the property that: 1) states 'sufficiently close' to those in the attractor lead eventually to states within the attractor. 2) states within the attractor lead immediately to other states within the attractor" (p. 22). On a topological surface, an attractor is a depression that drains everything in its vicinity.

There are varieties of such attractors. If the attractor is a "fixed point," it produces a single state. "Once the system is close to that state, it enters that state; and once the system is in that state, it doesn't leave" (p. 22). If the attractor is an ellipse or some other closed geometrical figure, the system will oscillate. It "leaves from one state, passes through a series of other states, then returns to the first state again, and so goes around the cycle again and again" (p. 22).

A *strange attractor* is a more complex region than those represented by points and circles. Strange attractors rule systems that exhibit continually fluctuating chaotic behavior but still create recognizable patterns. The strange attractor of the olfactory cortex, for example,

> has a large number of "wings," protuberances jutting out from it. Each "wing" corresponds to a certain recognized smell. When the system is presented with something new to smell, it wanders "randomly" around the strange attractor, until it settles down and restricts its fluctuations to one wing of the attractor, representing the smell which it has decided it is perceiving (p. 23).

This strange attractor, which is active in our smell discrimination, is an excellent intuitive model of the "mental" activity of self-organizing systems.

> Each wing . . . represents a certain pattern recognized-smell is chemical, it is just a matter of recognizing certain molecular patterns. In general, the states of a complex self-organizing system fluctuate within a strange attractor that has many wings, sub-wings, sub-sub-wings, and so on, each one corresponding to the presence of a certain pattern or collection of patterns within the system. There is chaotic, pseudo-random movement within the attractor, but the structure of the attractor itself imposes a rough global predictability (p. 23).

Within its attractor's structure, the system moves rapidly and easily from one part of the attractor to another, and finds movement outside the structure to be very difficult. In this way, the attractor imposes "a complex structural predictability" (p. 23).

In this way, the patterns of its strange attractor shape the structural dynamics of a complex system. The patterns of living systems have very highly organized strange attractors. The patterns *within* the attractor are closely related to the patterns of transition that are exhibited by their systems.

PATTERNS AND STRUCTURES

Goertzel analyses the mind as a network of regularities, habits, and patterns. "Each pattern takes the form of a process for acting on the other mental processes. And the avenues of access joining these processes adhere roughly to a specific global structure"(p. 13). A *pattern*, in general, is a representation of something that is simpler than what it represents. [Goertzel offers a mathematical definition in terms of algorithmic information, but mathematical definitions like this will be absent from this presentation of his thought.] A *fuzzy* set is "a set in which membership is not 'either/or' but gradual" (p. 18). A *structure* of an entity "is the [fuzzy] set of all patterns in that entity" (p. 15). The total amount of structure in an entity is its structural complexity.

Patterns and structures are essential to Goertzel's thought. The mathematical definitions and brief narrative expositions that he offers seem to be completely adequate for his purposes, but they are not sufficient in the less erudite confines of this book. For that reason, I expand upon his definitions with some phenomenological background supplied by Maturana, Varela, Bickerton, Luhmann, and others.

An organism is autopoietic. It reproduces itself and its self-reproductivity constantly in time and occasionally in space. It must retain this circularity or it dies. An organism is also more organized (from its perspective) than its environment, and it experiences itself as less confusing than that environment. It couples with its environment for its metabolic needs but retains its autonomy.

In this situation, the organism is faced with a tricky situation involving action and information. In order to survive, it has to adjust constantly to the changing complexity of both the environment and its own internal structure. To do this, it must have information that enables it to react appropriately. Before it can receive an environmental difference as information, however, the organism needs the ability to react to that difference. It does not receive all differences in its environment as information. For it, a difference is information only if that difference is something it can react to. In short, to survive, an organism needs information in order to react to differences, but it needs the ability to react to those differences before it can receive them. How is the organism to survive?

This is where a pattern plays its crucial role. The organism selects some aspect of the complexity facing it and reacts to that aspect as if it were the whole situation. (Faced with contradiction, it does something.) If its action harms itself, kills it, or hinders its self-reproductive ability, then that action is selected against. If, however, the action is successful, then the selected aspect comes to stand for the particular complex situation in which the selection was made. The selected aspect is a representation of something that is simpler than the thing it represents: it is an information *pattern*. The relatively simple pattern comes to represent a relatively complex element of the organism's environment.

The process is one of trial and error. Selections are repeated over and over, and each trial builds upon previous iterations. In this way, the organism adds to its stock of information for survival and growth and builds its reality.

These processes and the information they produce are intimately related. The processes find and produce survival-information. The information enables present and future survival and feeds back into the information-creating processes. Together, processes and information create a hypercyclic hermeneutic circle but, also, a circle that is open to new responses in novel environmental conditions.

Process and information are constantly influencing one another. It is to be expected, therefore, that they share some qualities. The information is generated in trial and error search processes that organize and reorganize it. Bits of information gathered in similar situations form fuzzy-sets whose elements are similar to each other. These elements have more interconnections with each other than they do with bits generated in quite different situations. The resulting trial-and-error organization is a "structurally associative memory."

The structure of constantly adjusting information is matched by a structuring of the information-generating processes. As the organism structures the complexity of its information, it also structures its processes by continually improvising and multiplying them. In the course of time, these processes group together and organize in functional levels so that upper level processes recognize patterns in the data presented by lower level processes. The priority structures represented in these levels organize both perception and motor activity. The resultant structure is what Goertzel calls the "perceptual-motor hierarchy."

Such a hierarchy of perception is found in the biological processes of organizing vision. Goertzel refers to recent findings about neural clusters of visual processors that organize in levels.

At the lowest level, in the retina, gradients are enhanced and spots are extracted-simple mechanical processes. Next come simple moving edge detectors. The next level up, the second level up from the retina, extracts more sophisticated information from the first level up from the retina-and so on. Admittedly, little is known about the processes two or more levels above the retina. It is clear, however, that there is a very prominent hierarchical structure, although it may be supplemented by more complex forms of hierarchical processing (p. 27).

Our multilevel perceptual and control mechanisms deal with questions of pattern.

The processes on level N are hypothesized to recognize patterns in the output of the processes on level N-1, and to instruct these processes in certain patterns of behavior. It is pattern which is passed between the different levels of the hierarchy (p. 29).

THE DUAL NETWORK

Goertzel defines "mind" as "the structure of an intelligent system" (p. 25). With this definition, he connects the mind to the physical world without placing it within the physical world. "Mind is made of *relations* between physical entities" (p. 25), but it is not itself a physical entity. It is a constellation of patterns emergent from interacting neurons and neurotransmitters. *Intelligence* is the ability of mind "to optimize complex functions of complex environments"(p. 25). For us, a "complex function" may be "anything from finding a mate to getting something to eat to building a transistor or browsing through a library" (p.25). To perform such functions we set certain *goals* and perform actions to achieve them.

Complex environments exhibit a troublesome measure of unpredictability. They cannot be well predicted in their details. One cannot know the name of the woman one is going to court. Still complex environments can be predicted in their future structures. One can expect, for instance, to find a woman who will meet certain expectations.

> In environments displaying this kind of unpredictability, prediction must proceed according to *pattern recognition*. An intelligent system must recognize patterns in the past, store them in memory, and construct a model of the future based on the *assumption* that some of these patterns will approximately continue into the future (p. 26).

Goertzel hypothesizes that there is only one type of structure that performs this kind of pattern recognition. He calls it the *dual network* of the mind.

> Every mind is a superposition of two structures: a *structurally associative memory* (also called "heterarchical network") and a *multilevel control hierarchy* ("perceptual- motor hierarchy" or "hierarchical network"). Both of these structures are defined in terms of their action on certain *patterns*. By superposing these two distinct structures, the mind combines memory, perception and control in a creative and effective way (p. 26).

The *multilevel control networks* have come about without any "master programmer" in an evolutionary process of directed trial and error. In this process, subnetworks of the control network that are not working effectively are (a) randomly varied, (b) swapped with one another, or (c) replaced with other subnetworks. Computer simulations indicate that "under appropriate conditions, this sort of process can indeed *converge* on efficient programs for executing various perceptual and motor tasks" (p. 29). This sort of process is used, for example, in predicting behavior in the stock market.

Structurally associative memory is a long-term memory model. In this model, connections between processes are determined, not by control structures, nor by any arbitrary classification system, but by patterned relations (p. 30). The process that sorts memories into association classes handles a continual influx of new information, and not at all perfectly. It works by approximation and "continually reorganize[s] itself" (pp. 31-32). In this way, it evolves structurally associative memories *by natural selection*.

Since there is no "memory supervisor," the reorganization process takes the form of directed, locally governed trial and error. It swaps and copies subnetworks. If its substitutions prove successful in increasing associativity, "then the new networks formed will tend not to be broken up" (p. 31). The reorganization process makes multilevel substitutions. "Large networks may be moved around, and at the same time the small networks which make them up may be *internally* rearranged" (p. 31). This local activity of trial and error enables radical changes because "the memory network is 'fractally' structured-structured in clusters within clusters . . . within clusters, or equivalently networks within networks . . . within networks" (p.32).

Separately, neither structurally associative memories nor multilevel control networks can develop intelligence. They do develop intelligence, however, by combining in a dual network that shares connections. They can coordinate because "the entities stored in the associative memory are distributed in an approximately 'fractal' way" (pp.32-33). A *dual network*, then, can be defined as:

> A collection of processes which are arranged simultaneously in a hierarchical network and a heterarchical network. Those processes with close parents in the hierarchical network are, on the whole, correspondingly closely related in the heterarchical network (p. 33).

Within the dual network, *substituting* networks within the memory network is equivalent to genetically optimizing the control network.

Only through the dual network, can the perceptual-motor hierarchical network do its job of guiding an organism. It relies on the analogical reasoning that is enabled by structurally associative memory.

> The purpose of each *cluster* in the dual networks is to instruct its subservient clusters in the way that it estimates will best fulfill the task given to it by its master cluster-and this estimation is based on reasoning analogically with respect to the information stored in its memory bank (pp. 33-34).

STRUCTURED TRANSFORMATION SYSTEMS

The brain that operates our dual network system is a massively *parallel* information processor. It has "a hundred billion neurons working at once, plus an unknown multitude of chemical interactions interacting with and underlying this neural behavior" (p. 35). Still it has to help us plan for the future, talk in sentences, simplify complex situations, write equations, and so on. In order to accomplish these sequential tasks, the brain turns itself into a virtual *serial* computer. In so doing, it hobbles itself to proceed one step at a time, thus sacrificing immense internal efficiency in order to enable our external functioning. Transformations from parallel processing to serial processing are operating every time we do something like putting our thoughts on paper. In this process, the gushing of our creative mind is forced into a thin trickle of sentences. Transformations of this kind create multilevel control structures by clustering elements into analogies within an overall blueprint.

Because our intelligence operates within a multitude of varying contexts, it requires more than a stack of blueprints. It needs "a transformation system structured in such a way that ordinary mental processes can serve as its blueprint-generating machine" (p. 40). Through its dual network, intelligence has such a system. It can induce relationships, reason by analogy, and search through its associative memory. It structures its blueprint generation in such a way that "the analogical reasoning mind can use it, in practice, to construct things to order" (p.

40). To function adequately this construction does not have to be infallible; it just has to work, approximately, most of the time.

Structured transformation systems make prediction possible by specifying original conditions, **A**, and the steps (dynamical rules) that are involved in moving from **A** to **B** in space and time. They make logical reasoning possible by supplying the step by step rules of logical deduction. In real life situations, they produce both simulations and logical reasoning in order to understand and predict the behavior of complex processes.

PSYCHOLOGY AND LOGIC

There is a history of dispute on the relationship between psychology and logic. Is logic part of psychology? Is psychology part of logic? Are psychology and logic completely separate? This is not a trivial set of questions. "The relation between psychology and logic is important, not only because of the central role of deductive logic in human thought, but also because it is a microcosm of the relation between language and thought in general" (p. 43).

The tendency nowadays is to consider psychology to be entirely alien to formal logic. It is assumed that Boolean logic stands independently on its own and that it applies non-controversially to practical situations. Such is not the case in the realm of philosophical mathematics.

Boolean logic, in its application to real situations, yields several paradoxes that have been noted in academic, logical circles. By considering these paradoxes, in terms of their everyday inutility, Goertzel sets a clean platform for "studying how the powerful reasoning tool that we call 'deductive logic' fits into the pattern of human reasoning" (p. 47).

Nietzsche's ideas on logic are pertinent to this discussion. Nietzsche held that logic is simply a particularly fancy manifestation of the will to power. "The will to make things equal," according to Nietzsche, is at the core of mathematics and logic.

> *The will to equality is the will to power* . . . the consequence of a will that as much as possible *shall* be equal. Logic is bound to the condition: assume there are identical cases. In fact, to make possible logical thinking and inferences, this condition must be treated fictitiously as fulfilled.
> The inventive force that invented categories labored in the service of our needs, namely our need for security, for quick understanding on the basis of signs and sounds, for means of abbreviation . . . (Nietzsche, 1968a, p. 277; as quoted by Goertzel, p. 58).

In another location Nietzsche says, "what is needed is that something must be held to be true-*not* that something *is* true" (1968a, p. 275).

There are two major points made by Nietzsche, according to Goertzel. First, "logic is a lie, but a necessary one" (Goertzel, p. 58). Also, a fact is an "interpretation which someone has used so often that they have come to depend upon it emotionally and cannot bear to conceive that it might not reflect 'true' reality" (p. 59). From the Nietzschean point of view, even the law of contradiction is "something cleverly *conceived* by human minds, in order to provide for more effective functioning in some circumstances" (Goertzel, p. 60). Goertzel embraces the Nietzschean point of view and makes it more concrete.

As answer to the questions posed at the beginning of this section, Goertzel says, "Logic . . . is a *special* psychological function of relatively recent invention, one with its own strengths, weaknesses, and peculiarities.   But it has neither meaning or utility outside of the context of the mind which maintains it and which it helps to maintain" (p. 61).

LINGUISTIC SYSTEMS

Goertzel describes language and meaning from a pattern-theoretic angle and stresses its systemic nature. He constructs "a general model of language as a special kind of structured transformation system" (p. 64). In this system, "syntactic rules form a transformation system, and semantics determines the analogical structure of this system" (p. 64).

In a syntactic system, "It is not the meaning of a word that matters, but only the way it relates to other words." For the purposes of *syntax*, a word is "a *relation* between other words"(p. 66). What makes a grouping of word/functions a *syntax* is a collection of constraints. "Constraints tell us which sorts of expressions may be put into which inputs of words" (p. 66). Examples of constraints are the rules for nouns, verb phrases, and so on.

Goertzel provides a definition of a syntactic system in mathematical notation that will not be presented here. His definition has the enormous virtue of making grammar accessible to mathematical exploration.

Semantics in its popular sense is the science dealing with the meanings of words. It lets us treat the meaning of any *entity* in a grammatical context. The idea of *meaning* in this abstract grammar transcends formal and natural human languages. The meaning of *any entity* is "*the fuzzy set of patterns related to its occurrence*" (p. 69). "In other words, everything which relates to the word 'Ice' is part of the meaning of the word" (p. 69). "A *semantic system* . . . is a collection of entities whose meanings consist primarily of patterns involving other elements of the system.

Something has to be added to syntax and semantics in order to constitute a linguistic system. There needs to be some way that syntactic and semantic systems are intimately linked functionally.   This linkage is abetted by the fractal composition of the dual system of mind. Goertzel proposes that this linkage is accomplished through a property of *continuous compositionality.*   By *compositionality*, he means that for most syntactic operations (call them **F**) there is a corresponding semantic operation (call it **G**) so that the meaning of **F** is close to the meaning of the **G** function that operates on it. In other words, the rules that hold for building up any syntactic structure are very similar to the rules that capture the meaning of that structure.

With these processes of syntax, semantics, and compositionality, Goertzel achieves his goal of constructing a linguistic system that, like his constructed model of mind, is a structured transformation system. With all of his key terms defined, Goertzel defines a linguistic system as:

1. a *syntactic* system, together with
2. a collection of *situations*,
3. so that relative to these situations the expressions of the syntactic system form a *semantic* system,
4. which is related to the syntactic system according to continuous compositionality (p. 81).

Finally, the *purpose* of language is to "communicate." What does this mean? Goertzel holds that *communication* is a process of doing something with language. It is a languaged way of molding the world. In light of this concept of communication, he reformulates his definition of language as follows: "A *linguistic system is a syntactic system coupled with a semantic system in such a way that the coupled system is useful for molding the world*" (p. 82).

Goertzel briefly identifies three linguistic systems other than Boolean logic and written/spoken language. Those three are perception, motor control, and social behavior. Each of these systems exhibits syntax, semantics, and compositionality. They are omitted from this discussion.

LANGUAGE, REALITY, CONSCIOUSNESS

Before he constructs his model of reality, Goertzel probes philosophically into the deep relations between language, logic, consciousness, and reality. He begins with the question: How effective is language in molding the world?

The Sapir-Whorf Hypothesis

According to the Sapir-Whorf hypothesis, reality is determined by the language that expresses it (linguistic determinism). "It claims that language is the main constructive force underlying the world that we see around us" (p. 90). Goertzel strongly endorses this hypothesis in the sense that "the structure of language is closely related to the structure of mind and 'subjective' reality'" (p. 90). He argues that, "When viewed in a sufficiently abstract way, linguistic determinism is a natural consequence of the structure of mind" (p. 90).

He notes that it has become acceptable to admit that language guides our categorization of the world, but he contends that it "guides our perceptions and cognitions in other ways" (p. 97). Besides, he asks, what can it mean when scholars make "a sharp distinction between 'our concept of reality' and 'reality'? What difference does this phenomenal/noumenal distinction make in practice?" (p. 97).

Consciousness

Implicit in this question is the underlying questions: What is consciousness? Where does it fit in? According to Goertzel, consciousness has a low limit (L) and an upper limit (U). Below L, exist unconscious perceptions. Above U, there are "perceptions that are in some sense beyond conscious perception: too abstract or general for consciousness to encompass" (p. 99). This idea is expressed by Jackendorff as follows: "consciousness corresponds to mental representations that lie midway between the most peripheral, sensory level and the most 'central,' thoughtlike level" (quoted on p. 99). In this perspective, consciousness is not the site of our most complex intelligence; instead, it exists in various sites midway between sensory awareness and the parallel processing of our dual network. "Consciousness is not in one place; it is rather associated with a collection of processes that occur in *intermediate* levels of the psychological hierarchy" (p. 100).

Consciousness is not intrinsic to the brain. It is a peculiar manifestation of the massive parallel processing that goes on in the brain. When it is faced with a modeling problem or some other need to take sequential steps, the dual-network multilevel *parallel* computer, which is our mind, can turn itself into a *virtual serial* processor. Consciousness is that virtual serial processor.

This interpretation of consciousness is closely related to that of Dennett: who also sees consciousness in terms of linear processing. He says, "Consciousness is a phenomenon 1) closely related with, 2) on the same levels as, and 3) dealing largely with the output of serial, linguistic processing" (Dennett, quoted on p. 102).

The joining of this theory of consciousness with a pattern-theoretic model of language and mind leads to an inevitable conclusion: "Language helps to determine the world we consciously perceive" (p. 102). Nietzsche puts this argument rather forcefully, as quoted by Goertzel (p. 103).

> Man, like every living being, thinks continually without knowing it; the thinking that rises to *consciousness* is only the smallest part of all this-the most superficial and worst part-for only this conscious thinking takes the *form of words, which is to say signs of communication*, and this fact uncovers the origin of consciousness (*The Gay Science*, p. 299).

Goertzel formulates his own theory of consciousness as follows:

> Consciousness works by mapping higher-level *thought processes into middle-level sensory data*. Consciousness consists of "fooling" the perceptual mechanism into thinking it is working with constructs built up directly from external sense data, when it is actually working with *transformed versions* of patterns from levels above

it. This explains what we mean when we say we are "thinking visually" about something, or "thinking in words." We mean that our self-perception uses the *standard perception routines* of the brain, which evolved for perception of data coming in from particular sense organs: eyes, ears, noses, taste buds, skin. Our ideas are mapped into pictures, sounds, perhaps even smells, and *in this disguise* they are grouped into wholes and "perceived." Then the perceptions obtained in this way give rise to higher-level patterns, which may be fed back down to the perceptual mechanisms, repeating the process and giving rise to the familiar *circularity of consciousness* (pp. 107-108).

Consciousness Produces Reality

According to Goertzel and Nietzsche, "the function of consciousness is to manufacture reality" (p. 108). Consciousness is a dynamic feedback process, according to Goertzel, in which patterns are cycled through higher-level (unconscious) cognitive processes and middle-level perceptual (linguistic and phenomenological) processes. "A pattern only acquires the presence, the solidity that we call 'reality,' if it has repeatedly passed through this feedback loop" (p. 109). Multiple passages through such a feedback loop tends to create imperviousness in perceptual patterns in specific middle-level perceptual processes:

> 1) Those processes that act to combine a group of different sensations from the same sense organ into a single cohesive entity-a "scene," "sound," "physical location," etc. 2) those processes which act to combine entities recognized by different senses (hearing, vision, touch, etc.) into a single, united form.
> Each time something is passed through these processes, it attains a degree of cohesion, a degree of resistance to being broken up. When something is passed through again and again and again, it achieves a superlative degree of cohesion and resistance-it becomes *real* (p. 109).

Consciousness processes sensations and patterns *serially* in this cyclical process that generates reality. It operates like the "serial processes of prediction, logical deduction, and syntactic sentence, percept-, or act-formation" (p. 109). Goertzel describes the similarities as follows:

> All of these processes involve a re-entry from higher to lower. Something is *built up*-a phrase, say, out of words; or a future, out of the present. And then it is *passed down* to the level where its parts came from: the phrase is plugged into a syntactic operation as if it were a word; the future scenario is treated conjecturally as a present and the mental routines for "present-world" manipulation are applied to it (p. 109).

Besides this similarity in their serial processing, there is a further structural connection between consciousness and deductive, serial reasoning.

> *In the context of the dual network, structured transformation systems require the interim assumption of reality every step of the way.* How could deduction work if one step were altered before the next were complete? How could prediction work if the one-week prediction were rearranged before the two-week prediction was done? . . . The reentrant processes involved in applying structural transformation systems require *reality* to be introduced at each step. And reality, I have argued, requires consciousness.
>
> . . . Language structures the memory which guides the structured transformation systems of deductive and predictive thought. But neither sentence formation nor deduction nor prediction could function without consciousness (p. 110).

Reification

All these structured transformation systems (consciousness, language, deduction, prediction, and so on) *reify* mental constructs. They create *imaginary subjects*, mental constructs that are treated as real things. They are constrained to such reification because of their need to impose a degree of solidity in the shifting dynamics of the mind. To work well, these transformation systems require a balance between rigidity and fluidity. Without rigidity, they cannot function. With too much rigidity, they grind on mechanically without benefiting their hosts. Our conscious serial processing provides the rigidity; our associative memories provide the fluidity.

Advancing to the realm of psychology, Goertzel explains the feeling of raw existence and the experience of the self in terms of the dual network. The feeling of "raw existence" . . . is simply the

> feeling of *subnetworks resisting the natural urge to shift*. It is the feeling of *solidity* resisting fluidity. And the feeling of "self-presence" is one level up from this; it is the feeling of *solidity which produces solidity*. "I am" means, "I, this mental process, make myself solid; I maintain my boundaries against the surrounding flux" (p. 111).

Goertzel does not equate consciousness with language. He considers consciousness to be something more basic and less complex than language. From the idea of consciousness as the iterative barrier-strengthening process he *deduces* its close connection with language and reason.

UNIVERSAL COMPUTATION

To this point, we have discussed Goertzel's static model of the mind as a dual network. He complements this static model with a dynamic model that is similar to the component-system model. Goertzel admires the theory of component-systems as it is presented by Kampis. He believes that Kampis describes complex system behavior accurately.

Goertzel believes, however, that Kampis errs in his conclusion that the behavior of component-systems is non-computable.

> Kampis's picture of complex system behavior is fundamentally right. Complex systems consist of components that act on one another to create new components. Thus they effectively violate the hierarchy of logical types . . . [But] the opposition which Kampis sets up between computation and component-systems, is in my opinion a false one (p. 130).

In response to Kampis's claim that component-system cannot be computed, Goertzel constructs a model using a rarefied mathematics that employs the concepts of stochastic computation and hyperfunctions. *Stochastic computation* is the kind performed by a computer that does "random coin tosses." Such a computer models situations that are impossible by the standard algorithmic programs that are envisioned by Kampis. *Hyperfunctions* expand set theory to include functions that operate on themselves. Goertzel believes that component-systems with their emergent structures can be modeled "in terms stochastic hyperfunctional iterations" (p. 130).

Goertzel constructs his computable model of self-generating systems with emergent properties. He describes its dynamics as follows:

> Self-generating dynamics is defined as a two-stage process. First, *universal action*: each component acts on each other component with a certain probability, yielding different new components with different probabilities. Then, *transformation*: these resultant components are transformed in some way, yielding a new collection of components. The results of transformation are then fed back into the first step, and used as fodder for universal action (p. 126).

According to Goertzel, a computer built upon this equation can model almost anything. It is a universal computer [given infinite time?]. Goertzel states that given a System at time t and an appropriately chosen T, the equation defining a self-generating system "can describe anything whatsoever. That is, after all, the meaning of a universal computation!" (p. 148).

THE COGNITIVE EQUATION

Goertzel aims to correct a deficiency in existing theories of cognition. As yet there does not exist a commonly accepted *law of cognitive motion* that simplifies cognitive dynamics in the way that Newton's laws of physical motion simplify classical dynamics. Goertzel generates just such a law, starting with the idea that the mind is a self-generating system.

This law is expressed in the *cognitive equation*. This equation is built on the following basic ideas.

> The reprogramming of processes by one another can be the causative agent behind the creation of new processes . . . This process of creation is unpredictable. Certain processes are more likely to arise than others, but almost anything is possible within the parameters imposed by the remainder of the network of processes that is the mind (p. 148).

The cognitive equation is the culmination of three other equations. They do not represent the interactions of the elements of the system; instead they represent the relationships between the patterns demonstrated by those elemental interactions.

These equations state that a self-generating system creates its successor in time by reiterating a moment-to-moment transformation. This transformative function is a hyperfunction, one that acts on itself. In this transformation, the system reproduces itself by incorporating essential patterns that exist in its Raw Potentiality. This Raw Potentiality is the unrestrained interaction that occurs between systemic and environmental processes that occurs when the system momentarily dissolves its components. In a self-reproductive process, the system breaks its component patterns, mixes them with environmental ones, reproduces its essential systemic processes, and incorporates compatible processes into itself. In so doing, the system makes itself substantial. In other words, "In a cognitive system, time is the process of structure becoming substance" (p. 149).

According to these equations, the interactions existing in the Raw Potentiality of a system at one moment generate patterns. The system selects its essential patterns plus compatible patterns that were generated with its environment. These patterns, both essential and newly integrated constitute the new reality of the system. "The system one moment later consists of the patterns in . . . this Raw Potentiality" (p. 149). In other words, "the entities which make up the system now all act on one another, and thus produce a new collection of entities which includes all the products of the interaction of entities currently existent" (p. 149). With the completion of this iteration, the patterns within this raw potentiality continue the process. In this way, they constitute the system as substance and not as mere ephemeral structure.

In operation, a cognitive system prevents itself from having merely ephemeral existence by the means described in these equations. As its momentary structure passes out of existence, the cognitive system creates a new structure that has all the components of the previous structure. It designs transformations that ensure that its components reappear in structure after structure after structure.

Goertzel concedes that he is representing "a bizarre type of dynamic-instead of acting on real numbers or vectors, it acts on *collections of hyperrelations*" (p.150). Nevertheless, he maintains that this dynamic "may still be studied using the basic concepts of dynamical systems theory-fixed points, limit cycles, attractors, and so forth" (p. 150).

The fourth and culminating equation is written below just to exemplify the nature of the mathematics employed.

Systemt+1 = F[R[Systemt] intersect St^ (G[R[Systemt+1]])].

This equation "says loosely that **System t+1** consists of the patterns which **System t** has recognized in itself" (p. 150). Goertzel says that this equation "brings abstract self-generating pattern dynamics down into the realm of physical reality" (p. 150).

The cognitive equation (or "the cognitive law of motion") expresses a self-generating pattern dynamic. This dynamic can be explored with the concepts of dynamical systems theory, such as limit cycles and attractors even though these concepts have not been well developed for pattern dynamics. "The general ideas of dynamical systems theory are applicable, but the more specific and powerful tools are not. If one wishes to understand the mind, however, *this* is the type of iteration, which one must master (p. 151).

Goertzel offers a brief informal description of the cognitive equation and the attractor it generates as follows:

> 1) Let all processes that are "connected" to one another act on one another. 2) Take all patterns that were recognized in other processes during Step (1), let these patterns be the new set of processes, and return to Step (1).
>
> An *attractor* for this dynamic is then a set of processes with the property that each element of the set is a) produced by the set of processes, b) a pattern in the set of entities produced by the set of processes (p.152).

BELIEF SYSTEMS

Goertzel claims that "the 'cognitive law of motion,' applied in the context of the dual network model, is adequate for describing the dynamics of mentality" (p. 166). To advance proof of this claim, he presents a mathematical theory of belief systems. In this section, the mathematics will be skipped; only key definitions, arguments, and conclusions will be presented.

Belief is "a mental process which, in some regard, gives some other mental process the 'benefit of the doubt.'" (p. 166). A *belief system* "is a group of beliefs which mutually support one another, in the sense that an increased degree of belief in one of the member beliefs will generally lead to increased degrees of belief in most other member beliefs" (p. 167). Belief systems with high levels of this mutual support are said to have high *systematicity*.

How do we select and coordinate our beliefs? We cannot routinely rely on logical analysis because correlating numerous beliefs logically is beyond our computing power. "Hence it is unreasonable to require that a system of beliefs be 'rational' in structure, at least if rationality is defined in terms of propositional logic" (p. 171). What we do use is a rough, approximate method that sorts out new potential beliefs in terms of how well they conform to our existing belief systems. In Goertzel's words:

> The mind confronts it [this new potential belief] with a combination of deduction, induction and analogy. It does indeed seek to enforce logical consistency, but lacking an effective general means of doing so, it looks for inconsistency wherever experience tells it inconsistency is most likely to lurk (p. 171).

Another way to express this thought is that fragmentary beliefs are not tested on an individual basis. They fit into a system of belief in some way or else they are discarded. "Belief is almost always systematic" (p. 171).

Belief systems are learning systems that satisfy the cognitive equation. They contain patterns of mental processes that support one another in a syntactic and semantic dual network structure. "Processes are grouped hierarchically for effective production and application; and heterarchically for effective associative reference" (p. 175). Thus

> A belief system emerges as a collection of mental processes which is closed under generation and pattern recognition-an attractor for the cognitive equation. In this way, a belief system emerges as a sort of "mini mind," complete in itself both structurally and dynamically (p.175).

There is "an enchanting paradox" involved in the utility and functioning of a belief system. "Only by attaining the ability to survive separately from the rest of the mind, can a belief system make itself of significant use to the rest of the mind" (p. 175). Because they are mini-minds, contrasting belief systems set up dialogues in our consciousness that enables the comparison and virtual activation of divergent perspectives and programs. They produce relationships among themselves that enrich the mind's ongoing self-reproduction through the cognitive equation.

CONTRASTING BELIEF SYSTEMS

To exemplify these ideas, Goertzel contrasts scientific research programs with paranoid belief systems. He examines why we consider one of them to be rational and the other, irrational.

Some belief systems, such as paranoid ones, are *too glib* in their generation of theories "New events never require new explanations . . . Let us call this property *conservatism*" (p. 181). Such belief systems generate theories to explain an event, which "never have much to do with the specific structure of the event . . . Let us call this property *irrelevance*" (p. 181). Conservatism, irrelevance, and resistance to test demonstrate a closed circular pattern of thinking which is unlikely to correct for changing situations and flawed inductions.

What is the structure of conservative and irrelevant belief systems? They are self-supporting, self-contained units. Like belief systems that we consider normal, irrational systems act are attractors of the cognitive equation. Their weakness "is that they gain *too much* of their support from internal self-generating dynamics-they do not draw enough on the remainder of the mental process network" (p. 183). They result from "*overly autonomous subattractors*" of the cognitive equation" (p. 183) that compete with and dissociate the operation of the overall mental attractor.

Severe dissociation of this sort is irrational. Less severe dissociation, however, can sometimes be an element in the work of genius. This fruitful dissociation may manifest itself as going against commonsense logic. In Galileo's day, for example, there were widely divergent descriptions of what was seen through the telescope. People who did not see what he saw generally contradicted Galileo's contention that he saw what was out there. He was seen as making preposterous statements that were based upon ideology. Still Galileo persisted in his delusions, and his dissociation is today's common sense.

How is Galileo's dissociated belief system different from a paranoid one? "Jane's belief system was conservative and irrelevant; Galileo's belief system was productive" (p. 184). Its productivity is validation for the rationality of Galileo's beliefs in spite of its dissociation from the common sense of his contemporaries. By believing and insisting on the reality of what he perceived and interpreted as being out there, Galileo started an industry of telescope sightings and theory construction.

> True, if it's not really out there then you're just constructing an elaborate network of theory and experiment about the workings of a particular gadget. But at least the assumption leads to a pursuit of some complexity: it produces new pattern. A conspiracy theory, taken to the extreme described above, does no such thing (p. 184).

A belief system is productive "to the extent that it is correlated with the emergence of new patterns in the mind of the system containing it" (p. 185).

Paranoid belief and theoretical fervor share a common disregard for test results. Scientific theory, however, maintains its fervor only in its bizarre, brainstorming, flighty, and exploratory stages-in order to generate and develop hypotheses. Scientific theory generally settles down in attempts to justify those hypotheses. In recognition of this fact, Goertzel offers the following fundamental normative rule: "During the developmental stage, a belief system may be permitted to be unresponsive to test results . . . However, after this initial stage has passed, this should not be considered justified" (p. 185).

TWO SURVIVAL STRATAGEMS

Belief systems can be monological or dialogical. Dialogical systems escape the clutches of conservatism and irrelevance. Monological ones, do not. Scientific systems in their early development are monological, but they progress to dialogicality; they can be called predialogical. "Dialogicality and productivity are roughly proportional" (p. 187).

Systems have two general survival stratagems: circularity and dialogue. Circularity in its extreme manifestations is monologicality, but usually it manifests itself as conservatism. Dialogue is open to and creative in accommodating challenges. "Dialogicality permits the belief system to adapt to new situations, and circular support structures permit the belief system to ignore new situations. In order to have long-term success, a belief system must carefully balance these two contradictory strategies" (p. 188). Goertzel suggests "precisely two properties common to successful belief systems:

1. being an attractor of the cognitive equation [and]
2. being productive, in the sense of creatively constructing new patterns in response to environmental demands.

A belief system cannot survive unless it meets both of these criteria (p. 192).

Dialogical thinking enjoys many advantages that monological thinking does not because the basic nature of thought is fundamentally interactive and parallel.

> Intelligence is achieved by the complex interactions of different agents . . . A dialogical system can solve problems by cooperative computation: by using its own processes to request contributions from outside processes. A monological system, on the other hand, cannot make a habit of interacting intensively with outside processes-if it did, it would not be monological (p. 189).

Monological belief systems subvert the tension between circularity and adaptation. They dissociate into semi-autonomous cells. "Lousy, unproductive belief systems are lousy precisely because they keep to themselves, they do not make use of the vast potential for cooperative computation that is implicit in the dual network" (p. 190).

SYSTEMATIC CREATIVITY

Goertzel believes that creativity results whenever a hierarchical structure and a structural associative memory are joined in continuous compositionality. Mind is his prototypical case. Language, because of the intimate relations between syntax and semantics, is also creative. Logic and belief systems, as subsets of language, are creative likewise. In addition to these structured transformation systems, Goertzel considers two other creative systems: evolutionary innovation and a body's immune system.

Evolutionary innovation is structurally similar to the productivity of belief systems. They have two basic attributes in common: sexual reproduction and structural instability. For evolutionary innovation, sexual reproduction "is systematic stabbing in the dark" (p. 195). Structural instability is evident in evolution because slight changes in genetic code "can cause disproportionately large changes in the appearance and behavior of the organism" (p. 195).

Belief systems develop in ways parallel to biological evolution. They engage in an abstract form of sexual reproduction when they dialogue with each other. They also require structural instability in order to be creative. "Belief systems are structured transformation systems that serve to systematically create new pattern via multilevel structural instability" (p. 199). Goertzel explains these parallels between evolutionary innovation and productivity in belief systems in mathematical detail.

The immune system is a self-organizing physical system that bears a marked similarity to belief systems. "Belief systems are to the mind as immune systems are to the body" (p. 200). This metaphor "holds up fairly well not only on the level of purpose, but on the level of internal dynamics as well" (p. 200). An immune system is an impressive structurally associative memory that operates with antibodies, antigens, long-lived "memory B-cells," and internal images. It functions in a state of chaos that enables it to react with dispatch to defend against both recognized and novel intruders.

In mental processes, fitness "corresponds with survival" (p.203). And "just as each antibody is some other antibody's antigen, each active belief is some other active belief's problem" (p. 203). Each active belief interacts constantly with other beliefs "thus creating a network of cybernetic activity" (p. 203). Belief systems function best when they maintain a certain chaos among their beliefs that enables "pseudorandom changes in the focus of attention" (p. 205). By maintaining themselves in a chaotic position and using conjunction (AND) as a default categorization for new beliefs and negation (NOT) as an override of the default categorization, belief systems "are capable of computing anything whatsoever" (p. 207). In addition, belief systems may organize our minds on their own without consciousness. "Belief systems themselves may in their natural course of operation perform much of the computation required for mental process.

MIND AND REALITY

Goertzel again takes up the "intimate relationship between language, thought, reality, self and consciousness" (p.213). He offers "the radical but necessary idea that self and reality are belief systems . . . [within] a formal model of the universe in which mind and reality reciprocally contain one another" (p. 213). In this way, he extends the scope of the cognitive equation to the point where it can be considered a *universal equation*. Then he confronts the paradoxes of our understanding of the physical world, especially quantum mechanics, and tries to resolve them with "the ideal that the world is made of patterns" (p. 213).

According to Nietzsche and Whorf, "external and internal reality are belief systems . . . [And] one of the main roles of consciousness and language is to maintain those belief systems" (p.214). A belief system is a special kind of linguistic system in which beliefs have meanings that "change roughly with the syntactic construction of the beliefs." (p. 214). In other words, the syntax of the belief systems creates its semantics. Now the semantics of the belief systems and the semantics of the general linguistic system overlap. These two abstract languages "intersect one another semantically, while being quite different syntactically" (p. 214) In particular, spoken language overlaps the belief system that we call "external reality."

> The belief system which we call external reality is a collection of processes for constructing three-dimensional space, linear time and coherent objects out of noise- and chaos-infused sense-data . . . If the mind had to apply conscious and/or deductive reasoning to every batch of sense-data it received, it would be paralyzed . . . For efficiency reasons, the mind instead applies certain commonsense beliefs about the way the world is structured, and automatically or semi-automatically processes sense-data in terms of these beliefs (pp. 214-215).

Our sense of self is also a belief system that is guided by our body-concept. "When we reason, we relate different ideas in a way that draws analogically on 1) the felt interrelations of our body, and 2) the relation of our body with various external objects. For example, the 'detached' feeling of logical reasoning is not unrelated to the feeling of separation between self and world" (p. 215).

Reality and self, being successful belief systems, are attractors of the cognitive equation. They are structural conspiracies that work in our everyday world and their efficacy reinforces our belief. They also have a strong element of internal conspiracy: one belief reinforces another automatically. Belief in one aspect of external reality reinforces or creates "belief in another aspect of external reality, and vice versa, even when those aspects of external reality have little or no support outside the belief system of external reality . . . Language is an accomplice in the conspiracy" (p. 217).

Goertzel's idea of collective reality builds upon these basic ideas. Collective reality is a hyperset that produces a purely sociological definition of reality. A thing exists in this reality hyperset if people recognize it as real *and* recognize that their collectivity also recognizes it as real.

The relationship between mind and reality in the hyperset perspective is expressed in the concept "universe," which is the set containing both mind and physical reality. "The universe may be understood as a collection of dual networks, linked at the bottom via certain 'connector processes'" (p. 225). Using this concept Goertzel discovers a "very close relationship between quantum measurement, pattern philosophy, and the cognitive equation" (p. 227).

DISSOCIATIVE DYNAMICS

Dissociation refers to a modularization of mental activity in which certain mental processes cut themselves off from the rest of the mind. Personality dissociation is dissociation that involves "significant portions" of the self/reality belief system. Partial dissociation is necessary and healthy. Large-scale dissociation can be horrible. "Partial personality dissociation is central to the formation of structurally conspiratorial belief systems, which are in turn essential to productive, creative logical thought. Or, in a formula: no powerful intelligence without strong internal conflict" (p. 242). Some measure of dissociation is necessary "to the evolution of useful logic-guiding systems within the neural network" (p. 242).

According to dual network theory, a perceptual/motor command hierarchy superimposed on a structured associative memory creates mind. Because of the fractal structure of this imposition, it is possible that part of this dual network might break off and function with semi-autonomy. This break-off is seen most dramatically in psychological states induced by traumatic stress, but it routinely occurs in its less severe forms. Only two things are required for dissociation to take place:

> In order to spin off and survive on its own, a subnetwork must first be complete in itself, in the sense of being a strong attractor of the cognitive equation. And second, it must have relatively few connections with the other parts of the mental networks-otherwise its autonomy would not last (p. 246)

A dissociated subnetwork bears some similarity to a conspiratorial belief system. It is an attractor for the cognitive equation, but it has more dual network structure and more hierarchical depth, more ability to actually make actions happen. "In other words, a dissociated subnetwork contains all the levels needed to do, whereas a belief system only *guides* other systems in doing" (p. 246).

A dissociated subnetwork of the mind is a subpersonality. It constructs a reality-image and a self-image that are accepted by the mind. Separate subpersonalities often share underlying parts of the self/reality system like speech patterns, posture, and memory. They all alter the common aspects of the self-reality system. "By altering the associative memory structure, each subpersonality affects the environment in which the other subpersonalities live" (p. 247). The situation is like "a community of psychokinetics, each one living a normal life, but also continually altering the physical world in response to the alterations made by the others" (p. 247).

As children we learn to set boundaries that fit different situations and we learn to adopt different attitudes and "personalities" with different people in different contexts. We are one person with our parents, another with our siblings, another with our teachers in school, and so on. As adults, we continue to develop new faces: a romantic personality, a work personality, a couch-potato personality . . . Each of these subpersonalities is an attractor of the cognitive equation.

In order to survive, each subpersonality must be rewarding in some way and must protect itself against being destroyed by other personalities. A family-first subpersonality, for example, may be destroyed by a workaholic subpersonality unless it finds joy in family life and works out a modus vivendi with the demands of making a living. This creative modus vivendi is a structural conspiracy that enables both work and family to thrive. It creates a new subpersonality attractor that maintains itself even in the face of fluctuating and hostile external forces.

Dissociation encourages mental productivity by encouraging "sexual reproduction" between subpersonalities, which results in strong and hardy structural conspiracies that thrive on the edge of chaos. Dissociation makes the analogical guidance system of logic vigorous and thereby supplies productivity to clever deductive rules. Dissociation, then, can be pathological, but it is necessary for creative logical thinking. "Without dissociation, ideological, paranoid and otherwise pathologically conspiratorial belief systems would be rare. But so would be productive belief systems; and hence, so would be creative logical thought" (p. 252).

CHAOTIC LOGIC

There is a cyclical dynamic at work between logic and the chaos that is enabled by dissociation. Goertzel calls the structural conspiracy of their joining: *chaotic logic*.

> Dissociated personality networks, and the structurally conspiratorial belief systems which they encourage, are attractors of the cognitive equation, supporting apparently chaotic dynamics. But without these strange attractors, the rich reserve of analogies required for deductive logic would never be created. Logic thrives on chaos. And, conversely, logic itself is a crucial tool of these belief systems and sub-personalities. It aids them in maintaining their attractor status . . . Chaos thrives on logic (p. 251).

Carl Jung and Friedrich Nietzsche, as examples, were creative people with severely dissociated personalities. They used that dissociation to synthesize aspects of their conflicting belief systems. "Jung and Nietzsche are two examples of creativity emerging from the synergy between the structurally conspiratorial belief systems of different subpersonalities" (p. 256).

From the structural conspiracies between subpersonalities that create works of note and even of genius, it is an easy step to the structural conspiracies between individuals and cultures that create collective reality.

> Subpersonalities are behooved to encourage structural conspiracies, so that the processes they create will not be destroyed by other subpersonalities. And by the very same reasoning, minds will do well to create reality structures which are structurally conspiratorial, so that the reality structures they create will not be disrupted by other minds. This suggests that the immensely conspiratorial nature of the reality belief system is partly due to its construction at the hands of competing individual consciousnesses (p. 258).

ARTIFICIAL INTERSUBJECTIVITY

Finally, Goertzel offers a program for constructing a new type of artificial intelligence that is based on his model of mind.   He calls it artificial intersubjectivity. His basic premise is: "The process of thinking is inseparable from the process of participating in the collective construction of a reality" (p.261). He believes that his universal network model "gives us a handy, elegant way of achieving this artificially intersubjective program. Namely, simulate a collection of dual networks connected at the bottom" (p. 261).   At the present time, Goertzel is testing, his theories in a project he calls "Webmind."   This project utilizes the internet to create a self-organizing social mind that can provide superior predictive power for processes like stock-market forecasting.

CONCLUSION

Goertzel began his book with the following stack of "intuitive equations" relating language, belief, mind, and reality:

Linguistic system = syntactic system + semantic system
Belief system = linguistic system + self-generating system
Mind = dual network + belief systems
Reality = minds + shared belief systems.

At the completion of the book, Goertzel assesses "how much and how little these systems-defining equations reveal" (p. 264). He says,

> The cognitive equation gives the flow of mind, and these equations describe attractors which direct this flow. To take the "flow" one step further, the system-defining equations are something like complexity-contoured continents, guiding the flow of the vast chaotic ocean that is pattern space. But yet they are not quite like continents, because they are themselves formed from the flow of the ocean itself (p. 264).

Goertzel realizes that he has taken just the first steps "toward understanding that most mysterious and most essential process by which logic interfaces with self-organizing habit . . . by which order synergizes with chaos to form the complex patterns of becoming that we call . . . mind" (p. 264).

CHAPTER 17

METAPHORS AND MAPS

ABSTRACT

Metaphors pervade our language. They provide our cognitive models of the world. Cognitive maps are models that we carry with us as guides for navigating our physical, social, psychic, and cultural worlds. Cultural cognitive maps are the carriers of sociocultural evolution.

METAPHORS

The Necessity of Metaphors

George **Lakoff**, in *Women, Fire, and Dangerous Things* (1987), summarizes his investigation into what categories tell us about our minds as follows:

> What is known about basic-level categorization suggests the existence of basic-level preconceptual structure, which arises as a result of our capacities for gestalt perception, mental imagery, and motor movement. The consideration of certain gross patterns in our experience-our vertical orientation, the nature of our bodies as containers and as wholes with parts . . . suggests that our experience is structured kinesthetically in at least a gross way in a variety or experiential domains.
>
> Cognitive models derive their fundamental meaningfulness directly from their ability to match up with preconceptual structure . . . In domains where there is no clearly discernible preconceptual structure to our experience, we import such structure via metaphor. Metaphor provides us with a means for comprehending domains of experience that do not have a preconceptual structure of their own (pp. 302-303).

In *Metaphors We Live By* (1980), Lakoff and Mark Johnson show how pervasive metaphors are in our everyday and scientific language. Metaphor and metonymy underlie our ontology, our spatial and temporal orientations, our ideas of ourselves, our ideas of causation, and the practical coherence of our world. They have a double function: they ground us in our bodies and they permit us to soar above them in realms of semantic reality.

We need metaphors. Without them, we could not make sense of our worlds. We do not even grasp many realities without the metaphors that give them context. Consider the use of metaphor in simple questions and answers like the following:

> Are you *in* the race on Sunday (race as CONTAINER OBJECT)
> Are you *going to* the race? (race as OBJECT)
> Did you see the race? (race as OBJECT)
> The *finish* of the race was really exciting. (finish as EVENT OBJECT
>   within CONTAINER OBJECT)
> There was *a lot of good running in* the race. (running as SUBSTANCE in
>   a CONTAINER)
> I couldn't do *much sprinting* until the end. (sprinting as SUBSTANCE)
> *Halfway into* the race, I ran out of energy. (race as CONTAINER)
> He's out of the race now. (race as CONTAINER OBJECT)
> (1980, p. 31).

In our metaphors, we find the cultural assumptions of our social perceptions and activities

> Metaphors have entailments through which they highlight and make coherent certain aspects of our experience.
> A given metaphor may be the only way to highlight and coherently organize exactly those aspects of experience.
> Metaphors may create realities for us, especially social realities. A metaphor may thus be a guide for future action. Such actions will, of course, fit the metaphor. This will, in turn, reinforce the power of the metaphor to make experience coherent. In this sense, metaphors can be self-fulfilling prophecies (1980, p. 156).

Julian **Jaynes**, in *The Origin of Consciousness* (1990), also argues the importance of metaphor for our describing and understanding ourselves. He examines the language of Homer to ascertain the roots of ideas such as soul, heart, spirit, mind, and emotion. In his extrapolation of these meanings, he draws upon and extends the work of Bruno Snell. Snell observed concerning "psyche," that Homer used it to designate a breath or a fluid that escaped a body at death. "It forsakes man at the moment of death, and . . . flutters about in Hades; but it is impossible to find out . . . the function of the *psyche* during man's lifetime (Snell, 1982, p. 8).

The words used by Homer have developed as metaphors to express our semantic pride in being intellectual creatures. Psyche, which we take to mean soul (or mind), meant "breath." *Phrenos*, which developed into our ideas of heart and courage, referred to sensations in the chest. *Thumos* is the word that lies behind our ideas of an emotional soul; it was originally just a word to describe agitation or willful motion. And *noos*, the word behind our exalted ideas of gnosis, referred to what one could see, what we metaphorically term a "field of vision." In brief, all the words we use to describe psychological reality are metaphors built upon bodily experience.

Non-Linguistic Metaphors

Are metaphors necessarily matters of language? Can physical, biological, social, psychic, and cultural structures be metaphors of each other? It would seem so. The structures of life are autopoietic. They have a common fractal structure of reproduction. Living things close to each other in the evolutionary tree show marked similarities.

If "the essence of metaphor is understanding and experiencing one kind of thing in terms of another" (Lakoff and Johnson, 1980, p. 5), then processes that relate to each other with internal and evolving consistency are essentially embodied metaphors. Organisms understand, experience, and select on the basis of shared metaphors (shared evolutionary histories) *and* on the basis of their own contingently evolved structures. It is to be expected that human knowledge, which grows autopoietically out of these universal processes, would make explicit these universal metaphors. It is not surprising that mathematics, the science of pattern, should miraculously model the physical, biological, social, psychic, and cultural processes of our worlds.

Mathematics As Metaphor

Mathematics itself is a marvelously complex and simple metaphor (organized complexity) of our world and lives. It is complex in the scope of the realities it encompasses and the contradictions that it manages. It is simple in its draconian selectivity. It repudiates Byzantine, too ornate, and luxuriant metaphors.

In the past, this selectivity led to a practical abhorrence of mushy, fuzzy, and hypercomplicated situations. Mathematical science could not digest matters of consciousness, intimacy, courage, intuition, and radical contingency. Today, as we have been observing, mathematicians are developing tools to manage even the complexity of human existence. With the aid of computer technology, they are working with ideas like dissipative structuring, chaos theory, fractal geometry, complexity theory, order from noise, synergy, information theory, transdisciplinary unified theory, autopoiesis, fuzzy logic, infinite sets, and chaotic logic. They may soon be modeling psychic and social behavior in ways that honor humanistic values.

There are factors other than computer technology that have made this humanistic extension of mathematics possible. Systems theory has provided an overarching intuition in which niches have been constructed by other perspectives and disciplines.   The disciplines of phenomenology, hermeneutics, and epistemology have sharpened our perceptions of limits and our consistency of knowledge. Anthropology and linguistics have clarified the origin and structure of our thoughts. Theories of self-reference have introduced a central new metaphor of emerging understanding. Neuroscience has located specific cognitive structures and cause-effect relationships that ground our speculations about consciousness in observable processes.

COGNITIVE MAPS

The General Evolution Research Group (henceforth called "the Group") is an assemblage of systems scientists and philosophers from around the world who are exploring the interconnections between science, evolution, systems thinking, and the ongoing development of our human world. In the words of Peter Saunders, one of the Group, they aim "to apply scientific findings to assist towards an understanding of how societies change and, in particular, to explore the role that the systems approach might make" (1993, p. 105). In *The Evolution of Cognitive Maps* (Laszlo and Masulli, 1993), the Group explores the potential of cognitive maps for explaining sociocultural evolution.

Mental maps were first postulated to explain the behavior of laboratory rats as they learned to run mazes (Tolman, 1932). Rats seem to remember a gestalt of their surroundings and guide their movements in the world on the basis of those gestalts (mental maps). A human cognitive map is "a model of the environment, a model in the systems science sense" (p. 4). It is an internalized analog of the environment.

Cultural Cognitive Maps

Cultural cognitive maps are conceptual descendants from the "collective representations" of Emile Durkheim who "pointed out that when a society described its world it was also describing itself" (quoted on p. 9). According to this conception, "[A society] maps its collective experience of the world and preserves the social structure resulting from collective experience, thus giving its cultural cognitive map an authority of its own" (p. 9). Cultural cognitive maps derive from individual experience and conceptualization, but they are the shared property of the group.

Sociocultural systems based on cognitive maps are distinctly different from the two poles of Luhmann's social analysis: interactions and society. They fill a lacuna in Luhmann's thought. They provide category for those things that he sometimes calls societal systems but for which an appellation of neither "interaction" nor "society" is appropriate. If "sociocultural systems" are the functional equivalent of "societal systems," then there is a hidden theoretic basis, even in Luhmann's thought, for the kind of action that most of us claim to see in our societies: the actions of individuals and/or groups that produce systemic change.

The Group is investigating the biological roots, the social manifestations, historical course, the possible future, and the usefulness of cognitive maps. They pose the hypothesis that cognitive maps occupy the place in sociocultural evolution that is held by the organisms in biological evolution. In this hypothesis, cultural cognitive maps are the essence of sociocultural systems. They "emerge, replicate, and change within the context of energy, matter, and information flows" (p. xi). They arise from the activities of individuals and groups. They unfold in the manner of dissipative structuring-with a memory. They possess a long-term, contingent history. In this hypothesis,

> [Societies are] emergent realities, systems in their own right . . . Every society has its own social cognitive map in the singular form . . . The flow of energy, information, and matter between members of a society constitutes a cycle with an integrity of its own . . . The idea of cognitive mapping [extends] to representations of society as a whole, made by the society as a whole, and reflected in the personal cognitive maps of the individuals making up the society (p. 7).

THE EVOLUTION OF COGNITIVE MAPS

The Group proposes two perspectives for viewing sociocultural evolution. One is neo-Darwinian evolution. The other is that of self-organizing systems.

The Neo-Darwinian Perspective

In the neo-Darwinian perspective, cognitive maps are cultural correlates of DNA in biological evolution. A cultural cognitive map "guides the replication of societal structures the way DNA informs the replication of biological structures, and it gives functional meaning to individual action" (p. 6). If cognitive maps do function as sociocultural DNA, then social development follows biological rules regarding basic templates, stochastic variability, and survival of the fittest. In a broader understanding of evolution, factors such as morphogenetic change, coevolution with the environment, adaptation to environmental change, and punctuated equilibrium would play important roles.

[In this connection, Liane Gabora (1999) has proposed a view of cultural evolution that arises from the school of Stuart Kauffman. In her proposal, *memes* (cultural memories) play the roles in cultural evolution that genes play in biological evolution.]

The Self-Organization Perspective

Prigogine's ideas of dissipative structuring are an integral part of the theory of self-organization.   Using Prigogine's ideas, several members of the Group offer historical explanations for the evolution of cultural maps.   They explore societal bifurcations in the processes of epistemology, cosmology, ecology, education, and societal change.   They show how accelerating processes reach unstable states and how small fluctuations trigger massive changes.

In the evolution of cognitive maps, turbulence is evidenced as widespread cognitive dissonance between a traditional cognitive map and cultural experience.   In times of turbulence, minor tinkering with the dominant cognitive map does not remove the dissonance.   In this condition, alternate cognitive paradigms receive a hearing.   Consequently, one of the alternative paradigms is likely to become dominant in a society.   Naturally, the influences that shape the selection of that paradigm become matters of great importance.

Cultural History

The first representations of the world presumably were produced by simple life-forms as accommodations to their worlds that promoted their survival.   At a higher rung on the evolutionary ladder, some insects, as we know from studies of honeybees, communicate their vision (cognitive map) of the world and its salient contingent particulars within their colonies.   Birds and mammals teach intricate cognitive maps to their offspring by leading them into new terrain and displaying behavior that is to be mimicked.   Monkeys teach each other new ways of coping with the world and pass those new ways on to their young in processes of sociocultural evolution.   Birds and primates (and other animals) have proto-languages in which they communicate information that is crucial to their survival or is helpful in their social situations.

Early language among humans was almost certainly of the protolanguage variety.   As it is explained by Durkheim (and Habermas), the first words arose as a response to, and a remembrance of, awe-inspiring events in the social life of early humans.   Over time, portrayals of these events became the myths and rituals of a society.   They were the basis of communal life.

In the course of history, cultural cognitive maps based on myth failed to adequately interpret the complexity of life as tribes mixed and joined in towns.   Town society required the more adaptable cognitive maps that religion provided.   Over time, the abstraction of religion increased, notably in the faiths of Christianity, Judaism, and Islam.   Because of their abstractness, these faiths had a portability and generality that was sufficient in the West until the cultural revolutions of the 14th, 15th, and 16th centuries.   During that period of crisis, an even more abstract cognitive map was needed.   The West found the abstract symbols it needed in the scientific paradigm of Galileo, Descartes, and Newton.   In the 19th century, the cultural myth of science was enhanced by the belief in progress as epitomized in Darwin's theory of natural selection.

In the latter half of the 20th century, the belief in science and progress has waned. Today, we are in a postmodern period. We no longer possess an overarching map. We live in an age without "metanarratives" (Lyotard). In this period of disorientation and cultural chaos, we face two alternatives. We can resign ourselves to living in a world without a center, and perhaps rejoice in its diversity, as do Luhmann and others. Or we can attempt to find and/or create a new center as do Habermas and members of the Group.

An Emerging Paradigm

The Group proposes that self-reflection and self-organization will shape the emerging paradigm. This paradigm will integrate versions of self-organization that bridge biological, social, psychic, and cultural spheres. Within this paradigm, one can even explain how the paradigm will come to be.

Cognitive maps shape social evolution. Societies reduce their uncertainty about their lives using cognitive maps such as primitive. Relying on their cognitive maps, they face life with expectations and develop complex adaptations and meanings. Because they can test response to proposed actions in the contexts of their maps, societies can take risks and accelerate their creative development.

Societies can process environmental information within the context of their cognitive maps "according to a variety of programs created when individuals invent new symbols or redefine old ones. It is thus possible for societies to anticipate different future possibilities through manipulation of their symbol systems" (p.15). With their cognitive maps in place, societies anticipate future possibilities, weigh alternative futures, and move quickly to implement their selections. They become adept at making satisfactory selections. An unfortunate side effect of this effectiveness is that societies tend to harden their maps into dogmas that are resistant to change.

The internal and external complexity of social systems increases constantly. Within bounds, the ingrained cognitive maps of a society adapt well to changes. Eventually, however, the change-induced complexity overwhelms the ordering capacity of a society's cognitive map. When that happens, the system experiences instability. In this crisis of confidence, part of the population will try to enforce old verities, part will just muddle along, and a small part may have a new solution.

When a society reaches a bifurcation, the influences of small dissident groups within it are amplified. One or more dissident group, or perhaps a single dissident individual, will gain an audience. In the newly unstable system, their dissident views gain a hearing. They may generate a consensus that becomes dominant in a new period of more or less stable progress.

New cognitive maps are emergent phenomena that change societies and the people in them. They are discontinuous from previous maps and are unpredictable. They are dependent upon individual and small group initiative. Their influence is amplified by the acceptance of a populace. "The study of cultural cognitive maps suggests that people can change, sometimes dramatically, as a result of their creative acts" (p. 17).

*   *   *

The individual contributors to *Cognitive Maps* document the workings of cognitive maps in cultural evolution. The chapters by Artigiani, Treumann, Laszlo, Masulli, Allen, Lesser, Kampis, and Campanella have special relevance for understanding the nature of the emerging systemic paradigm.

ROBERT ARTIGIANI

Artigiani considers cognitive maps that are currently far from equilibrium in the realms of epistemology (postmodernism) and cosmology (20th century physics). He shows how dissipative structuring is working in both realms.

There is, according to Artigiani, a close correlation in this century between the histories of postmodernism and physical science. Within both communities, there was a period in which theorists expressed convictions that they no longer had anything new to say. In their self-congratulation and self-doubt, the belief was abroad that their fields were complete.

In this situation, literature and science became self-reflective. Einstein, Bohr, and Heisenberg incorporated the knower into their science. Saussure, Barth, Foucault, and Lyotard, among others, demonstrated the radical subjectivity of our knowledge. Both science and philosophy began to incorporate knowers and authors into their methodologies. Both are in the process of creatively coping with the chaos that they have uncovered.

Postmodernism

Architects, artists, poets, authors, and philosophers became self-reflective in this century. They became aware that they had nothing original to say. All the stories had been told. All points of view had been exploited. They envisioned more of the same with only peripheral additions. In their self-reflection, they began to produce works that portrayed the circularity of their creations. Escher portrays their mindset in the sketch where two hands are drawing each other and being drawn at the same time.

Ferdinand de Saussure laid theoretic ground for postmodernism with his semantics of language. "Words . . . are merely arbitrary symbolic conventions whose meanings are determined by other, equally arbitrary words" (quoted on p. 33). In self-reflection, artists and philosophers alike came to recognize their work as artifice. In addition, authors like John Barth, came to a realization of the exhaustion of literature. For them, the limits of the novel were reached, "there are simply no spaces left for modern literature to explore and no stories left for it to tell" (p. 34).

Barth reacted to this situation, by continuing to write while restricting himself to telling old tales, "condensing an entire tradition into its method of expression" (p. 35). He said that he wrote "novels that imitate the form of the Novel, by an author who imitates the role of Author" (Barth, 1984, p. 72). He folded literature back upon itself, "added a new hierarchical level to the structure of modern culture and exposed the difference between linguistic description and the environment" (p. 35). In so doing, he exposed the meaning of "description" by making the difference between descriptions and reality obvious. "Barth's novels about novels, therefore, made the process of cultural evolution possible" (Artigiani, p. 35).

Across the board, postmodernist architects, artists, and philosophers reveal the artifice of our expressions. They create a chaos with a distinction obvious within it: the difference between reality and its expression. In this chaos with its difference, new meanings emerge, "altering behaviors and the relations defining society" (p. 35).

Among the general population, postmodernism is seen as a catastrophe of faith in which the old paradigms that gave meaning to life are derelict. Still, there are values to be found in uncertainty and cultural relativity. Awareness of relativity makes dictatorships based upon unquestioned right and might less tenable. Awareness of our fallibility lessens the imperative we might feel to force our beliefs on other people. Because of the diversity of viewpoints and strategies within our postmodern cultural cognitive maps, we are able to make rapid responses to environmental change. Postmodernism provides the flexibility and quickness of response that is necessary for survival in the information age.

The Fallen Ideal of Classical Physics

Artigiani pursues parallels in the history of postmodernism and the history of 20th century science. He explores

> the possibility that Post-Modernism represents an evolutionary response to the crisis in contemporary culture common to both science and the arts. The crisis results from a mismatch between the cultural cognitive map and its environment, a mismatch leading to a sense of exhaustion resulting in an awareness that the society has lost contact with reality (p. 36).

Artigiani develops his thesis in three stages. First, he shows, from a survey of turn-of-the-century science, that science was experiencing a general sense of exhaustion. Second, he shows that quantum theory is a self-reflexive science that turns a critical eye to its own problematics and epistemology. Third, he interprets Prigogine's theory of dissipative structures as the "reconceptualization of physics" that grounds physics in the reality of its evolutionary processes.

The bare thread of Artigiani's argument is anchored on Mach's statement that the goal of science "was to produce 'a mimetic reproduction' of nature in thought" (quoted on p. 37). Because of the anomalies that are uncovered in relativity and quantum theory, scientists gave up the ideal of a mimetic reproduction. Einstein, Bohr, Heisenberg, and the great scientists of the early 20th century were appalled by the uncertainty and contradictions of quantum theory. They saw that their science, even while it worked magnificently, was cut off from everyday perceptions and common sense. In the words of Heisenberg, "The atomic physicist has to resign himself to the fact that his science is but a link in the infinite chain of man's argument with nature, and that it cannot simply speak of nature itself" (1955, p. 15). Science had to renounce its claims to realism. Bohr said regarding his science that "we are suspended in language," and that "physics concerns what we can say about nature" (quoted by Artigiani, p. 41). In this chaos induced by science's self-reflection, Heisenberg claimed "the renunciation" of the classical goal of mimetic reproduction "will be final . . . We are unable to reach full understanding no matter how far we travel" (quoted by Artigiani, p. 44).

In retrospect, Artigiani is able to say that the Copenhagen Interpretation of quantum physics revealed the difference "between classical science and the natural environment" (p. 45). In bringing this difference to light, "the Copenhagen Interpretation of quantum theory . . . performed the quintessential Post-Modernist act: it . . . deconstructed science and, unintentionally, paved the way for its 'reconceptualization' by a succeeding generation" (p. 45).

This deconstruction revealed the arbitrary nature of scientific codes and created a fluid communication structure. In this fluidity "intransigent realities become signs of more complex, unexamined realms" (p. 45). One particular intransigent reality came to the fore: the contradiction involved in a science that claims to be "both complete and incapable of describing the scientists who produce it or of postulating existence without scientists to observe it" (p. 51).

Self-Reflective Nature

The science of dissipative structures, which Artigiani takes to be postmodern science, describes a situation in which nature observes itself in a manner analogous to that used by a scientist in a laboratory. "That is, as science observes nature, nature is involved in continuous observation of itself" (p. 51). The core elements of dissipative structuring are processes that experience perturbations within their symbiotic environment, leading to bifurcations, and emergent structures that result from successful bifurcations. These elements correspond to instrumental readings (perturbations), laboratory conditions (environment), alternative organization possibilities (bifurcations), and measurements (emergent structures). In this conception, the "scientific method" employed by nature to evolve the scientist, is the same method that the scientist uses to comprehend nature.

Natural observation is not an idea that is easy to accept. We are accustomed to thinking that all observation is human observation and that information is recorded in the minds of scientists. We assume that nature and science behave differently. Still, in the flux of our postmodern uncertainty, we entertain the idea that nature and science follow the same laws.

In natural observation, evolving systems experience perturbations, react to them, and record their successful reactions as emergent structures. "Emergent structures thus act analogously to the pointer readings of scientific experiments. Both record information resulting from observations" (p. 53). The emergent structures actually result from the mutual observations of their component entities. "Each component of an emergent structure . . . [is] both captured and capturing" (p. 53). When these "components observe one another, each defines the rest" (p. 53); the components co-define each other and co-constitute the emergent reality, without which they would not exist. Therefore, observation transforms potentiality into actuality in natural nonequilibrium conditions as well as in laboratory experiments.

Mimetic Reproduction?

At the end of the 20th century, the new science of emergent reality fulfills Mach's injunction that science should produce "a mimetic reproduction" of nature in a totally unexpected way. It demonstrates that the *method* of science is the method of nature. "By proposing new models and metaphors Post-Modern science mimics the processes by which nature self-organizes" (p. 55).

RUDOLF TREUMANN

Treumann reflects on the role of rationality and irrationality in the growth of individuals and societies. His model for rationality is reductionistic science. His model for irrationality is art. Both science and art give form to experience and thought. Both provide the necessary *form* that makes *inform*ation and communication possible. Art provides form without the necessity of words; it can be precise in many ways, but it is always ambiguous. Science uses precise words in order to make explicit the underlying laws of nature. In some ways, art is more basic than science because it gives form to unconscious and chaotic processes. Science, however, is more practical because it provides the unambiguous information we need to build machines and bridges.

Artistic Form

Communication is impossible without the forms that embody it. The necessary forms for communication are provided by the creation of symbols. It is the position of Treumann that science uses symbols, but art creates them.

> In the self-referential definition, art is understood as the elementary or, in the literal sense of the word, the primitive (primeval) cognitive activity which creates symbols. There is no learning, no understanding, no communication without symbols. But symbols have to be generated by some cognitive process. They are not just found or plucked from midair . . . For something which is new and not symbolized, it is important to find an appropriate symbol. At first there is only a feeling that something exists which requires symbolization. Before this symbolization is done, we do not even know what exists (p. 82).

In the concepts of Peirce: Science engages in deduction and induction, but the artistic impulse accomplishes the abduction.

For instance, the idea of chaos was introduced into Western music by the classical composer Haydn for whom "the world was a perfect world, created by the Lord as perfect as it ever could be" (p. 95). The conservative Haydn, in his oratorio, *The Creation*, confronted the problem of expressing the chaos that existed before creation. He solved his problem by creating music that never resolves into a melody or a harmony. He composed the most painful chord he could imagine, a flattened sept-chord. Haydn had invented atonal music. It is said, that Haydn was overcome while hearing the first choral performance of the oratorio. He fainted and fell from his chair.

The symbol of chaos introduced by Haydn and other artists provided grist for the philosophers and scientists of chaos. As part of society's cognitive map, chaos interacted with other symbols within the map. It was shaped and enlarged in literary and philosophical elaborations. From its early conceptualization as the opposite of order, it developed into the idea of a complement of order, and has even become mathematicized as the science of deterministic chaos.

Information and Time

Classical physics was like classic music in its feeling of completeness and good order. It made no space for the really interesting things in life like chaos and evolution. For centuries it was

> blind to change and evolution and . . . described the world in terms of unchangeable states, where transitions from one state to another were outside the interest of physics, and . . . [were] described by reversible equations. In this respect physics . . . described a dead and uninteresting world, a world without change, a world with

no room for creation and where creativity was an alien intrusion, suspect and not understandable in terms of physics (pp. 89-90).

In contrast to the classical dynamics of physical forces, the new science of thermodynamics deals with quantities of information. In the new theory, "evolution and creation can be described by the theory of deterministic chaos, which is nothing else but the theory of information dynamics" (p. 90). Information dynamics treats time as it is quantified by Shannon's definition, and as it is present in Kolmogorov entropy. According to the Kolmogorov equation, only chaotic systems generate information; stable, periodic, and stochastic systems do not. Information creates time; chaotic systems generate information.

> Physical time is generated if and only if there is an increase of information . . . All regular or stable periodic systems to do not generate time and are dead systems, while chaotic systems always generate time   . . . It is chaotic systems which contribute to physical time, without them there would be no time at all, no evolution at all, and the world would be a dead world (p. 93).

Treumann points out that "this definition of physical time is very satisfactory" (p. 93) because it confirms our intuitions "that time intervals when nothing happens are extremely boring, time seems not to flow, and such time intervals disappear in retrospect . . . On the other hand, time intervals where many things happen flow very fast and appear in retrospect long and filled" (p. 93). Since the accumulation of information is the process by which time is generated, "there is a continuous but nonetheless nonuniform creation of time" (p. 93).

Science Mimics Nature

There is a delightful surprise for science in this theory of information dynamics, because as chaos in science "rises from determinism, from the impossibility to avoid errors in determining any quantity" (p. 94), chaos in nature arises because "nature is itself supplying these errors, because nature is itself measuring all the time" (p. 94). The measuring of science mimics the measuring of nature. As a result, science's deterministic description of the world is not only an idealization made by human minds; it is also a correct idealization "because it corrects itself and generates the chaos we are confronted with daily" (p. 94). The method of science is the method of evolution and, not surprisingly, it produces ever-more-precise representations of life.

Deterministic chaos theory makes creation and creativity understandable from the basic principles of physics.

> Creativity turns out to be something understandable as the action of a chaotic mind. A mind lacking chaos is pathological because it moves and thinks only in well-defined and fixed modes of thinking and has no possibility of leaving the vicious circle which has captured it.   Creativity requires some kind of chaotic,

unpredictable thinking, some kind of unexpectedness and surprise, some kind of risky activity of the mind. Only under the conditions of such a risky activity will new ideas be generated and will there be evolution in our recognition of the world (p. 94).

Perhaps the riskiest type of thinking is done in the arts. In their sensitive grasping for symbols, artists dwell in the chaos of creativity. They provide the irrationality that is necessary for long-term successful thinking.

PETER ALLEN AND MICHAEL LESSER

Allen and Lesser present an evolutionary paradigm in which evolution across species is generated by selection on the basis of learning. They develop their argument in the following stages. (1) The world is all detail. (2) Our cognitive maps are reduced descriptions of that world on the basis of some classification scheme. "Models, equations systems, descriptions in language, any cognitive map, must lack the particularity and diversity of the world it represents" (p. 120). (3) On the basis of such cognitive models, "the future of the system can now apparently but wrongly be predicted by the simple expedient of considering the behavior of the model" (p. 121). (4) If the future of a system were correctly predicted, in the manner, for instance, of a Newtonian clockwork mechanism, then evolution could be extended linearly into the future as an inevitable ascent. (5) "This view is defective" (p. 121).

Average and Non-Average Detail

Models reduce reality to average predictable detail, but what actually occurs depends "on local, nonaverage detail--the accidents of history" (p. 121). Models based on average detail are capable of functioning but not of evolving. It is the nonaverage detail, which is missing from the model, that provides the evolutionary change in the real complex world. There are then *two* types of changes that determine the future of a system: those brought about by the typical behavior of average components, and the qualitatively different ones brought about by non-average components and conditions.

Selection

It is the presence of these two kinds of changes that makes the concept of evolution meaningful. It is the dialogue between them that creates selection.

The process that results in what we call selection emerges from a dialogue between the describable average dynamics and the unpredictable, exploratory nonaverage perturbations around this average which result from the inevitable occurrence of nonaverage events and components. These perturbations search or explore a stochastic process that generates information (pp. 121 & 123).

On the basis of this definition of selection, Allen and Lesser present "the concept of evolutionary drive, which shows that evolution selects for populations with the ability to learn, rather that for populations exhibiting allegedly optimal behaviors" (p. 123). In general evolution, selection for learning takes the form of "diversity-creating mechanisms in the behavior of populations" (p. 123) such as genetic variation or cognitive processes. In the mathematical model presented by Allen and Lesser, "evolution selects for imperfect reproduction even though at any moment it would always be better not to make errors" (p. 125). If it made no errors the model would fall into a positive feedback trap that allows for neither "improvement in birth rate nor increased positive feedback" (p. 128). Furthermore, it turns out that evolution does not represent "the discovery of preexisting niches. The niches are themselves the creation of the evolutionary process" (p. 130).

Cartesians and Stochasts

In evolutionary human systems, long-term success requires two apparently contradictory kinds of behavior:

> First, the ability to organize one's behavior so as to exploit the information available concerning net benefits, a rational ["Cartesian"] approach . . . Second, the ability to ignore present information and to explore beyond present knowledge, an apparently irrational ["Stochastic"] approach . . . The Cartesians make good use of information, but . . . the Stochasts generate it! . . . Over a longer period the best performance will not come from the most rational but instead from behavior which is a complex compromise (pp. 131-132).

In summary, "ignorance permits learning, but learning creates new ignorance. More generally, nonaverage behavior generates adaptation, which in its turn generates new nonaverage behavior" (p. 133).

ERVIN LASZLO

Laszlo considers the ecological dimension of cognitive maps. In the history of the race, the paradigm of man as the dominator of nature arose with the origin of agricultural civilizations and religions. This paradigm has served humanity for good and ill for millennia. In our day, the paradigm of dominance is particularly hurtful to our world and us. As a result, the paradigm is starting to wane, but its demise needs to be hastened.

According to Laszlo, "a new holistic map of man and world is now emerging, in sharp contrast to the previous domination map" (p. 63). The domination map was anthropocentric, Eurocentric, atomistic, materialistic, economically competitive, dualistic in its mechanical objectifying of the body, and hierarchical. The emergent holistic worldview holds values diametrically opposed to the dominance paradigm.

The new paradigm is making inroads into mainstream society, into boardrooms, legislatures, and scientific conferences. It has the makings of a powerful political movement. It delivers a convincing picture of a desirable life and future, and a striking contrast to the ecological disaster portended by the paradigm of dominance. Hard-nosed scientists have entered the fray, and not only in the realm of environmental impacts. They are shaping the contours of basic science into new physics, new cosmology, new sciences of nature, life, and complexity that

> elaborate and specify a cognitive map of a self-evolving, dynamic, and essentially unified reality in which life and mind, and human beings, and human societies are integral elements . . . In these days mathematical physicists, rather than just poets and philosophers, are attempting to specify the contours of the emergent holistic map" (p. 65).

IGNAZIO MASULLI

Masulli believes "that we are now at the threshold of a great transformation: "the passage from an industrial society, already mature, to another kind of society, which for the sake of convenience we shall call 'postindustrial' but whose characteristics are as yet undefined" (p. 67). He sees quantitative and qualitative challenges that civilization has not faced before. These challenges come so fast that we seem to organize our societies and lives around crises rather than around some shared rationality. The old rationality was built around recurring advances in mechanical and social technology. In the recent past, the positive feedback engendered in these recurring cyclic advances "Has led to a series of transformations and emergencies that ultimately turned out to be too difficult to control" (p. 68).

Subduing the Earth

The landscape in which we live has been radically transformed. We have largely abandoned the man-nature exchange that is our evolutionary heritage. Today we live mainly within a "milieu of artifacts and technological manufactures . . . [that provide] a substitute planning realm (planning, comfort, design)" (p. 68). Our synthetic landscape involves us with human artifacts like schools, industries, politics, overtime, and round-the-clock entertainment. It draws us up and down, hither and yon. It denies us a stable human point of view.

The unbridled technology and exploitation of the earth, which is the hallmark of our mature industrial society, brings us face to face with the limits of the industrial paradigm. "The paradigm of industrial civilization has been, and still remains, that of promise and an expectation of indefinite growth" (p. 68). This paradigm is encountering problems from nature itself, from a sense of social justice, and from the demographics of population growth. These problems can hardly be addressed in isolation; they are all related in the system that is our modern world.

Alternative Paradigms

Because of the excesses of rampant technology and the social chaos that they engender, the old paradigm is not as solid as it used to be. People are actively exploring alternative paradigms. In the muddle of cognitive maps that are being tried, the future paradigm will emerge. Masulli hopes that the future rests with the wise who will initiate beneficial changes in tendencies and values. He lists some of those changes:

> from attitudes that formerly were predominantly aimed to counter death to attitudes in favor of life; from attitudes opposed to disease to attitudes in favor of health; from preoccupation with controlling death towards a control of births; from self-repression to self-expression, and from behavior guided by external constraint to a self-disciplined behavior (p. 72).

Bifurcating Paradigms

Masulli explores how we might change into a more hopeful paradigm. He also examines interrelationships between cognitive maps and the working regimes of social systems. He offers the following hypothesis:

> When significant alterations occur in cognitive maps, such as to affect the formation of new patterns of thought and action, this phenomenon must be connected with the substantial transformations that have come about in the working regime of social systems-transformations that involve the birth of new structures (p. 170).

He finds that these transformations follow "the bifurcation processes that occur in the evolution of complex systems" (p. 170). At bifurcation points, fluctuations become abnormally high and enable "choices" among various possibilities. The activities of systems beyond these points retain their relations with previous history, but they are unpredictable. In human history,

> The ability of innovators . . . can be seen to consist in their taking advantage of the conditions of non-linearity that characterized the previous working regime of the

system. When they succeed in interpreting the drives, the needs, and the
contradictions, the new leaders forge a new type of organization (p. 172).

The Industrial Revolution

Masulli illustrates the interactions between cognitive maps and social change
through the complex example of the Industrial Revolution, which "was a deeply
traumatic process, precisely because it involved simultaneously far-reaching social
change and a profound alteration in cognitive maps" (p. 173). From his discussion,
he notes how new cognitive maps on the part of the industrial and agrarian elites
created social structures of exploitation that were superimposed on the old cognitive
maps of agrarian society. This superimposition created widespread psychological
and economic distress. This distress made possible new cognitive maps, new social
movements, and eventually new social structures oriented toward social welfare.

These new developments occurred very rapidly as cognitive maps and
social structures acted upon each other catalytically in a conflict that eventually
resulted in a synthesis. This mechanism is "well known in the dynamics of complex
systems. In effect, we have here an instance of cross-catalysis in which the two
different products, or groups of products, act upon one another as catalysts for their
own synthesis" (p. 178). The resulting synthesis was not accomplished on the
economic plane alone but all across the sociocultural spectrum.

Finally, the idea of progress was integral to both the capitalistic and
socialist movements of the nineteenth century. Progress was the basic pattern that

> right from the start . . . constituted a basic pattern capable of informing highly
> contrasting, even conflicting, modes of thought and action. While the agrarian and
> industrial bourgeoisie made use of this pattern to further their own ends and
> legitimize their hegemony, the same pattern also inspired the socialist movement,
> in which the social objectives were diametrically opposed (p. 178).
>
> . . . In this case it was this relation between cognitive maps and social change
> that enabled contrasting objectives, interests and ideas to interact and stimulate an
> evolutionary leap (p. 179).

Progressing Beyond Progress

The paradigm of progress constituted the historical limits that enabled evolutionary
progress in the past. In the present, however, this paradigm with its historically
encumbered limits inhibits further evolutionary progress. Masulli sums up our
predicament as follows:

> The destiny that mankind must follow in its effort to emancipate itself from
> subjection to nature is simultaneously mankind's own destiny. In the end, man, in
> whose interests the enslavement, the objectivization, and the disenchantment of
> nature were brought about, is himself so oppressed, objectivized, and disenchanted

with himself that even his own efforts at emancipation work against him by consolidating the illusory context in which he is prisoner (p.74).

The stage is set for a major bifurcation.

GEORGE KAMPIS

Kampis presents a model of mind based upon referential information, component systems, and material causality. He shows how utterly temporized information leads to enduring cognitive maps.

Referential Information

Non-symbolic referential signs are the interactions of mental components prior to their encapsulation into symbols. These signs are called "referential" because they interact with each other and the entire learning system in a process of multiple-context internal determination. They have the transient nature of molecular signs. They are without external, process-independent machinery or "scaffolding." They constitute their own ever-evolving meaning in their relationships. For that reason,

> Classical information theory, with its traditional metaphors of 'communication,' 'entropy,' and the like is irrelevant to them . . . We have to look at the living flesh of the referential signs, endowed with selectively realized properties in the given system . . . In a nonsymbolic system . . . the information cannot be separated or abstracted from its carrier because the carrier is active (p. 140).

Kampis claims that "referential and not nonreferential [symbolic], information is the key to understanding what the mind is and what it does. In other words, referential information offers a key to the mode of existence of cognitive maps" (p. 141).

If we restrict knowledge, as we usually do, to nonreferential (symbolic) information, then we must admit that language cannot express all the contents of our cognitive maps. In reflecting on things like our innate sense of balance, we realize our non-linguistic prowess. In examples such as this, we become aware of information processing that performs work. This intermediate processing is not yet "knowledge," but it contains information and processes it. This processing is nonsymbolic thinking.

Kampis describes the origination of cognitive maps around three propositions. *Proposition one*: The mind is a processor of referential (nonsymbolic information). This proposition has just been discussed

Utterly Temporized Information

Proposition two: When processing nonsymbolic (referential) information, the original information content disappears and new information is created. This proposition gets to the core of the utter temporality of component systems: the constant turnover of their elements. If the original information is sensory output, for example, it cannot be carried over to perception because it is not symbolic. It simply comes and passes. "The sovereign information content of the last element of this chain is the one that the mind 'sees'" (p. 142).

In the long chain of transformations that take place from external stimuli to what we know, each link processes its predecessor for information.

> Every transition leaves us with a new world defined by the arbitrary new information content. What we sense and what we know have little in common; in the purely logical sense they have got nothing to do with each other. They occupy opposite ends of a long sequence of transitions which, according to its own internal laws and without the use of any external information, melts the original information and reshapes it to its own likeness. The beginning and end are not unrelated, of course, but . . . there is no a priori formal transformation that could link the two (p. 142).

The Nature of Cognitive Maps

Proposition three is based upon the following argument. If "the existing information content disappears at each and every step of the processing" (p. 143), but the information of cognitive maps remains stable, then "the transformation that leads to that must be very special" (p. 142). The nature of that transformation is the content of the third proposition: "Cognitive maps are self-reproducing assemblies of referential information carriers. They are 'fixed point' solutions of a self-modifying process" (p. 145). In this conception, a single self-reproducing cycle is the smallest stable unit of nonreferential information; that is, such a cycle constitutes a simple cognitive map characterized by "the first steps towards symbols: permanence and relatability" (p. 145).

> Cognitive maps can be conceptualized as intricate, hierarchically and heterarchically organized replicative superstructures built from permanent linkages of many elementary mental models, that is, of many elementary cycles. The mind appears to be a continuously changing collection of such superstructures (p. 145).

Mind, then, structures itself in relation to its environment by replicating according to the principles that have been elsewhere described by Csanyi and Kampis.

> Replication is not a property of the components but . . . of the component assemblies. It is all [the] components of a cognitive map together that determine

what can be perceived by the mind. And this is highly selective, according to the conditions and logic of replication. Only what conforms to these conditions can enter the system as a further component—others will be discarded already at the nonsymbolic level (pp. 145-146).

Thus, the component system restricts its replication to components that are compatible to its essential map.

Cognitive maps are judged by the success or failure of the mental models that they engender. They experience a selection pressure, on an evolutionary time scale, against organisms that "prefer biologically useless or disadvantageous modes of information processing and cognitive map construction" (p. 146). In human societies, "another selective force is imposed by communication. Personal cognitive maps that cannot be communicated to other humans or those which are not accepted by other humans will be eliminated from the mind of average persons. Culture and society are born at this point" (p. 146).

Symbolic Information

When we turn to symbolic (nonreferential) information, we have information that is communicable according to meanings and procedures. It would seem that this kind of information is ideally suited to interpersonal communication, and in some ways it is, but the process is not at all straightforward. First of all, "Nonreferential information is communicable only within the system in which the relevant symbolic universe is available" (p. 146). Such availability is present

> only within the mind of the same person, where the replicative units that together span the symbolic system are all available. To communicate between different minds, these units are not suitable, for they are not available and not communicable themselves; only behaviors, not contents, are accessible by communication partners (p. 146).

Within the mind, one operates with cognitive maps, meanings, and mental models that are under one's control. Internal language enables one to communicate, interrelate, and control one's thoughts. Spoken language, however, is not communicable under the same open rules of internal language because it is

> already a translation of mental language to action, and all actions correspond to referential rather than nonreferential information-because they themselves perform work . . . Spoken (expressed) language is no longer symbolic, it is an object that has lost its denotational value when it has left the mind" (pp. 146-147).

"Not symbolic," in this sense, means "not a natural sign; that is, one that can be understood without conventions for understanding the sign."

This interpretation flies in the face of our usual understanding of communication based upon the channel metaphor. It does not totally trash that metaphor, however. It just places it at the end of a rather complicated process of communication whose intermediate steps need to be noted.

In conversations, individually isolated cognitive maps are brought into a kind of match in some areas as people indicate their point of reference and its associated referential information. Through processes of intuition, trial, and error, the participants of conversations "begin to bring forth identical actions in identical situations. That is, the speakers behave as if their mental models were identical" (p. 147), in certain circumscribed situations.

In this process, the principal function of language is to supply names that activate similar mental models, superstructures, and replicative structures in the persons who use it.

> Words and other linguistic units can act as keys, which can replace or signify environmental situations. In this way, a human organism can stimulate itself, and this is what we may call thinking. The verbal keys, if expressed, may also stimulate others by evoking their personal linguistic keys. In doing so, a cultural context is born as a new organizing quality (p. 147).

A Global Cultural Computer

Culture works amazingly well given the complicated way in which it is evoked. It works like a huge computer that processes global conversational information. We, as individuals, are born into this higher level [global] reality and spend our whole lives attached to this computer" (p. 147). It works so well that we almost never notice that it "does not transfer nonreferential information but only evokes personal inference mechanisms to reconstruct it" (p. 147). Notice of its complicated underpinnings arises "only in case of stressed situations such as socialization problems or mental illnesses, where the system is not properly bootstrapped" (p. 147). The system also has troubles sometimes in "communicating emotions, new thoughts, and everything that is not already circulating and providing a stable context in which the illusion of understanding may develop, in the system whose products we are" (p.147).

It appears that, "from the information-theoretical viewpoint, we spend our whole life communicating with 'aliens'" (p. 148). It is amazing that we communicate with each other as well as we do. It took evolution a long time to make it possible. "Only our four billion years of common history that built in us many similarities by hard-wiring and the feedback mechanisms of societal life save us from paranoia, solipsism, and mental chaos" (p. 148).

MIRIAM CAMPANELLA

Campanella examines the uses of cognitive mapping in political activity. She compares two paradigms of decision-making: the traditional *rational* mode and the emerging *cognitive* one.

The Rational Paradigm

In the rational paradigm, one goes along with the assumptions of common language in which "we speak of 'occurrences' not as unstructured happenings but rather as structured and conscious choices among alternatives" (p. 240). Within the bounds of this assumption, rationality requires "consistent, value-maximizing choice within specified constraints" (p. 243) and strategies of optimal choice and cost-benefit analysis in terms of a goals-means economy.

The rational model of decision-making ignores certain limits to rationality. It presupposes, for example, that there is a unitary actor in international and national decision-making who can maximize values in a zero-sum game. Many situations, such as nuclear threat, are not zero-sum games. The actual actors are integrated people and not the mere economic actors of the theory. Also no political actor acts in an organizational vacuum. In addition, time constraints and fluid situations render dispassionate and leisurely contemplation of rational alternatives impossible and, therefore, require shorthand rules for decision-making.

The Cognitive Paradigm

The *cognitive* paradigm centers on the real-life strategies that we use to learn and make decisions. It considers the human actor "not at rest but in action" (p. 242). It "emphasizes unexpected factors, pertaining to the subject and to the environment of the action, and these vanish from view when operating under the assumption of rationality and consistency" (p. 242). It assumes that people form a simplified cognitive map of their world in order to cope with the confusing complexity of their lives. They use this map to make choices that are not optimal but are nevertheless satisfying.

In the cognitive approach, individual confront situations of high complexity and ambiguity with a simplified and flexible set of beliefs and information. They use these cognitive filters "to overcome (or succumb to) the complexity, the ambiguity, the uncertainty, etc. of the surrounding world" (p. 243)?

To understand this cognitive filter we must go "beyond rationality." We must investigate "the elements upon which rationality is built" (p. 243). We must perform a cognitive analysis of the decision-making process by focusing on the pre-rational knowledge of the agent.

Two Viewpoints, Two Strategies

In political science, the cognitive researcher considers the decision-making process *from the outside as an observer*. From that viewpoint,

> The agent [who is being observed] does not use his/her rational calculation abilities, but a different mix of direct and indirect experience on the basis of which to build the learning process. Although there is rational reasoning on the part of agents, there is evidence that they use their own cognitive map and not optimal rationality (p. 243).

From the viewpoint of the agents, however, "what is selected is not the result of a cognitive (limited) map, but what appears to be the 'one best way'" (p. 144). The viewpoint that produces knowledge of cognitive mappings is that of the third person observer of systemic interactions. The viewpoint that produces the conviction of rational choice is that of the first person agent living his or her autopoietic existence. The outside and the inside viewpoints produce different conceptions of how the decision-making process works.

Campanella summarizes the applicability and effectiveness of these two paradigms in the following paragraph:

> The cognitive paradigm considers more clearly the ever-changing situation of the actor-in-action, while the rational paradigm demands steady-state situations. If the rational paradigm is able to resolve problems under selected conditions of space and time, the cognitive paradigm copes with "fuzzy gamble situations" (p. 244).

The use of cognitive maps makes possible simulation techniques for decision-making. By using these simulation models, we can better understand the decision-making process and improve it because we come to see the interdependent network of relations in which each actor is entangled. As a result of the success of the cognitive model, "interdependence and cooperation are becoming the new creed of international policy making" (p. 249). Another result of the ascendance of the cognitive approach is a growing realization that the direction of the world's future does not issue from a separate steering capacity. It comes instead from the interaction of the organizations and nations that are co-present in decision-making.

SUMMARY

Metaphors expand our lived bodily experience and allow us to think abstractly about the world. Cognitive maps are our internal models of our world. Cultural cognitive maps are our shared sociocultural perceptions organized on the basis of paradigms. Leaps of sociocultural evolution develop when paradigms are forced far from equilibrium by social changes.

This chapter concludes the expansive part of this investigation. Along with previous chapters, it provides core insights into the thoughts of about thirty major systemic thinkers. The following chapters synthesize the thoughts of those thinkers. They deal with the topics of design, social theory, and knowledge.

PART FOUR

FIVE EMERGING SYNTHESES

FIVE EMERGING SYNTHESES

The preceding chapters indicate the spirit and horizons of this inquiry into present-day systems thinking. They engage seminal thoughts about systems and social processes from a variety of eminent thinkers. These summaries can be thought of as virtual interviews that were conducted with their authors. The following chapters present syntheses of what these thinkers have to say collectively about design, the social world, communication, cognition, and epistemology.

In this introduction to the remaining chapters, a brief description of the method used to generate the syntheses is in order. More complete descriptions are available in previous publications (Bausch, 1999, 1999b). From the works summarized above and a few others, I precipitated about 1000 key statements. I divided these statements into definitions, consigned to a lexicon, and into "standards," which propose norms for the practical, ethical, and theoretical use of systems theory in understanding and directing social processes. Then I further divided the standards into five topical categories. I used those groupings to construct composite renderings in the areas of design, social theory, communication, cognition, and epistemology. The following chapters are further refinements of that work.

THE WORK OF CONDENSATION

When the summaries contained in the previous chapters were complete (in rough form), I had no set ideas about how to identify the categories of social processes nor schemas for synthesizing them. I had to devise methods of condensation, analysis, and synthesis. The two methods were "systemic content analysis," and an adaptation of the Interactive Management methodology.

Systemic Content Analysis

Systemic content analysis is a method for reducing the complexity of textual information that is similar to the method of reducing phenomenological reports that Giorgi calls "empirical phenomenological analysis" (1974, p. 83). In using this methodology, one processes the material gained from interviews on a topic, such as anger, in five steps. I applied those five steps to the virtual interviews with systemic authors that I generated in reading and condensing their works.

1. I composed and reread the summaries.
2. I read them carefully and highlighted the passages that stated basic principles and clear applications. Then I precipitated out those statements that were most emphatically highlighted.
3. I eliminated redundancies of phraseology by selecting those statements in the thought of an individual author that were more prototypical of his or her thought. That is, statements that were well-phrased and more precise than others that had a roughly equivalent meaning.
   a. When the statements of an individual author were reduced by the procedure described in (3), these statements were mixed with the similarly reduced statements of the other authors under consideration. The result was a collage of about 1000 statements
   b. This jumble of statements was sorted into suitable categories on the basis of a content analysis that sought to encompass what those statements said and to partition them into coherent wholes. This process eventually produced six categories: a lexicon, and five groups of standards that dealt with the practice and ethics of design, the structure of the social world, communication, cognition, and epistemology .

Condensation/Analysis Diagram

This part of the condensation/analysis is portrayed in the following diagram. This diagram indicates six steps: summary (A), precipitation (B), textual analysis (C, D, E), and analysis/synthesis (F). In this diagram, all of the numbers are approximate, except those in the last column that state the number of standards that were utilized in the syntheses.

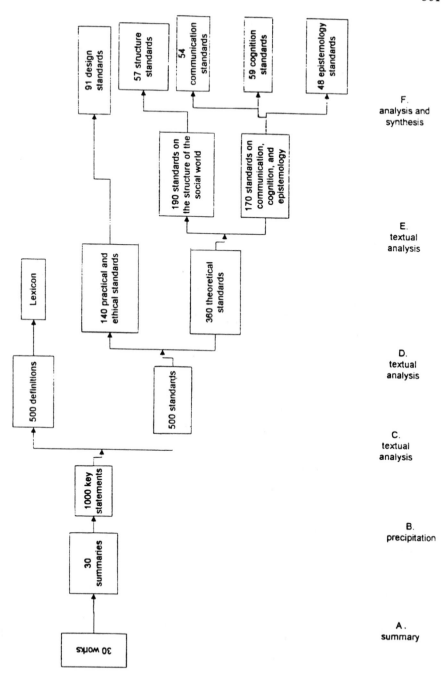

Condensation-Analysis Diagram

The final column of this diagram identifies five areas of summary/synthesis. They are the topics of the following chapters. They concern:

- the practice and ethics of design
- the structure of the social world
- communication
- cognition
- epistemology.

SYNTHESIS METHODOLOGY

Two additional steps adapted from Giorgi's methodology were used in the work of synthesis

4. To develop an eventual synthesis, I reflected upon the statements, gathered into categories, which were still expressed essentially in the language of their authors, in order to develop a rounded expression of what systems theorists have to say about particular social processes. Within the categories, the standards proposed by the various authors were compared and subjected to varieties of imaginative variation to remove redundancy in the manner of (3) above, but with the added criterion of achieving diversity of authors and viewpoints and not just diversity in meaning. In this way, I elaborated the meaning of the statements by relating them to each other and to the sense of the whole.
5. When the comparisons and imaginative variations were performed in (4), above, I integrated the insights and positions of these various authors on specific topics concerning design, the social world, communication, cognition, and epistemology.

This method was employed throughout the process of synthesis and is the sole method used in the chapters that deal with communication, cognition, and epistemology. In the chapters dealing with design and the social world, this method was augmented by the methodology of the CogniScope^{TM}.

The first task in creating the synthesis indicated in (5) above was to compose the elements of these complexes into fragmentary mosaic portraits around specific topics that relate systems theory and social processes. The second task was to draw a composite portrait for each of the five general topics. In this book, I report a refined version of that portrait.

Interactive Management

In an ideal situation, a synthesis like this would be conducted in open discussion by the authors themselves. They would decide among themselves how systems theory applies to social processes. In such a situation, there would be real dialogue. Such an ideal was impossible in my situation.

Even a situation which had all these authors in the same room would hardly be "ideal," because these authors would encounter the same barriers to communication that arise in any interdisciplinary group meeting: different cognitive maps, different languages, different meanings for the same words, differing perceptions about the relative saliency of standards, power inequalities, and so on. Even if the authors were in the same room, they might break into small conversations to explore small clusters of ideas, or they might break up in disgust. In order to generate an approximation of "the ideal speech situation" in this large group, a major innovation in the design of the conversation is required.

Such situations are the focus of John Warfield and Alexander Christakis. In their collaborative work over the past 25 years, they have developed a system of Interactive Management and a process called the CogniScope^{TM} that facilitate conversations concerning complex situations among people of diverse backgrounds. I used the CogniScope^{TM} in the following two chapters to organize the clutter of standards in the areas of systems design and social theory.

Open dialogue would seem to be the exact opposite to the situation of the lone inquirer. His or her work would seem to be monological. As Ulrich, Goertzel, and others have pointed out, a monological situation limits our ability to make adequate sense of complex "messes" because it limits possible solutions to only one perspective.

The lone investigator of this book is not, of course, limited to a monological situation. I sat in on multiple dialogues between the texts of the authors under discussion. These virtual dialogues are minimally filtered through me. They are dynamic interactions between the thoughts of their authors.

In using systemic content analysis and the CogniScope^{TM}, I enabled virtual dialogues between the authors themselves via the proxy of their recorded statements. These methods open channels of multiple parallel (mostly unconscious) processing in my brain and those of my congenial critics. Moreover, the CogniScope^{TM} incorporates major portions of this kind of parallel processing into its programmed deep logic, thereby allowing focused attention to relatively simple paired comparisons.

In complex situations, which require attention to multiple sources of information, the rush of data submitted to the mind for processing overwhelms our ability to adequately handle it. The methods used in this synthesis slow down and process the data rush.

As Goertzel points out, we handle a degree of this rushing complexity by creating virtual dialogues within our minds among various opinions on a subject. In the CogniScope™ sessions, Goertzel's notion of an internal virtual dialogue was expanded to create computer-assisted, external virtual conversations.

Advantages of These Methodologies

The combination of these methodologies enabled me to draw out common threads of discourse from a diverse array of systemic thinkers, to juxtapose them, and weave them into coherent narratives that honored their diversity of viewpoints. In particular the CogniScope™ generated an enhancement pattern that was far beyond my unaided skill even were I granted an unlimited time to work with the imponderables that it organized. The CogniScope™ also contributed clarity and efficiency in organizing the plethora of standards that dealt with the structure of the social world. In addition, it presented the logical framework for making sense of the standards that pertain to communication, cognition, and epistemology.

These two novel methodologies offer synthesis-generating tools that are not available in the ordinary armamentarium. As such, they can be extremely valuable in an age when social science is woefully inadequate in synthesis. They can help us generate acceptable metanarratives for the information age.

CHAPTER 18

THE PRACTICE AND ETHICS OF DESIGN

ABSTRACT

This chapter presents the results of a virtual dialogue about the practice and ethics of design. The dialogue was between statements acting as surrogates for their authors. The virtual conversations were accomplished through a modified method of paired comparisons. Each statement was, in effect, compared on a one-to-one basis to every other standard (cf. Bausch, 1999, 1999a).

These paired comparisons were made in three iterations. The first set of comparisons inquired whether a certain pair of standards were talking about the same general thing. The second set questioned whether one standard was more salient than another standard. These two preliminary sets of paired comparisons set the stage for the third set that inquired whether the implementation of standard **X** would substantially enable the implementation of standard **Y**, and vice versa. This third set of paired comparisons yielded an enhancement pattern that exists among those standards. This pattern indicates how implementations of those standards influence one another.

I do not present the processes that lead to the enhancement pattern here. I begin with the enhancement pattern that provides the structure for this chapter. It will be presented after some preliminary discussion of basic ideas.

PRACTICE-ETHICS-DESIGN

Designing lets us take some control of our present and our future. *Design* in its general meanings is defined as a verb, "to conceive and plan out in the mind"; and as a noun, "a mental project or scheme in which means to an end are laid down" (Webster, 1983, p. 343). *Systems design* is a method to guide change, "a decision-oriented disciplined inquiry" (Banathy, 1996, p. 9). *Social systems design*, which is the focus of this chapter, has received numerous definitions (27 are listed in Banathy, 1996, pp. 11-13). Churchman (1971), for example, states that, "Design is primarily a thought process and communication process, transferring ideas into action by communication" (quoted by Banathy, 1996, p. 12). In m own formulation social systems design organizes expectations for collective action. The general idea of a *social system* is that of a self-organizing system built around shared expectations.

The design that is envisaged in this investigation not merely *systematic,* that is, it does not merely involve steps or phases in logical and linear arrangements. It is rather *systemic.* It involves specific techniques, generalized rules, and universal principles of design. In Banathy's words, the systemic approach "reflects a dynamic, open, and learning-focused approach to conducting systems design" (1996, p. 16).

In the context of this chapter, the *practice* of design refers to the practical guidelines that are followed by effective designers. The *ethics* of design refers to norms (or standards) that should be adhered to in the designing process and the qualities that should be present in the resulting design.

OVERVIEW OF THE ENHANCEMENT PATTERN

The enhancement pattern on the following page was generated from 91 unstructured standards for designing in an intense six-hour CogniScope™ session. During that time, the computer, programmed with Interactive Management logic, allowed Dr. Christakis, Diane, and myself to make three sets of paired comparisons among those standards. The enhancement pattern shows the relative influence that various standards have on each other.

There are six levels of influence in the graphic schema of this enhancement pattern. Level VI is the deepest driver of the process of design; that is, it has the most influence on the processes in the levels above it. Conversely, Level I has little or marginal influence on the processes in the levels below it. Each level from VI up through II influences the processes in the levels above it.

The activities at Levels VI, V, and IV are the deep drivers of the process of design. These activities ground everything in Levels III, II, and I. Solid attention to Comprehensiveness (Level VI), Velocity and Innovation (Level V), and the lack

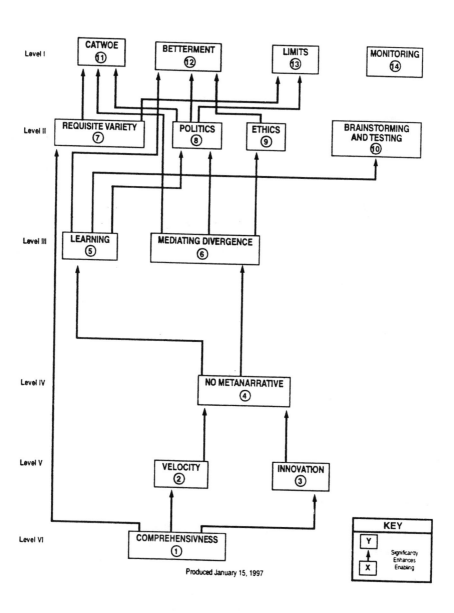

Enhancement Pattern among Design Standards

of Metanarrative (Level IV) provides a secure foundation for a Design that is ethical and practical. Neglect of these foundation activities puts all subsequent designing into question.

Neglecting to involve sufficient stakeholders, points of view, and relevant considerations (Level VI, Box 1) leads to the ethical and practical calamities described by soft and critical systems theorists. Neglecting the limits of human information processing (Level V, Box 2) leads to numbness, "groupthink," inefficient design sessions, and lousy design (cf. Warfield, pp. 37-42). A narrow focus on results and a neglect of creativity (Level V, Box 3) can lead to designs that merely repeat old patterns and fail to produce the innovations that keep an organization vital. Assumptions of consensus regarding lifeworld concerns (metanarrative; Level IV, Box 4) are almost always counterfactual in our postmodernist era. Every successful design has to be aware of dissent and the limited parameters of consent in which it can function.

The activities of Level III, Mediating Divergence and Learning, point to group activities and the goals of design. Divergence in the group, if it is properly managed, creates the likelihood of productive (creative) thinking. A balance of circularity and openness in a system's design keeps on the edge of chaos and stimulates its creativity.

The activities of Level II (Requisite Variety, Politics, Ethics, Brainstorming and Testing) and Level I (CATWOE {customers, agents, transformation process, worldview, owners, and environmental impact}, Betterment, Limits, and Monitoring) are also important. Failure to carry them out can doom a design and its implementation. Still, the activities lower in the enhancement pattern significantly enable the activities above them; and the enabling, for the most part, is not reciprocal.

**LEVEL-BY-LEVEL
AND BOX-BY-BOX ELABORATIONS**

Level VI; Box 1: Comprehensiveness

Authors express this most basic requirement of systems design in various ways. Ulrich states that systems design should look beyond the horizons of systems science into social, ideological, ontological, and epistemological concerns (cf. Ulrich, 1991, p. 247). Churchman says that designs should be as comprehensive as they can be. Flood urges us to mix metaphors in order to obtain a rounded appreciation of situations (cf. Flood, 1990, pp. 80- 81). Luhmann recommends that we welcome and channel the noise that generates social systems (cf. Luhmann, p. 214).

In terms of operational practicality, comprehensiveness demands attention to the social, ideological, ontological, and epistemological parameters of the design situation. In terms of expression, it requires the employment of metaphors that fit the various parameters and explicit recognition of the biases involved in those metaphors. In terms of the orientation of the design situation, it requires an acceptance and adaptability to noise. This noise is generated by expressive stakeholders, divergent views and philosophies, and an awareness focused on the break-offs of any design.

If comprehensiveness is neglected, important considerations may be ignored. Narrow interests can be thrust upon people unknowingly. Relationships can be warped by inappropriate metaphors. Emergent solutions will lack sufficient noise.

Level V; Box 2: Velocity

Two authors stress that information can be received only if it is presented in measured doses. Warfield says, "The rate of presenting information for processing to the human mind should be controlled in order to avoid its overload during the Design Process." Bickerton reminds designers that a difference is information for an organism only if the organism has an ability to react to that difference.

The overriding principle of this box derives from Bateson's dictum: "Information is a difference that makes a difference" (Bateson, 1972, p. 315). Because of our psychological limitations as human subjects, we cannot simultaneously process more than about seven items at one time. If we are consistently presented with more than seven items for consideration, we go into information overload (cf. Warfield, pp. 45-46). In our overloaded condition, differences no longer constitute information for us. Therefore, the rate of presenting information in a design situation needs to be controlled. The use of computer assistance (one of Warfield's Principles) greatly enables the achievement of the standards of this box.

If the velocity of "information" (more precisely, the velocity of differences) is not controlled, the purported receivers of that "information" will not get it. Instead, they will behave precisely as Goertzel, Campanella, and Varela say that they will. They will either file some part or perhaps none of what is presented into their belief system (without questioning it); they will adapt the "information" or ignore it so that it does not disturb their cognitive map; or they will distort or ignore the "information" in a purely pragmatic way. If truly rational consideration (that is, consideration on the basis of reasons, arguments, and facts) is to be given to the decisions involved in design, the velocity of "information" must be regulated. The "bounded rationality" (Warfield, p. 45) of the human organism must be respected.

Level V, Box 3: Innovation

Numerous authors urge the importance of creativity in the design process. Masulli urges us to take advantage of the conditions of non-linearity that may be present in the working regime of a system. Luhmann urges us to recognize and ride the dynamics of interpenetration. Banathy says that we have to conceptually leap out of an existing situation. Churchman asks us to design societies so that people of the future will be able to design their lives in ways that express their own humanity. Allen and Lesser remind us that

> in evolutionary human systems, long-term   success requires two apparently contradictory kinds of behavior: First, the ability to organize one's behavior so as to exploit the information available concerning net benefits, a rational ["Cartesian"] approach . . . Second, the ability to ignore present information and to explore beyond present knowledge, an apparently irrational ["Stochastic"] approach . . . The Cartesians make good use of information, but . . . the Stochasts generate it! . . . Over a longer period the best performance will not come from the most rational but instead from behavior which is a complex compromise (pp. 131-132).

Finally, Treumann and Lakoff alert us to how metaphors create social realities for us. A good new metaphor is a guide for future action.

Bifurcations are sources of creative solutions in messy situations. They come about when systemic processes move far-from-equilibrium. They are both dangerous and wondrous. They can destroy a system or they can raise it to a higher, emergent level of functioning. They arise from the untapped potential of the past. To tap that potential, one orients oneself toward a goal and leaps into the unknown. The direction of the leap might be Churchman's orientation to future generations or Luhmann's orientation to the system/environment difference.

Operationally in the design process, bifurcations come about through the dynamics of interpenetration as described by Luhmann. The people around the table, because of their face-to-face situation, constrain their behavior, and they experience the urge to create some kind of (undetermined) accommodation. As they mix their ideas and cognitive maps in conversations, "aha experiences" occur and agreements are reached. These bifurcations occur as the participants mix rational "Cartesian" and irrational "Stochastic" behaviors under the pressure to do something.

Designers can catalyze bifurcations and abet design dynamics by encouraging design participants to bring their inchoate thoughts into expression. They can encourage the generation of symbols and metaphors that give shape and visibility to structures and processes that would otherwise go unnoticed. Innovative brainstorming and "heartstorming" (Frantz, 1993) of this kind are an integral part of any process that transcends an existing situation.

Level IV; Box 4: No Metanarrative

Luhmann states the importance of this box most directly. He says that we should assume that no consensus exists about norms of rational action or rational values. Numerous other authors allude to this principle. They detail ways to deal with the lack of a metanarrative. Churchman urges us to avoid the environmental fallacy of mistaking the part for the whole. Checkland urges the creation of several root definitions in accord with divergent worldviews. Ulrich says that we should understand how metaphors build biases into systems thinking (cf. Ulrich, 1991, p. 260). He also says that we should use normative grounds to justify the inevitable break-offs in systems design (cf. Ulrich, 1991, pp. 106-107).

Luhmann urges us to incorporate as many conflicts as there are. For that reason, we should welcome and accentuate differences in designing sessions and incorporate them as minority positions in the design. In a design situation lacking in consensus, planning and consensus-formation are two sides of the same coin. We should create believable and credible models (Mitroff); and we should carry on all discussions and decision-making openly and without coercion (Checkland).

Warfield offers a comprehensive list of activities that should be incorporated in the designing process. The following principles guard against a false presumption of consensus.

1.   Groups should do designing.
2.   Design should articulate the dimensions of the problem.
3.   Design methodology should provide constructive group capabilities for generating and structuring ideas, for designing alternatives, and for doing tradeoff analyses.
4.   The group design should not demand infeasible behavior from participants. It should promote full participation; and should provide opportunity for focused group dialog in structuring, designing alternatives, and doing tradeoff analysis.

The standards in this box place systems design squarely in the era of postmodernism. They cycle upon each other in the following way. When we do not presume consensus, we look for ways to manage disagreement. And vice versa, when we accommodate conflicts we are more likely to recognize a lack of consensus. The mindset required for dealing with a lack of consensus requires:

5.   alertness to the biases contained in the metaphors that lie behind any systems design,
6.   awareness that systems design cannot move forward without consensus-formation,
7.   caution against presuming that any partial representation of the situation constitutes the whole of the situation, and
8.   openness to differences and conflicts and the way that we accord them full consideration and representation in the design.

The lack of consensus on values, priorities, norms of truth and behavior, expectations for the future, and the use of competing technical semantics, requires direct attention. That attention is provided by the standards in boxes 5 and 6. Box 5 deals with how to build a learning organization that can deal with the un-certainties of the future. Box 6 offers ways to manage divergence, lack of consensus, and uncertainty in the design situation.

Level III; Box 5: Learning

Some authors emphasize that learning should take place in the designing process. Other authors stress that the end-product system should be able to learn.

Designing Systems That Learn

Allen and Lesser stress the end product. "Design, like evolution, should select for populations with the ability to learn, rather than for populations exhibiting allegedly optimal behaviors." Mitroff says that the design model should be adaptable; its structure and parameters should be easily updated or modified. Banathy urges us to include processes that enable the system to learn from its experience and improve its design over time. One such process is the empowering of stakeholders to modify the design whenever they wish. With such processes in place designed systems will function after the design team has departed.

Goertzel says that productive designs encourage the emergence of new patterns in the mind of the system containing it. For that reason, designers should design dialogical systems that can solve problems by cooperative computation (Goertzel, p. 189). Gell-Mann tells us that all complex adaptive systems, acquire greater complexity through the competitive evolution of their schemata. Kauffman reminds us that design, like selection, is unable to maintain optimal fitness. Designers should settle on some medium fitness, but a fitness that is metastable on the edge of chaos.

Kampis says that adequate models must respect the spontaneity (non-formality) of living systems at all levels of their material functioning. Ackoff says that idealistic designs are not utopian; they are "the most effective ideal-seeking system of which designers can conceive" (Ackoff, p. 107). Jackson says that we should not judge the success of a social theory by criteria of social control (cf. Jackson, 1991, p.121).

Learning in the Designing Process

Other authors talk directly about the conduct of the designing process. Ulrich urges us to be utopian. We should change from today's biological inspired systems concept . . . to a systems concept that inspired by practical philosophy's emancipatory utopia of a community of autonomous and responsible citizens (cf. Ulrich, 1991, p. 260). Banathy asks us to introduce design experimentation in order to resolve issues that emerge in the course of the design. Goertzel urges us to create competition between various positions because this competition creates an evolutionary pressure.

All the standards in Box 5 stress the key element that enables a social system to survive: its ability to learn. They do not derive from the old mechanistic and organic metaphors that stressed equilibrium and homeostasis. Many are in the humanistic tradition of Churchman. They derive in part from the neuro-cybernetic metaphor that was pioneered by Stafford Beer. Some derive, via Luhmann, from the noise theory of Von Foerster. Others derive from the complexity metaphor.

These standards reflect a shift from the standards of a previous era that relied on mechanical or organic models of social systems. The previous standards urged the optimizing of technical operation and the maximizing of social control in an effort to maintain equilibrium or homeostasis. The previous standards still hold for systems, and parts of systems, that are relatively stable and predictable, but they have little utility in situations of human complexity and rapid change.

The new standards deal with the long-term viability of the designed system. They rely upon neuro-cybernetic, self-organizing, and emergent models of social systems. They incorporate years of experience in designing and recent conclusions from cybernetics and complexity theory.

It can be argued that design processes that include the humanistic concerns of Boxes 1, 2, 3, 4, and 5 will design social systems that are able to adapt, learn, and evolve. Social systems, so designed, do not aim for effectiveness in terms of social control or optimal fitness. They balance technical and human concerns, Cartesian and stochastic behavior, different points of view, conflicting strategies, conformity, and a respect for spontaneity. They exist, as a natural result of the democratic processes that create them, on a delicate balance of conflicts, on the edge of chaos, where they learn and evolve on the basis of their own dynamics. Such social systems adapt to novel situations and evolve new patterns of behavior.

Incorporating Conflict

Behind such learning ability lies some incorporation of conflict, competition, cooperative computation, and decentralization of authority within the system. Conflicts and competition create evolutionary pressure for cooperative computation (group effort focused on situation assessment and strategic policy). They create dialogic thought, emergent ideas, and synergistic action. They keep an organization alert to oncoming bifurcations and ready to make the best of them. The key to cooperative computation, of course, is an alert and educated workforce or, in the larger society, a sophisticated citizenry.

In capitalist countries, for example, sophisticated consumers and entrepreneurs do not require micromanagement by designers and governments. Entrepreneurial organizations develop a kind of immune system. As they incorporate conflicts, they create stores of toxins and antitoxins that exist in a stable instability at the edge of chaos. Their metastability enables them to react creatively to threatening situations (infections). Because they (and all social systems) are constantly reproducing themselves, they can readily produce creative ways to cope with dangers and breakdowns.

Also key to the establishment of a learning organization is a clear understanding (by all of the stakeholders) of the dynamics of the situation that is under design. Such understanding would include (a) a clear explanation of the feedback loops that exist among the components of the social system, (b) the identification of key leverage points where those loops can be controlled, (c) the establishment of criteria of management and measurement concerning those leverage points, and (d) ongoing monitoring to attain those criteria.

Learning systems of this kind are idealistic because they respect individual differences and the conflicting values of their participants; in other words, they dare to be dialogical. They are utopian in the operational sense because they embody communities of autonomous and responsible participants. They are not utopian in the sense of envisioning some impossible ideal; they are simply effective ideal-seeking systems.

Level III, Box 6: Mediating Divergence

The standards of this box stress the crucial importance of maintaining both expert argumentation and democratic participation in the designing situation. They concern the tasks of mediation, the orientation of designers, and the ways the designing process should proceed.

The Essential Tasks of Mediation

Ulrich offers an overview of the tasks that need to be accomplished to successfully mediate divergence. He says that we should go beyond the actual state of practical philosophy. We should develop ways to mediate the divergent requirements of cogent argumentation (on the part of the involved) and democratic participation (on the part of the affected; Ulrich, 1991, p. 253).

To this end, we should build our expectations around the cognitive paradigm that centers on our real-life strategies (Campanella, p. 242). We should go beyond mutual acceptance of problem definitions. We should structure the design and its follow up design so that groups achieve mutual appreciation of their cognitive maps.

Habermas stresses that we should consider how to implement the normative meaning of democracy in theory and practice. To that end, we should remain self-reflective with respect to particular and all positions or approaches (Flood, 1990, p. 165). We should also recognize that fragmentary beliefs are not tested on an individual basis. They fit into a system of belief in some way or else they are discarded (Goertzel). Finally, we need to recognize there are two types of changes that determine the future of a system: those brought about by the typical behavior of average components, and the qualitatively different ones brought about by nonaverage components and conditions (Allen and Lesser).

In a design situation that is genuinely participatory, "a key task . . . is to enable the designing community to use the appropriate technical perspective, to understand and respect the personal perspectives of its members, integrate those perspectives, and forge an operational perspective that is acceptable to all" (Banathy, p. 180). Because of the accustomed dominance of the technical perspective, special effort is needed to reconcile the requirements of expertise and democracy. This effort can be abetted by attitudes that recognize the individuality of cognitive maps, elicit consent based upon belief-system conviction, honor value perspectives, respect insecurities, and value dialogue.

The Orientation of Designers

In conducting the designing process we should forego aiming at an objective solution to the problem of boundary judgments, but aim at a critical solution instead (cf. Ulrich, 1991, p. 107). We should recognize and compensate for imbalances in power relationships (Jackson, 1991, p. 130). We should amplify expression and resolution of insecurities (Luhmann). We should refrain from automatically turning practical questions into technical questions (Habermas). We should not confuse non-technical rationality with irrationality (cf. Ulrich, 1991, p. 251).

Designers must specify the boundaries of their design and the degrees of feedforward and feedback that they will accept from the design's environment (Banathy). They should design systems to include dialogicality and circularity. "Dialogicality permits the belief system to adapt to new situations, and circular support structures permit the belief system to ignore new situations. In order to have long-term success, a belief system must carefully balance these two contradictory strategies" (Goertzel, p. 188). These designs should complement the "monological" utilitarian dimension with the communicative dimension of rational practice (cf. Ulrich, 1991, p. 251).

To achieve successful counterpoint between expertise and democracy, however, the polar opposites that exist in the design process need to be honored: the rational and the cognitive paradigm, the technical and the practical approach to problems, the typical and non-average response to situations, and the circularity and dialogicality of creative thinking. These objectives are enabled by the implementation of the more explicitly practical standards in this box. They are also advanced by design processes (such as Interactive Management and the CogniScope^{TM}) that remove blockages of power, linguistic confusion, mental overload, and boredom. In addition, they are achieved because explicit group awareness is focused on them, and the dynamics of interaction therefore generate new emergent solutions.

Ways of Proceeding

Procedurally, the process should orient group members to their underlying values and create a dialogue that leads to a mutual appreciation of cognitive maps. The process should contain specific provisions for learning that both highlight the different ways that group members judge the salience of design options and lead to a consensual accommodation about relative salience. The design process "must incorporate specific provisions for human learning that offer the strong possibility of diminishing significantly the variability in perception of relative saliency of design options" (Warfield, p. 182). Because fragmentary beliefs are tested on the basis of their conformity to existing individual belief systems, designer/stakeholders need to be self-reflective about their decision making. They should not automatically assume that they are acting rationally.

If environmentalists, loggers, politicians, and forestry administrators are trying to work out a design, the relative saliency of several issues needs to be hashed out in open (but structured) debate. Some of these issues are: saving the spotted owl, preserving old forest, protecting the ecology, making profit for contractors, keeping people employed, tax revenues, long-term economic viability, the need for lumber, getting re-elected, clear administrative guidelines, and the need to get along with one another amicably and without violence. The judgments about relative saliency in the paired comparisons of the CogniScope^{TM} methodology are an excellent means to this end. This kind of attention to both the utilitarian and human dimensions of a situation enables stakeholder/designers to develop consensual/critical definitions of boundaries and determinations about the kind of feedback and feedforward they will accept.

In design situations that are structured along the lines of Interactive Management, power imbalances are recognized and largely neutralized for several reasons. First, when decisions are made on the basis of simple paired comparisons, in which human memory is not stressed, information is considered (therefore differences are not mindlessly processed), and judgments are made on the evidence. If the velocity of the "information" is not controlled, participants are forced to use the expedient that they usually use. That is, they slip bits and pieces into their cognitive maps and lose the thread of the argument. Subsequent discussion then becomes emotional and terminates either in confusion or unreflective groupthink. Second, in enforcing separate roles, the influential people (managers, experts, dignitaries) and their organizational or cultural subordinates exist on an equal footing as rational makers of well-defined decisions. Because the powerful people relinquish the role of group-leader to the facilitator, they lose many of their perquisites to shape the discussion, introduce its metaphors, and regulate its decision-making process.

Level II

Many of the standards on levels II and I make the standards in the previous boxes more explicit and operational. They become more important as the processes of completing the design and implementing it come into play. Some of the standards, however, have an ideal quality, especially those regarding ethics and betterment.

Regarding these latter standards, a peculiar realization emerges. The desires to act ethically and to better the human condition are basic motives for entering into the design process in the first place. They appear, however, to have only a light influence in the enhancement pattern. One reason for this anomaly is the largely non-operational character of these standards. They indicate the civic virtues and responsibilities of the designer/stakeholders and the tentatively progressive attention to human concerns that should guide their deliberations.

To an extent these civic virtues are operationalized by the standards in the deeper levels of the enhancement pattern. It is difficult to see how many of the ethical standards, proposed in box #9, would be present at the beginning of the designing process. But there are reasons to believe that some such virtue would exist at the end of a process that incorporates the standards on levels VI through III. Because of their efforts to show mutual respect, attain mutual understanding, cooperate to attain shared goals, and stretch their capabilities to learn and trust, the participants make meaningful human connections in a setting that brings out the best in them.

Level II; Box 7: Requisite Variety

Warfield restates Ashby's principle of requisite variety. The complexity of the design must match the complexity of its target. If the design is too complex or not complex enough, the designer will be frustrated. Mitroff expands the same idea. He says that the model should be complete. It should describe all of the important elements of the system and its relationships. It should be simple. It should depend on few assumptions, have few parameters, and have straightforward relationships.

Luhmann points to an element that must be included in this requisite variety. We should make explicit the criteria that guide the process of negotiating inclusion and exclusion on topical, temporal, and social grounds. Flood points out another element. We should recognize how non-sovereign power relations (Foucault) subjugate knowledges and prescribe which knowledges are operative in a society (cf. Flood, 1990, p. 42).

These standards require that explicit criteria be in place for negotiating inclusion and inclusion on topical, temporal, and social grounds (Ulrich's attention to break-offs could be repeated here). They require explicit attention to the influence of power relations.

To attain these intermediate objectives, the standards in boxes 11 (CATWOE) and 13 (Limits) need to come into play. In addition, many important criteria that have been distilled from the design literature need to be addressed. These requirements for good design have sometimes been enumerated as follows:

- socially desirable
- culturally acceptable
- psychologically nourishing
- economically sustainable
- technologically feasible
- operationally viable
- environmentally friendly
- generationally sensitive.

Obtaining consensus on the relative importance and meaning of these process/product criteria is a major task in the designing process. Also important is the defining of the objectives and operational procedures that further them.

Level II, Box 8: Politics

The standards in this box deal with conflicts that arise between parties who make decisions on the basis of different belief systems. Because of the presence of differing beliefs, we should consider status questions as normative questions in disguise (cf. Flood, 1990, p. 82), and we should use metaphors critically (cf. Flood, 1990, p. 100).

Luhmann would have us recognize the side effects and backlash that our designs occasion, and make appropriate adjustments. We can lessen backlash by including conservative and liberal political considerations in the parameters of the design and in the plans for its implementation. Any agreement made by disparate parties in such a situation is a temporized action that has different meanings for the people reaching agreement. Luhmann would also have us understand that a societal system tries to act rationally from within its world horizon (cognitive map); that is, rationally in terms of information-for (Kampis).

Much of our thinking during the designing process is set on automatic. Varela puts it this way: When we are functioning on the plane where we are making and implementing decisions, knowledge is pure pragmatism. Goertzel says that we do not routinely select and coordinate our beliefs on the basis of logical analysis. To accommodate this conservative bent of our thinking, we must respect continuity with the past (Bridges, p. 32).

Luhmann points out a way to overcome the wasteful unconscious conflict of competing cognitive maps. He asks us to focus on the effect that our decisions have on our environment. He says that planners should use the system/environment difference as a schema for acquiring information (Luhmann, p. 474). By focusing on system-environment interactions, we can control a designed system's environmental impact. We do that by checking the resulting environment's repercussions upon itself (p. 475). In this way, we differentiate actions that can be absorbed by the environment without repercussions, from other actions that do produce repercussions.

People and organizations ordinarily act rationally within the world as they perceive it. They do not, however, enjoy the perspective of the detached observer; they are autopoietically enmeshed in their situations. They do the things that work for them. They do not routinely sort out and select beliefs on the basis of logical analysis; they select beliefs that conform to their existing beliefs. When they make agreements, they commit themselves to a temporized action that may have different meanings for each party involved.

In the actual practice of design, awareness of divergent belief systems leads to explicit attention to politics. Designers acknowledge that the status that they assign to parameters is derived from political decisions. They use metaphors with an awareness of their political impact. They recognize that their decisions will arouse inevitable side effects including backlash. Recognizing the need to obtain consensus, they attend to conservative concerns of social and political continuity and liberal concerns for public opinion, parliamentary discussion, and legitimate binding decisions.

Level II, Box 9: Ethics

Banathy requires that "The product of the system design should include explicitly stated ethical standards as guidelines of system behavior" (Banathy, p. 181). Churchman requires us to make plans and decisions that serve the future of humanity. Other authors offer multiple values and perspectives for evaluating the rightness and wrongness of designs. They include:

- self-realization,
- social responsibility,
- ecological responsibility,
- evolutionary responsibility,
- honoring multilevel values, and
- caring for the whole system.

All of these perspectives are to be honored in the design situation (Banathy). Habermas adds another requirement for responsible designs. They should specify empirically the feedback and control mechanisms that are supposedly at work to keep the system directed toward it goals.

Ethical conduct during the design process "implies balance among personal and moral action, rational and technical inquiry, and just organizational behavior" (Banathy, p. 179). The designing group should be frank, earnest, and on its good behavior. It should work with the realization that:

1. Good designs are the result of good designing.
2. "No one has the right to design social systems for someone else."
3. "A design decision is ethical if the stakeholders (the designing community) give their informed approval."
4. "Participants are not coerced" (Banathy, p. 180).

The stakeholders of the designing group should risk being confrontational, but they should treat each other with civility. The "stakeholders" include everyone who is affected by the design, including its possible victims or people who may incur collateral damage.

The values listed in this box and elsewhere in the enhancement pattern operationalize to some extent other values, like the fostering of freedom, compassion, and social justice, that could be added to a list of ideals. Concern for these values vastly increases the probability that a design will serve the future of humanity.

Level II, Box 10: Brainstorming and Testing

"During the developmental stage, a belief system may be permitted to be unresponsive to test results . . . However, after this initial stage has passed, this should not be considered justified" (Goertzel, p. 185). This standard simply states that irrational, stochastic, and idiosyncratic behavior is encouraged during the brainstorming and hypothesizing stages of design formation, but that the designs must be tested against reality in the later stages of their development.

Level I; Box 11: CATWOE

A key component of the Soft Systems Method of Checkland requires that every root definition in a design model should identify its Customers, Actors, Transformation process, Worldview, Owners, and Environmental constraints. Attention to these key elements informs every element of the design process. For that reason, CATWOE could be located in several boxes of this enhancement pattern. It is placed here as a kind of checklist that ensures a well-rounded design.

Level I, Box 12: Betterment

In the design world of wicked problems, the aim is not to find the truth but to design systems that enhance human betterment and improve human quality (Banathy, p. 31). This standard indicates that design, unlike science, does not aim for truth; instead it tries to better the human condition. In other words, technical correctness and optimization are not the prime criteria for evaluating social systems and their design. The criteria for judging social systems design are: the respect that they show for individuals and divergent viewpoints, their learning ability (viability), and their cultural, spiritual, and economic impacts.

Level I, Box 13: Limits

Warfield urges us to discover the limitations of design in any particular design situation (cf. Warfield, p. 177). He specifies seven critical factors in the design process. If all seven of these critical factors are attended to the designer has some assurance of success. If one critical factor is ignored, one opens the door to failure. "The seven factors are: leadership, financial support, component availability, designer participation, documentation support, and design processes that converge to informed agreement" (Warfield, p. 180).

Mitroff discusses the often-present reality of limited resources. He says that the systems model should restrict itself to a knowledge base that is available and economically feasible. Banathy echoes this thought. He says that no design should incorporate unknown or unusable technologies (Banathy). Warfield also recognizes the need for economy. He says that a designer should fit the design process to the design Target and design situation. The whole panoply of methodology of generic design science does not have to be applied to every design system (Warfield).

These standards enumerate explicit limits that need to be examined in a situation to determine beforehand if a design is feasible. The designing situation must contain adequate leadership, financial support, component availability, designer participation, documentation support, and effective design processes. The design processes should be adequate to, but not excessive of, the requirements of the design target and the design situation. In addition, the model that design produces should be economical in its requirements of an easily accessed knowledge base, and it should not incorporate unknown or unusable technologies.

Level I, Box 14: Monitoring

Evaluative/monitoring processes should be introduced in the course of the design inquiry that introduce corrections in the design whenever indicated (Banathy, p. 192). Processes for evaluating the performance of a design should be a part of the design. Follow-up processes need to be scheduled and performed to monitor the success of the design and introduce corrections when needed.

It is peculiar that this standard, important as it is, is not enhanced by any of the standards below it, and does not enhance any other standard. This peculiarity is probably due to an insufficient description of the preceding standards. In particular, a more complete explanation of what is entailed in a learning organization (as is indicated in the commentary on box #5) would spotlight the importance of monitoring. If that consideration had been present at the time that this enhancement pattern was generated, monitoring would be cycling in the learning box.

SUMMARY EXPLANATION

Every one of the 91 standards for design are important. Many of those standards approach the same aspect of designing, but approach it from a different angle. Some standards have more ramifications, however, than others. The deep drivers in the enhancement pattern have greater impact than the standards of the higher levels. Therefore, the deep drivers have more practical and ethical importance than standards above them, even though they may seem less "practical" or less "ethical." The lines and arrows in the enhancement pattern indicate the influences that exist among the boxes. Thus, for example, Box 6 influences Boxes 8, 9, 10, and 12.

On the basis of their relative influence as it is portrayed in the enhancement pattern, designer/stakeholders may profitably pay explicit attention to its deep drivers at the onset of the design process, to Level III during the formative stages of the design, and to Levels II and I as the design nears completion and implementation begins.

DESIGN AND THE IDEAL SPEECH SITUATION

Habermas voices serious objections concerning the application of systems theory to social processes. He finds a technocratic bias and a mechanistic characterization of society in systems theory that reinforces the power of elites by denying the importance of normative considerations and participatory democracy. Systems theory, as he understands it, undermines democracy by reinforcing the impersonal play of market forces (1989, p. 345). It contributes to the colonization of the lifeworld by promoting exclusively systemic considerations, to the detriment of other, more humanistic, concerns.

The stakeholder/designer enhancement pattern makes it clear that systems theory-as it is applied to design in the schools of soft systems, critical systems, and interactive management-incorporates practices that enable participatory democracy. The standards for design advocated by these schools bolster lifeworld concerns in small group, production-line, boardroom, community organization, national polity, and global policy forums where design and decisions are made. They explicitly address Habermas's concern for the "rightness" of communicative action by their requirements of inclusion, mutual respect, and efficiency.

Soft systems' thinking demonstrates inclusion and mutual respect by developing root definitions (and ensuing designs) for each Worldview that is present among the stakeholder/designers. Critical systems thinking takes explicit notice of the imbalance of power relations that often skews soft systems thinking to favor establishment and management concerns. In particular, Ulrich's attention to the break-offs of design emphasizes that boundary decisions are value decisions in disguise. His suggested programs, for instructing the public about the politics and break-offs of design are operational procedures that empower less influential stakeholders.

Banathy's exposition of a design culture furthers this empowerment thrust by advocating academic attention to the ways that we can create the future. The General Evolutionary Research Group backs his advocacy. It maintains that evolution has become conscious of itself in our age and that cultural cognitive maps working in a bifurcation scenario hold the keys to our future.

The concrete applications of the science of generic design (Warfield), as practiced in interactive management and the CogniScope^{TM}, produce a close approximation of an "ideal speech situation." If the CogniScope^{TM} method, for example, is properly employed, the concerns that are most critical to democratic decision-making, those expressed in the deep drivers of the enhancement pattern (Levels VI, V, IV, and box #6 of Level III), are given efficient and careful consideration.

In a well-run designing process, comprehensiveness (Box #1) is achieved by including representatives of all the stakeholders (customers, victims, actors, owners of the process) and available points of view in the design conversation.

Velocity control (Box #2) is handled with computer assistance to enable multiple rational discussions with limited focus that are joined in chains of long and deep logic. Limiting the scope of a discussion (to paired comparisons, for example) allows discussants to carefully consider their reasons and feelings about an issue.

The interactive management process avoids several possible major miscarriages of democratic decision-making:

- It prevents unreflective cognitive processing. When data is dumped upon people, it is not information; it is just differential overload. This kind of data is ignored, accepted, or opposed on the basis of a merely emotional response. When the velocity of data is controlled, rational consideration is possible.
- The limitation of velocity enables rational decision-making. Therefore, it limits the otherwise overwhelming power of cognitive processing on the basis of cognitive maps. When decisions about the meaning, saliency, or enhancement of small things are made in a calm and unhurried environment, then quick calculations on the basis of prejudice and power relations are replaced with rational decisions that produce limited agreements.
- The process prevents the 'groupthink' that occurs in overly long and disorganized meetings. When people lose track of what is going on and feel

pressed by their biological and social needs to agree with something in order to go home and remain friends, they agree to things without knowing what they are agreeing to.

- It reduces power imbalances on the basis of position and/or expertise among the designer/stakeholders. The influence of positional power is lessened; the power of cognitive maps is weakened; and issues are judged on their merits. When everyone is fully informed about the decisions they are making, the probability for rational and synergistic decisions is greatly increased. In addition, the presences of an independent facilitator, a neutral recorder, and a computer acting as group logic-keeper remove perquisites of power from the control of dominant interests. The power influences from outside the group, in its environment, however, such as money, force, and violence, cannot be controlled by the group process. They can only be negotiated around and influenced by political activity (box 8).
- The limitation of discussion to many rather minute decisions does not produce mere atoms of agreement, because of the computer-assistance. The computer arranges those decisions in trees and webs of long and deep logic that surpasses the logic available in the individual cognitive maps of the discussants.
- Innovation (Box #3) is encouraged not only by the well-tested brainstorming and Nominal Group techniques, but by the rationality encouraged by Box #2, which brings active minds into synergy.
- The lack of metanarrative (Box #4) is acknowledged in the whole interactive management process. It is brought to the focal awareness of the designing group especially in the paired comparisons that rank goals, objectives, and standards on the basis of their saliency.
- The CogniScope™ manages diversity (Box #6) not only through the respectful attention to diverse and divergent opinions, but also through the processes already described, which make differences explicit *and* alleviate them.

SUMMARY

In summary, systems design, as the authors under consideration propose it and attempt to democratically practice it, is not at all the enemy of lifeworld concerns. It is, to the contrary, a stalwart advocate of practical participatory democracy. The spread of systemic design- and decision-making, on the model of the enhancement pattern, holds great promise for producing both more practical/ethical social systems and more comprehensive scientific theory.

Even neuro-cybernetic methods of systems design (for example, Beer's Viable Systems Model) are not the enemies of lifeworld concerns. By enabling systems to monitor leverage points and feedback loops, they provide an early "heads up" to on-coming bifurcations. The cybernetic methods that have been developed in industrial, commercial, and information technological applications are, in principle, available for the design of our future lifeworld. In practice, the most notable effort in this regard was the cybernetic intervention in Chile from 1970-73

(Espejo, 1990; Beer, 1979). The merely partial success this project (or of the Forrester projections published by the Club of Rome) should not deter more sophisticated ventures that build upon the successes of cybernetic interventions in business management..

A process of decision-making that exhibits the standards incorporated into this enhancement pattern, especially those of levels VI, V, IV, and III, exhibits traits of an "ideal speech situation." The sessions with the CogniScope™ described in this chapter, had they been done with real (not virtual) authors, would have obviated (in the manner already described) many of the power relations that might otherwise have arisen. They would also have reduced the limitations of memory, language, recording, and logic (addressed by Warfield). What would one call a conversation about a complicated, fast-changing situation among people of different disciplines, power-bases, and worldviews, if it worked? Would one call it an approximation of an "ideal speech situation"?

The next four chapters apply important theoretical concepts of systems theory to social processes. The next chapter examines how systems theory views the social world. It concludes that the social world is a pattern of processes in continual reproduction. The subsequent chapters treat some special social processes: communication, cognition, and epistemology.

CHAPTER 19

THE STRUCTURE
OF THE SOCIAL WORLD

ABSTRACT

Habermas, Luhmann, and the General Evolution Research Group offer models of society based upon different premises. Habermas presents a model of society based upon the allegedly universal pragmatics of language. Luhmann portrays society as a temporized entity, that is, as a system whose events disappear from moment to moment, whose only enduring structure is its constant self-reproduction in the autopoiesis of communication. The Group's model also portrays sociocultural groups as temporized entities, but it postulates that they reproduce and evolve through the dynamics of their cultural cognitive maps.

In addition to models that relate specifically to social systems, several models represent self-organizing systems of which society is seen as a special example. Csanyi describes the nature and formation of component-systems. Maturana and Varela describe autopoietic systems from a predominantly biological viewpoint. Goertzel creates a mathematical model of mind and belief systems. Kauffman describes the basic mathematics of complex systems. These authors extend the concepts of non-equilibrium thermodynamics, hypercycles, complexity, component-systems, and autopoiesis to social reality.

In an abstract sense, the concepts mentioned above are metaphors and models even in the fields in which they find their primary designees. They are doubly metaphorical in their application to social processes, but as metaphors they are no more exceptional than the familiar appellations of social processes as "homeostatic," "orderly," or "dialectic." The validity of these metaphorical extensions is to be judged on the basis of how well they explain social reality.

Judgments of their validity must await consideration of how they explain social processes and the utility of their explanations.

METHODOLOGICAL CONSIDERATIONS

As previously explained, I precipitated key statements from the writings of the authors under consideration. I reduced the statements concerning the structure of the social world to 57 prototype standards. I sorted these fifty-seven into dimensions with the aid of the CogniScope™ and portrayed them in an Attributes Field [see preceding pages]. The computer assistance in this first sorting was of crucial importance. It allowed me to place some preliminary order into the confusing welter of statements that differed in perspective, context, and language.

Using the graphic potential of this field, I was able to organize those fifty-seven statements into four Attribute Profiles: Discourse/Lifeworld, Autopoiesis, Component-Systems, and Reproduction. These profiles contained the ideas of the various authors in much their own words. In these profiles, I combined statements across the boundaries of dimensions, perspectives, and authors. If a statement of Luhmann buttressed a statement of Habermas, for example, it was placed within the same profile.

Subsequently, I used techniques of imaginative variation, which I borrowed from phenomenology, to construct narratives for these profiles. The end products are a lifeworld/discourse narrative, an autopoiesis narrative, and a component-system narrative. In this process, the reproduction profile was eliminated as it was incorporated into those other narratives. The processes of generating standards, dimensions, Attributes Field, and Attributes Profiles are fully described in a previous publication (Bausch, 1999).

CLUSTER #1 --
TRADITIONAL

- **(Standard 1)** Social systems are open systems that perform functions on inputs and outputs. They are stable things that manipulate things external to themselves; or they are some "whole" equipped with static and variable "qualities" in which the static qualities are the essence of the whole (structural functionalism).

- **(Standard 37)** The semantics of cognitions and norms become ontologized in our use of the words "is" and "ought" (Luhmann).

CLUSTER #2 --
EVOLUTION
THRU DISCOURSE

- **(Standard 2)** The norms and structures of society devolve from the nature of discourse (Habermas).

- **(Standard 3)** Language and the social order depend, often counterfactually, upon the truth, truthfulness, and rightness of discourse (Habermas).

- **(Standard 4)** The structures of the lifeworld have evolved through the linguistification of the sacred (Habermas).

- **(Standard 5)** The rules and structures of the lifeworld are conventional but not arbitrary (Habermas).

- **(Standard 7)** Media such as money and power are institutionalized in the lifeworld (Habermas).

- **(Standard 11)** Human social structures are built upon the social structures and accomplishments of non-human species (Bickerton).

- **(Standard 16)** Models based on grammars afford an underlying account of the ways couplings occur among symbol strings. Thus an entity such as an autocatalytic set of symbol strings announces its functional integration! The members of the system collectively make one another (Kauffman, p. 371).

CLUSTER #3 --
PRESCRIPTIVE/
IDEAL

- **(Standard 6)** The ideal resolution of the tension between lifeworld and systems requires that systems fulfill conditions for the maintenance of sociocultural lifeworlds (Habermas).

- **(Standard 34)** The expectations of person, role, program, and value increase the range of choice of members of society. In modern society, freedom and constraint increase together; the greater the types of constraints, the more choice about how to respond to them (Luhmann).

- **(Standard 35)** Progress for autopoietic systems involves increasing their levels of acceptable insecurity. It requires modalizations: cognitive and normative methods for dealing with disappointment (Luhmann).

Attributes Field for the Structure of the Social World

CLUSTER #4 -- REPRODUCTION

- **(Standard 8)** The interaction of cognitive maps and cultural change follows the patterns of bifurcation processes (Masulli).

- **(Standard 13)** Theories of dynamics, thermodynamics, and complexity cannot explain the spontaneous innovations of life and thought (Kampis).

- **(Standard 14)** Component-systems do not endure as "things"; they are genesis machines that string together momentary presences (Kampis).

- **(Standard 17)** Historical, social, psychological, and evolutionary processes exhibit many similarities. They are all loci "of law, accident, design, selection, ever unfolding and transforming in novel functionally integrated forms" (Kauffman, p. 403).

- **(Standard 22)** Autopoietic systems have to make selections from the complexity of their environments (Luhmann).

- **(Standard 23)** Autopoietic systems manage selection process through a system of graduated expectations which they develop through processes of trial and error, memory, and adaptation (Luhmann).

- **(Standard 24)** Social systems reproduce themselves using a strategy of expectations (Luhmann).

- **(Standard 25)** A social system uses its moment-to-moment dissolution to create greater organized complexity (Luhmann).

- **(Standard 26)** Expectations are crucial functions of autopoietic systems. They constrain a system's search for possible new connections and thereby shape its structure of self-reproduction (Luhmann).

- **(Standard 48)** The mind is a structured transformation system (STS) which occurs as the brain transforms its massive parallel processing into a virtual serial processor in order to cope with linear tasks. It functions as a blueprint-generating machine (Goertzel).

- **(Standard 49)** Language is an STS functioning in the social realm. It is a serial processor governed by the dual network of syntax and semantics (Goertzel).

CLUSTER #5 -- COMPONENT SYSTEMS

- **(Standard 9)** Societies as component systems have (1) physical components that are constantly assembling and disassembling, and (2) energy fluxes running through them that can excite some of their components (Csanyi).

- **(Standard 15)** The coherence and endurance of its pattern of information is the essence of a component-system (Kampis).

- **(Standard 18)** Complex adaptive systems acquire greater complexity through the competitive evolution of their schemata which can be embodied in atomic, molecular, neural, or communicative media (Gell-Mann).

- **(Standard 44)** Mind is a patterned process of relations between physical entities, but it is not itself a physical entity (Goertzel).

- **(Standard 45)** Mind has two components: a perceptual-motor hierarchy and a structurally associative memory (Goertzel).

- **(Standard 54)** Component-systems have two survival stratagems: circularity and dialogue. To be successful, they must use both of them (Goertzel).

- **(Standard 56)** The cognitive equation, in the context of self and reality as belief systems, becomes a universal equation.

Attributes Field for the Structure of the Social World (cont.)

CLUSTER #6 --
SELF-ORGANIZATION

- (Standard 10) Societies, as autopoietic unities, do not require teleonomic concepts of information and function; such concepts are applied to these unities when we observe what they do in their temporal and spatial environments (Maturana and Varela).

- (Standard 12) Societies are third-order structural couplings formed by the coordination of social behaviors (Maturana and Varela).

- (Standard 28) Through risky sharing and mutual adjustment of expectations, a structure of mutual expectancy is differentiated out in the domain of double contingency (Luhmann).

- (Standard 39) Structural formation is not preformed in principle, nor does it occur according to objective historical laws that establish how state A is transformed into state B. Instead, the decisive point is the translation of problems in system formation into differences. If a decisive point is reached, order emerges out of chance events in the course of time (Luhmann).

- (Standard 43) Language, media of dissemination, and symbolically generalized communication media interdependently ground the processing of information and increase what can be produced by social communication. This is how society produces and reproduces itself as a social system (Luhmann, p. 162).

- (Standard 50) A self-generating system is a process in two stages: universal action and transformation (Goertzel).

- (Standard 51) According to the cognitive equation, mind at time t-1 commingles its elements with elements from its environment. Mind at time t recognizes and maintains the patterns that it discerns in that commingling and turns these patterns into its time t elements which are again commingled with environmental elements (Goertzel).

- (Standard 52) Belief systems are learning systems that obey the cognitive equation (Goertzel).

CLUSTER #7 --
INTERPENETRATION
SYSTEMS

- (Standard 19) Living systems, psychic systems, interaction systems, and society itself are the autopoietic systems that inhabit the social world (Luhmann).

- (Standard 20) Autopoietic systems exist as environment to each other; they influence each other through interpenetration (Luhmann).

- (Standard 21) Autopoietic systems of the same type communicate under the conditions of double contingency (Luhmann).

- (Standard 32) Conceptualizations based on the thing, and especially on that special thing, the "human being," no longer suffice (Luhmann).

- (Standard 55) Self and external reality are belief systems that mutually generate one another as the inside and outside of experience. Together they constitute the universe. (Goertzel).

- (Standard 57) A thing exists in collective reality if people recognize it as real and recognize that their collectivity also recognizes it as real (Goertzel).

Attributes Field for the Structure of the Social World (cont.)

332

Chapter 19

CLUSTER #8 – MODALIZATION

- **(Standard 27)** In social situations, expectations have to be reflexive (Luhmann).
- **(Standard 29)** Progress into insecurity is made efficient by the use of "symbolic abbreviations representing highly complex expectational situations" (Luhmann, p. 306), whole categories of do's and don'ts (Luhmann).
- **(Standard 30)** By making its expectations ambiguous and tentative a system takes some control of its situation (Luhmann).
- **(Standard 31)** Psychic and social systems give their expectation endurance by attaching their expectations to something that is not an event. They attach their expectations to names; then they factually order those names, establishing connections and distinctions (Luhmann).
- **(Standard 33)** Person, role, program, and value provide a graduated scale for assessing the expectations that are put on human behavior. They enlarge the range of expectations that we can safely have in the social arena, especially by allowing us to function together without requiring value-conformity of one another (Luhmann).

- **(Standard 36)** Modalization refers to the way one responds to disappointment. It is insurance in the form of risk-assessment and contingency-planning (Luhmann).
- **(Standard 38)** Through cognitive and normative modalizations of our expectations, we form generalized contingency plans that enable us to harbor highly improbable expectations (Luhmann).
- **(Standard 40)** Norms are derived from society's need to attain security for its expectations. They come into demand and are generated as counterfactually necessary (Luhmann).
- **(Standard 41)** Norms have a certain independence from actual existing laws and customs, just as those laws and customs have a certain generalized independence from the events to which they refer (Luhmann).
- **(Standard 42)** Symbolically generalized media, like "truth, love, property/money, power/law; and also, in rudimentary form, religious belief, art, and today standardized 'basic values'" (p. 161), help condition the selection of communication so that it can achieve acceptance (Luhmann).

CLUSTER #9 – DUAL NETWORK

- **(Standard 46)** The hierarchical network supplies the perceptual/motor engine (Goertzel).
- **(Standard 47)** The heterarchical (memory) network provides the memory and the capability of analogical reasoning (Goertzel).
- **(Standard 53)** Belief systems use the semantics of language, but they add their own special syntax (Goertzel).

Attributes Field for the Structure of the Social World (cont.)

COMMENTS ON THE ATTRIBUTES FIELD

The presence of different perspectives in the same dimension of the attributes field requires some explanation.

Cluster #2 presents the expected Habermasian emphasis upon the evolution of society through discourse, but it points backward to the primitive organizing power of/ communication among non-human species and even among polymers. The confluence of these ideas opens the possibility that the theory of communicative action might be incorporated in an overall theory of evolution as communicative action.

Cluster #3 contrasts the prescriptive ideals of Habermas and Luhmann. Habermas wants to resolve the tension between lifeworld and system by giving the lifeworld priority. Luhmann wants to maintain the ideal of systemic independence. This difference in prescriptive ideals mirrors the differences between a consciously ordered society and a society that trusts the interplay of impersonal forces. It also contrasts the necessary components of a productive belief system: logic and chaos. Habermas positions himself on the side of logic, practical rationality, and reconstructive science. He prescribes an ideal speech situation, in which people make decisions on the basis of the best argument, that is, on logical and rational grounds. He derives practical rules and norms from the nature of discourse that are quasi-transcendently necessary in the Kantian sense; that is, they are norms that are necessary for the very existence of communicative action. He prescribes a societal ideal in which lifeworld concerns for justice, democracy, and humanism control the impersonal forces of economic and bureaucratic power.

Luhmann positions himself on the side of indeterminacy, noise and *laissez faire*. He trusts the chaotic interplay of systemic social conflicts that "draws in chance straight-away, creates sensitivity to chance, and when no value consensus exists, one can thereby invent it. "The system emerges *etsi non daretur Deus* [even if God doesn't exist]" (1995, p.105). With the noise of incorporated contradictions, psychic and social systems maintain a delicate homeostasis. They increase their arena of freedom, and generate solutions for novel situations (cf. Luhmann, 1995, p. 306).

Cluster #5 deals with component-systems not only in the thought of Csanyi and Kampis, but also in the thought of Gell-Mann and Goertzel. It relates component-systems to cognitive maps in the thought of Masulli and Treumann. It also forges a connection between Luhmann's autopoiesis and component-systems.

The standards of **cluster #6** are closely related to those in the previous cluster. In these standards, authors proclaim that self-organization is endemic to chaotic systems and they detail several of the ways that chaotic systems manage their survival and evolution. The authors represented in this cluster agree that systems self-organize. Nevertheless, they have various interpretations and names for self-organization: autopoiesis, component-systems, self-modifying systems, self-generating systems, chaotic dynamical systems, complex adaptive systems, and nonlinear dynamic systems.

Cluster #7 deals with systems and how they communicate with each other. The systems include the four autopoietic systems of Luhmann: vital, psychic, interactions, and society. They also include the lifeworlds of Habermas, the belief systems of Goertzel, and the cultural cognitive maps of the Group.

Luhmann's four systems are autopoietic. Lifeworlds, belief systems, and cultural cognitive maps are component-systems. The relationship between autopoietic and component-systems requires some clarification. They may indeed be different names (and perspectives) that describe the same vital, psychic, and social realities. The likelihood of this confluence of ideas is enhanced by the concept of a belief system, which is defined as a subset of language (Luhmann's society) that shares its semantics but uses a different syntax (Goertzel).

According to Luhmann (1990, pp. 3-5), society is coterminous with human communication. As an all-inclusive entity, it cannot enter into communication with psychic and interaction systems; it can only interpenetrate (or structurally couple) with them; and them with it. For this reason, society is somewhat insulated from individual or collective influences. It is a self-referential system that develops autonomously just like any other autopoietic system.

If collective belief systems, cultural cognitive maps, and lifeworlds, however, do indeed constitute component- and/or autopoietic systems, then "society" assumes its everyday denotation. Then there are a multitude of "societies" and not just the overarching autopoietic "society." If a collective belief system is a component-system, then the interaction between it, individual initiative, and concerted group effort becomes the kind of rational and cognitive interaction that we normally call communication. In this eventuality, the realms of science, ethics, and politics (the traditional lifeworld concerns) are radically dependent on lifeworld activities of individuals, interactions, and concerted group effort.

Cluster #8 is all Luhmann except for one entry by Habermas, which states that strategic activity in the lifeworld generates systems of law and money that expeditiously handle conflict resolution and economic activity. For Luhmann, modalizations of expectation are the structures of the social world. Habermas appropriates only the media modalizations of money and power from Luhmann's systems theory, and he uses them only as they standardize and depersonalize communicative action.

THREE NARRATIVES

In this chapter, I omit the mechanics of how attribute profiles were generated and manipulated. I present only the three narrative explanations of the social world. The discourse/lifeworld narrative tells how society develops from communication. The autopoiesis narrative tells how systems structure their reproduction. The component-system narrative weaves together themes from thermodynamics, complexity and information theory, cognitive maps, and the cognitive equation.

The Discourse/Lifeworld Story

Habermas is the chief proponent of communicative action. He proposes that the norms and structures of society evolve, in part, from the nature of discourse. He argues that language and the social order depend, often counterfactually, upon the truth, truthfulness, and rightness of discourse.

In the tradition of Mead and Durkheim, Habermas finds the constituting origins of societal structure in the awesome (holy) events that a societal group experiences early in its history. When the group remembers those experiences and re-enacts them, it constitutes itself as a people. Over time, re-enactments become myth, ritual, and taboo. In the course of history, myth morphs sequentially into religion, into theology, into philosophy, into science, and even into postmodernism. Ritual transforms into the structural processes of familial, tribal, national, and global lifeworlds. These processes eventually generate systemic media that coordinate social transactions. Taboos become the conventional but not arbitrary rules that govern social interactions. Habermas calls this process the linguistification of the sacred.

Other writers share the view that the social world is generated through evolving processes of discourse. Maturana and Varela hold that societies are third-order structural couplings formed by the coordination of social behaviors. In this process, individuals and societies co-constitute each other in human and animal groups. Bickerton agrees. He proposes that human social structures are built upon the social structures and accomplishments of non-human species. Kauffman traces the evolution of order all the way back to the random grammars of polymeric strings. He says, "Models based on grammars afford an underlying account of the ways couplings occur among symbol strings. Thus an entity such as an autocatalytic set of symbol strings announces its functional integration! The members of the system collectively make one another" (Kauffman, p. 371).

Goertzel supports the discourse/lifeworld model within his overall conception of the mind. He proposes that the mind self-organizes into a dual structure of perceptual/motor hierarchy and associative memory (a structured transformation system: STS) in constant iterations of the cognitive equation. He finds similar STSs to exist in the social world. For him, language is an STS functioning in the social realm. It is a serial processor governed by the dual network of syntax and semantics. Belief systems are also STSs. They use the semantics of language, but they add their own special syntax. An individual belief system becomes collective if people share it. In other words, a thing exists in collective reality if people recognize it as real and recognize that their collectivity also recognizes it as real. Furthermore, self and external reality are belief systems that mutually generate one another as the inside and outside of experience. Together they constitute the universe. The cognitive equation becomes a universal equation.

The cognitive equation is an abstract formulation for constant internal communication. This communication is externalized in language. It is manifested in the semantic structures that organize our social lives.   Goertzel proposes a mathematical formulation of how discourse creates our lifeworlds.

Luhmann's ideas are not wholly outside the discourse/lifeworld context. He holds that autopoietic systems exist as environment to each other; they influence each other through interpenetration. He also holds that autopoietic systems of the same type communicate under the conditions of double contingency.

Habermas acknowledges that lifeworld discourse is inadequate to explain the complexity of modern societal interaction. He acknowledges that the systemic media of money and power are institutionalized in the lifeworld. In his overall schema, these media generate a tension between lifeworld and systems.   The resolution of this tension requires, for Habermas, that systems fulfill conditions for the maintenance of sociocultural lifeworlds.

The Autopoiesis Story

There is a basic tension in the understanding of social autopoiesis as originally proposed by Maturana and Varela, and as interpreted by Luhmann. Maturana and Varela consider societies to be third-order structural couplings formed by the coordination of social behaviors. For Luhmann, societal systems are not structural couplings of individuals. They are independent autopoietic systems constituted by communication. The overall description of autopoiesis, however, with the one exception just mentioned, is generally consistent in the renderings of Maturana, Varela, and Luhmann.

All self-organizing systems maintain themselves by means of expectations. Kauffman conceives of polymeric "agents" creating models of one another in an attempt to predict one another's behavior. In a similar way, Luhmann states that autopoietic systems, in their ongoing reproductive processes, make selections that maintain and expand their meaning (their organized complexity). They manage this selection process through a system of graduated expectations that they develop in processes of trial and error, memory, and adaptation.

Expectations are crucial functions of autopoietic systems. They constrain a system's search for possible new connections and thereby shape its structure of self-reproduction. Because of their expectations, social systems reproduce themselves in a manner that embodies constraint and openness. In turn, this modulated self-reproduction provides invariance in time for a system by constraining its selections to those that have survival potential. In their ongoing performance of this reproductive function, expectations provide a sense of self. They also provide flexible patterned structures that equip social systems to perform structural changes upon themselves without lapsing into either rigidity or chaos.

In social situations, expectations have to be reflexive because autopoietic systems of the same type communicate under the conditions of double contingency. In this domain of double contingency, a structure of mutual expectancy is differentiated out through risky sharing and mutual adjustment of expectations.

Participants in this domain lessen their risk by making their expectations ambiguous and tentative. In this way, they take some control of their situations.

Psychic and social systems give their expectations endurance by attaching them to names; then they factually order those names, establishing connections and distinctions. By using these names for "highly complex expectational situations" (Luhmann, p. 306), societies create norms, whole categories of do's and don'ts. In this way, they facilitate their progress into mutual respect and attain security for their expectations. They generate these norms in the press of difficult social bargaining even when the norms are counterfactual to actual societal conduct.

Autopoietic systems increase their levels of acceptable insecurity in order to progress and increase the range of choice of their members. They make the insecurity bearable by using the expectations of person, role, program, and value. These expectations increase constraints on individuals *and* give them freedom. In modern society, freedom and constraint increase together; the greater the types of constraints (in the form of expectations), the more choice is had by societal members about how to respond to them.

In dealing with their many expectations, societies and individuals develop generalized contingency plans that enable them to harbor improbable expectations. Using cognitive and normative strategies for dealing with disappointment, they create social structures that intertwine risk and security. Their cognitive openness to new information embodies their acceptance of risk. Their normative rejection of unacceptable and dangerous information expresses their regard for security. Normative behavior reflects the necessary circularity of living systems. Risky behavior reflects the asymmetry that is necessary for growth.

Structure is subsidiary and consequent to the way self-referential systems maintain themselves in time. Structural formation is not preformed in principle, nor does it occur according to objective historical laws that establish how state A is transformed into state B. Instead, the decisive point is the translation of problems in system formation into differences and conflicts. If a decisive degree of turbulence is reached, new structures emerge.

Systemic media have emerged in this way to meet society's need for more rapid and universal communication. Symbolically generalized media, like "truth, love, property/money, power/law; and also, in rudimentary form, religious belief, art, and today standardized 'basic values'" (Luhmann, p. 161), provide acceptable forms for easy communication. Language, mass media, and symbolic media (like money) interdependently ground the processing of information and increase what can be produced by social communication.

Habermas agrees with Luhmann to this point. He says that the evolving complexity of strategic activity generates systems of law and money in order to expeditiously handle conflict resolution and economic activity. He would not, presumably, agree with the following assertion, that the generation and maintenance of media of communication are how society produces and reproduces itself as a social system (cf. Luhmann, p. 162).

The Component-System Story

Csanyi says that the laws of energy conservation and irreversible thermodynamics express the necessary but *not sufficient* conditions for the origin of living systems. Kampis adds that theories of dynamics, thermodynamics, and complexity cannot explain the spontaneous innovations of life and thought. To explain component-systems and life, the circularity of reproductive existence working through material implications must be added to the principles of far-from-equilibrium thermodynamics.

Component-systems do not endure as "things"; they are genesis machines that string together momentary presences (Kampis). In this, they resemble autopoietic systems as described by Luhmann. He says that social elements are action/events of minimal duration, whose embodied relationships also have minimal duration. Therefore, autopoietic systems do not endure as structures of elements because they remain the same in that way only for the duration of an event. They endure through their structures of self-reproduction. For them, the classical dichotomy between static and dynamic systems is voided.

The constraints upon change are themselves changeable; they are the autopoietic structures of reproduction. These reproductive structures maintain the relationship over time between a system's relations and the elements of those relations (Luhmann). Goertzel elaborates on this idea by saying that component-systems are not mere blueprints of their living structure; they are also blueprint-generating machines. He explains how this works in the cognitive equation: Mind at time $t - 1$ commingles its elements with elements from its environment. Mind at time t recognizes and maintains the patterns that it discerns in that commingling and turns these patterns into its time t elements. At time $t + 1$, mind again commingles its elements with environmental elements, and so on.

Kampis says that the coherent endurance of its pattern of information is the essence of a component-system. Other authors say the same thing in different words. Luhmann says that structure maintains the relationship over time between a system's relations and the elements of those relations, and structure exists on a level of active patterned information. Treumann says that the evolution of cognitive maps depends on their quantity of information. Gell-Mann says that complex adaptive systems acquire greater complexity through the competitive evolution of their schemata embodied in atomic, molecular, neural, or communicative media. Goertzel bases his whole systemic view of mind and reality on patterns of information.

According to Goertzel, belief systems are learning systems that obey the cognitive equation. They have two survival stratagems: circularity and dialogue. To be successful, they must use both. In this, he is in agreement with Luhmann who says that social systems reproduce themselves in a manner that embodies constraint and openness. He is also in agreement with Masulli who says that the interaction of cognitive maps and cultural change follows the patterns of bifurcation processes. In the same vein, Treumann says that the evolution of cognitive maps is an instance of deterministic chaos that is explained by information dynamics. The agreement between Goertzel, Luhmann, and these members of the Group is made obvious if we realize that a society's daring to dialogue (be open) is the act that makes it asymmetrical and pushes it far from equilibrium. It is in this way that a self-generating system is a process in two stages: universal action and transformation (Goertzel).

REFLECTIONS ON THE NARRATIVES

Lifeworld

The lifeworld perspective can be expanded in at least two ways. First, we can look to the evolutionary beginnings of communication and language. With Kauffman, Kampis, and others, we can trace these beginnings to the grammars embodied in polymer communication. With Maturana, Varela, and Bickerton, we can trace the development of communicative and/or representational power in the animal kingdom.

Second, in the ongoing development of the lifeworld, we can expand beyond development based on rational decision-making, which is emphasized in Habermas's account, especially in regard to the ideal speech situation. We can give explicit recognition to the cognitive paradigm that Campanella describes (1993):

> The cognitive paradigm emphasizes unexpected factors, pertaining to the subject and to the environment of the action, and these vanish from view when operating under the assumption of rationality and consistency. The cognitive paradigm is centered on human cognition. The actor is not at rest but in action . . . The proponents of the cognitive approach and the cognitive mapping model consider the computational capabilities of an agent as a core problem that differentiates their approach from the classical rational paradigm . . . The cognitive paradigm considers more closely the ever-changing situation of the actor-in action, while the rational paradigm demands steady-state situations. If the rational paradigm is able to resolve problems under selected conditions of space and time, the cognitive paradigm copes with fuzzy gamble situations (pp. 242 -244).

By employing this version of the cognitive paradigm and its underlying concepts of cultural cognitive maps and collective belief systems, we open new avenues for explaining development. As societies accommodate the autopoietic viewpoints of individuals, they modify their collective belief systems. They progress in a stochastic way, but they retain the guiding forces of idiosyncratic (autopoietic) reason and reality testing. In this way, individual idealism and risk-taking influence collective belief systems and reconstitute socio-cultural identities.

If we use this cognitive approach, we restore theoretical vitality to the lifeworld story by incorporating the autopoietic viewpoint into it. One no longer sees society using more or less objective criteria for practical decision-making. One realizes that rational decision-making as an exception rather than the rule. One sees discourses pulsing with the push and pull of dissonant cognitive maps. One expects decisions to be made on the basis of a logic that is interior to an individual or group, on the basis of subjective and parochial belief systems. Such decisions are made often on a feeling of rightness that does not always rise to conscious consideration. With this cognitive approach, one generates a multi-centered perspective that strives for an inside understanding of how individuals and collectivities rely upon their ideals, pride, and strivings for autonomy-and how they are working out their survival.

Autopoiesis

The idea of autopoiesis, as a closed cycle of self-reproduction in which systems survive and progress by structurally coupling with their environments, is a major catalyst of much present-day systems thinking. This idea sparked Luhmann to his conception of society as an autopoietic system of communication. In Luhmann's version, society is not the third-order autopoietic unity that is described by Maturana and Varela. It is a first-order autopoietic system that is formally defined as a system of communication. The advantage of this choice of Luhmann is that it allows him to streamline his richly articulated explanation of social systems. The disadvantage is the removal of real organisms, real people, and their material interactions from "society."

A further theoretical split between Maturana/Varela and Luhmann follows from this decision of Luhmann. Maturana and Varela carefully craft a model of biological phenomenology in order to maintain the inside, autopoietic viewpoint. Luhmann switches between the autopoietic viewpoint and the viewpoint of the detached, objective observer.

With the idea that autopoietic systems use expectations to select from the complexity of their environment, Luhmann reconstructs the interior phenomenology of autopoietic systems. In this description he relies on our own experience of complexity, information, and meaning. In this sense, he portrays a view from the viewpoint of an autopoietic system.

In a similar way, Luhmann reconstructs the difficulty that autopoietic systems have in constructing reliable expectations of each other from our own experience. When we try to predict each other's behavior, we discover that we are often notoriously unsuccessful. From our frustration, we conclude that the actions of the other party are contingent; that is, they can go one way or another. Upon reflection, we conclude that our actions are also contingent. Because of double contingency, we grant independent free action to others and to ourselves. In Luhmann's conception, we mutually create ourselves and others as persons. Luhmann generalizes this experience to apply to all autopoietic systems with the parable of the black boxes, which was told earlier.

With the problem of double contingency and its resolution, Luhmann provides an interior standpoint for viewing the organic growth of the modalizations in the lifeworld. Modalizations are strategic forms of risk assessment and contingency planning. They allow systems to make their expectations ambiguous and tentative in highly complex situations by using "symbolic abbreviations." In other words, they are concepts and names that we create to give permanence and security to the ever-reproducing events that underlie all physical, organic, psychic, and social phenomena. The name/modalizations that are identified by Luhmann include "person," which has already been discussed. They also include: role, program, and value, especially the values of norms and cognitions. These modalizations are structures of the lifeworld.

To this point, Luhmann follows the consequences of his adopted autopoietic viewpoint taken from Maturana and Varela. In his grand scheme of four classes of autopoietic systems (vital, psychic, interaction, and society), however, Luhmann departs from a phenomenological approach and adopts the position of the objective, dispassionate academic observer. He simply declares that these are the only kinds of autopoietic systems that exist.

This claim creates a chasm in Luhmann's social world. One half of his social world consists of "atoms" and "molecules" of communication (i.e., interactions which are temporally and spatially very limited). The other half is the universe of communication (that is, society that is coterminous with human communication). The whole interior spectrum of communication (the lifeworld) is absent. In his ensuing theory, the social system differentiates itself according to its rules of evolutionary logic into self-sufficient subsystems, such as monetary systems or systems of positive law. It develops without direction from the lifeworld forces of individual and collective initiative.

Luhmann does allow interpenetration (structural coupling) between lifeworld systems and society as a whole. He explains how interaction systems and society mutually influence each other. He does not, however, seem to leave room for collective belief systems and organized groups that transcend individual interactions. He does not believe that belief systems and organizations can shape their societies.

Lifeworld and Autopoiesis

In the theories of lifeworld and autopoiesis as I have presented them, a subtle double switch occurs, which needs to be spotlighted. Habermas describes communication as a dispassionate academic observer. He develops his theory of communicative action with careful attention to detail; he provides structure for his theory by reconstructing the thought of Weber, Marx, Mead, and Durkheim. Through his method of scientific reconstruction, he gains distance and a certain mediated objectivity for his conclusions.

Luhmann, like Maturana and Varela before him, explains the origins of the social world from the viewpoint of a participant making selections from the complexity of its world. In his model, he builds social structures upon the never-finished project of resolving double contingency. Luhmann later adopts the position of a theoretician as we have just noted. Luhmann jumps from the involved-participant perspective to the all-encompassing viewpoint of the "objective" observer. He switches from the internal perspective of an autopoietic system facing an uncertain world to an objective theorizing perspective that prescribes a developmental logic for autopoietic systems.

The addition of the cognitive paradigm and cognitive maps to the lifeworld perspective works an inverse transformation of viewpoints in the lifeworld model. When the rational actors of the lifeworld become self-reflective, they recognize how they reason within the limitations of their cognitive maps. With this recognition, they can evolve meaningful strategies for designing emergent social systems that honor diversity. As designer/stakeholders, they actively influence the development of their societies. They do this as participants from the internal (autopoietic) perspective.

The lifeworld model and Luhmann's autopoietic model have crossed in the night. The lifeworld model was objective (descriptive and reconstructive); it has become autopoietic. Luhmann's model started with an autopoietic viewpoint; it has become objective (prescriptive and somewhat idealistic). Somewhere in that crossing, there are lessons to be garnered.

The Component-System Profile

The theory of component-systems, which was originated by Csanyi, championed by Kampis, and modified by Goertzel, provides the theoretical backdrop for much of the theorizing about cognitive maps that is produced by the General Evolution Research Group. It employs the tools of information theory. It dialogues with the autopoietic and enaction theories of Maturana and Varela. It incorporates elements of the thought of Prigogine, Eigen, Kauffman, and Gell-Mann. In short, it is centrally located among the theories of self-organizing systems.

The founding example of a component-system is Csanyi's explanation of the development of life from pre-existing chemical processes. Given acceptable temperature parameters and the presence of chemicals that interact with each other, polymers automatically form from previous monomers (simple molecules). In the correct environment, these polymers are constantly assembling and disassembling. When the action of one polymer affects the assembly of another, and the action of the other cycles back to affect the assembly of the first, a supercycle develops that mimics the theorized beginnings of life on earth. A component-system has an energy flux running through it that can excite its constantly assembling and disassembling components. To this extent, component-systems conform to theories of non-equilibrium thermodynamics, such as Prigogine's dissipative structures.

Component-systems also require closure in a manner that is roughly equivalent to the manner that hypercycles become compartmented, in Eigen's thought, in order to produce a living cell. In addition, component-systems consist of physical elements that interact in a way that is a priori opaque to observation, formal description, and prediction. They are completely temporized because they constantly undergo disassembly and assembly. Therefore, they are not enduring "things" but genesis machines. In Luhmann's terminology, their structure is secondary to their mode of reproduction.

In the thought of the Group, cognitive maps and cultural change follow the rules of non-equilibrium thermodynamics, probabilistic chaos, and information dynamics. Cultural cognitive maps develop in the cyclical flow of energy, information, and matter between members of a society. In our embodied situation, every act of communicative action (in the broad sense that includes instrumental and strategic action, and even social interactions such as work or the systemic action of steering media) embodies a pattern of information. Societies hold themselves together by the coherence and endurance of their embodied patterns of information.

In the thought of Goertzel, component-systems are modeled by the cognitive equation and the dual network. For him, mind is the prototypical component-system, but many other social realities are also component-systems that obey the cognitive law of motion. Language is a major component-system. It has several subsets that share its semantics, but employ their own syntax. Some of those subsets are propositional logic, mathematics, and belief systems of all sorts. Two of the principal belief systems are *self* and *reality*.

Belief systems equate approximately to cognitive maps. Collective belief systems equate to cultural cognitive maps. Other terms in use by authors in this research mean roughly the same thing. Some of those words are "virtual world," "worldview," "lifeworld," and "common sense."

Component-System and Lifeworld

The theories of belief system and lifeworld are largely compatible, but any synthesis between them would have to overcome several obstacles. Some of the similarities are: Both belief system and lifeworld describe the results of interpersonal and cultural activity. They both explain the origins of social structures. They both express respect for individual and collective freedom.

There are also modest differences. Belief systems, as defined by Goertzel, function within the cognitive paradigm. They are relativistic. They do not establish firm rules of truthfulness, justice, freedom, and equality. Therefore, they do not fit into the pragmatic rational paradigm, advocated by Habermas, that aims to justify norms of social justice.

There are, then, some difficulties to be anticipated in integrating the theories of component-system and lifeworld. But the difficulties do not seem to be insurmountable. An extended dialogue between them would foster cross-fertilization of ideas. The emergence of breakthrough thinking in such a dialogue is inevitable given the laws of complexity. The specific ideas that might emerge are, however, mostly unpredictable-such is the nature of emergent ideas.

The general idea of a component-system/lifeworld synthesis was broached in the sketch of a lifeworld/autopoiesis synthesis. Component-system dynamics-including bifurcations, the influence upon bifurcations of tiny initial conditions, the mutual binding of components, emergent patterns, the dual structure of learning systems, and the cognitive equation-give greater articulation to the processes of lifeworld rationalization that are described by Habermas. In turn, the real life concerns of Habermas-freedom, norms of truth and social justice, participatory democracy, and establishing lifeworld control over mechanical social system processes-provide fertile areas for developing the social cogency of component-systems thinking.

Habermas's project-to establish firm, pragmatic, rational norms for social justice that honor and protect claims of truthfulness, freedom, and equality-can be augmented by arguments from both autopoiesis and component-systems. For example, evolutionary redundancy and the behavior necessary to overcome double contingency offer strong support for Habermasian norms of social justice.

Component-Systems and Autopoiesis

The theories of component-systems and autopoiesis in the biological realm consider the same realities, but do so from contrasting viewpoints. Autopoiesis attempts to reconstruct the ongoing experience of organisms from the interior "performative" (cf. Habermas, 1996, p. 18) standpoint of the organism itself. Component-system theory attempts its explanation from the viewpoint of the "objectivating" (cf. Habermas, 1996, p.18) observer.

The performative-objectivating distinction between these theories holds to a *lesser* extent in the psychological and social realms, because psychic and social systems, with their reflective abilities, incorporate both viewpoints within themselves. The broader contexts discussed in this chapter blur the subjective/objective distinction even more.

In some ways, the autopoietic social systems of Luhmann are more formal than the component-systems of Kampis and Goertzel. Luhmann frames his ideas with an acute attention to the necessity of setting clear formal boundaries to social systems. The boundaries of component-systems are defined by material implications. They resist formal description.

Because of the contingency of material implications, the "total information content" of material systems "is distributed over time. At no point in the system's history is all information manifested, nor is it expressible" (Kampis, 1991, p. 443). Material systems can be properly described only by referring to their entire history of semi-contingent bifurcations. "Only the complete history of the system defines it as a complete unity. We are identical with our cumulative history, and not with our transitory states" (p. 443).

Many component-systems can be described only as fuzzy sets. This is a handicap if we desire to perform operations of propositional logic with them. Their fuzzy nature does not, however, lessen their explanatory power. In their recognition of material implications, component-system theorists make room for precisely the lifeworld influences that Habermas insists upon (and which many good sociologists record).

The discourse/lifeworld profile uses the idea of communicative action to build the lifeworld. The autopoiesis profile explains life and social processes as reproductive continuity with changing elements. The component-system profile begins with replicative chemical processes and concludes with Goertzel's universal equation. Laszlo's contribution of the "interconnected universe" integrates the ideal of evolution to the nature of quanta and the necessity of interconnection.

The idea of reproduction is basic to all theories of evolution, be they Darwinian, neo-Darwinian, or systemic (in the manner of the theories we have been discussing). Reproduction is implicit in the discourse/lifeworld profile. It is explicit in autopoiesis and component-systems. The centrality of reproduction in autopoietic and component-systems derives from their temporized existence. Such systems do not exist as structures independent of their reproductive activity. Their structure *is* their pattern of reproduction.

CONCLUSION

Three principles underlie the structure of the world in the systems view that has been developed in this book: interconnection, temporality, and reproduction. The result of these principles is the emergence of new systems. Without consistent *interconnection*, the self-organization proposed by the authors considered in this research would be impossible. As Laszlo demonstrates, through his application of reconstructive science, *pure* chance cannot have evolved the universe as we know it. Laszlo contends that interconnection as physical memory enables evolution without recourse to the deus ex machina explanations of teleology and teleonomy.

Temporality is radically indicated in our understanding of sub-atomic physics, in which electrons and positrons instantaneously generate and exterminate one another. This temporality is evident in all areas of reality. All physical objects are products of constant circular processes. Living systems reproduce themselves moment-to-moment. Psychic systems undergo reiterations of the cognitive equation (Goertzel) as they constantly reshape themselves in response to novel situations. Social systems are reshaped with every act of communication.

In a totally temporized space, constituted by constant creation and dissolution, there is no ongoing structure in temporal atoms (events). Continuous structure involves temporal and spatial patterns of intermeshed processes. It requires energetic, sustained, and patterned interaction. As some processes influence the occurrence of other processes, they constitute a chain of reproduction that is a component-system.

The fundamental quantity in this new physics is *information*. In information theory, "the quantity of information is the number of binary (yes/no) choices which have to be made to achieve a unique selection from the possibilities" (Checkland, 1981, p. 315). This mathematical measurement of information leads to theoretic quantifications of complexity of all systems, be they physical, organic, psychic, or social. A different definition describes information in terms of what it is, and not just how it is measured. In this definition, information is a pattern of interconnected processes.

In isolation, differences do not even exist. In isolation, every event is just a nameless nothing. Information as the pattern and organization of inter-connections supplies the difference within unity that is the universe. Information, as new patterns of interconnections, is the evolution of the universe in temporized space.

In the views of Luhmann, Maturana, and Varela, the structure of autopoietic systems is provided in the fidelity and flexibility of their *reproduction*. In the estimation of Goertzel, reality (as we perceive it) and particular realities are patterns that have gained a feeling of solidity. This feeling arises because those patterns are continually present in the self-reproduction of our belief systems. Their reliable presence in our circle of cognitive reproduction produces our assumption of unchangeable things, hence, our reality. In Goertzel's words, "self and external reality are belief systems that mutually generate one another as the inside and outside of experience. Together they constitute the universe."

The openness of reproduction within the confines of circularity is the dynamo of evolution. This truth is prominent in the theories of autopoiesis, component-systems, and complexity. These theories describe the growth potential of a long-maintained sensitive balance between openness and circularity (stochastic and Cartesian behavior, chaos and logic, conservatism and liberalism, and so on).

The sciences of self-organization can lend support for Habermas's norms for communicative action-truth, truthfulness, and rightness. The argument goes as follows: Truth, truthfulness, and rightness have been upheld as ideals during all of recorded history with few major exceptions. These norms came about largely because they enable successful communicative action. They gained strength because the were continually tested and refined. Aberrations almost always proved unsatisfactory in the long run. These norms are strong. They continue to survive and gain nuance in spite of multiple affronts. As a result, these norms have achieved the redundancy of being practically certain.

In the systems view that emerges from this research, self-organizing processes produce the organized complexity of the universe. The activity of these processes has produced the universe both as it might be out there and as we know it. These evolutionary processes have generated us. They are operative in our discussions and in all of our interactions.

The three narrative explanations of the social world developed in this chapter have remarkable underlying similarities and possibilities for mutual enrichment. The precise contours of their likely synthesis have not yet developed. Some of its essential contents, however, can be grasped in the underlying dynamics of those profiles. A rough synthesis of those dynamics has the following elements.

- The world, our systems, and our selves are temporized entities. They exist in an unstable stability, in which their reproductive process is their structure. Social structures, such as person, role, family, corporation, or nation, have minimal temporal duration as *things*. At every next moment, these structures are materially different; that is, they are different *things*. These structures exist in extended time as *things*, therefore, only as metaphors. What remains over time is their ongoing autopoietic reproduction.
- The reproduction of our world evolves more complex structures through its openness to energy and information, in which an organization, for example, might develop inner differentiation or an alliance (structural coupling) with another organization.
- The relationship between the circularity and openness of self-generating systems can be explained in terms of pattern recognition. As one generation mixes its patterns with its experienced environment, subsequent generations use the patterns it recognizes in that mixing as components of itself. This process is succinctly stated in the "law of cognitive motion." Goertzel offers a brief informal description of this law and the attractor that it generates as follows:

    (1) Let all processes that are "connected" to one another act on one another.
    (2) Take all patterns that were recognized in other processes during Step (1), let these patterns be the new set of processes, and return to Step (1).

    An attractor for this dynamic is then a set of processes with the property that each element of the set is (a) produced by the set of processes, (b) a pattern in the set of entities produced by the set of processes (1994, p. 152).

Motorola, the electronic and information giant, offers an example of the cognitive equation. In its observation of its own internal patterns as they interact with emerging patterns, Motorola decided to morph into an electronic, information, *and bio-tech* corporation.

- The patterns referred to by Goertzel are equivalent to the functions described by Csanyi and Kampis.   Patterns and functions influence each other's behaviors.
- Systems undergoing circular autopoiesis and/or evolutionary bifurcations utilize the energy and information that flows through them. In autopoiesis, the system takes in more energy than it leaves out and uses the difference for the reproduction of its essential processes. In bifurcations, a system absorbs

higher than usual amounts of energy that create turbulence and instability. In its far-from-equilibrium state, it engages in deviant forms of reproduction that often result in higher degrees of systemic complexity and adaptability. In autopoietic theory progressive bifurcations are called "structural couplings." This is an apt term because a component-system generates more complex organization by coupling patterns from its previous generation with patterns from its environment.

- This evolutionary progress combines the two principal notions of "information" in systems thinking. Information #1 is a difference in a system's environment that is important for it (that it can react to). Information #2 is the measure of the complexity in a system that has incorporated information #1. Information #2 corresponds to Luhmann's notion of "meaning," but in the ultra-simplified Boolean language of information theory.

The above statements are some of the more basic principles in the new synthesis. The following statements are subsidiary principles.

- Evolving systems (organic, psychic, social) reproduce patterns in themselves that have the nature of a dual network: a perceptual/command hierarchy and a heterarchical structurally associative memory. The reproduction depicted in the cognitive equation produces patterns that are deposited in either part of this dual network and fractally reproduced in the other part. The result of this sharing of patterns is aptly called "continuous compositionality" (Goertzel).
- A pivotal ideal-typical case in human social life and evolution occurs when two people or social systems encounter each other and discover that they are not able to predict one another's behavior (the situation of double contingency).
- To deal with this double contingency, psychic and social systems modalize their expectations.
- As people and societies modalize their expectations, they generate the structures of the social world and constantly revise them in continual reproduction.
- Minute influences at the time when a component-system bifurcates have enormous influence on the subsequent behavior and evolution of the system.
- We can monitor the rise of turbulence and the imminence of bifurcation by a kind of "systemic seismology." In the toy systems of information theory, such increased turbulence can be monitored and bifurcations can be predicted. In the application of neurocybernetic systems models to organizational situations, we can monitor the activity of feedback loops interacting at leverage points and prepare to make early use of imminent bifurcations. In this situation, breakdowns are seen as opportunities for growth. In very big cultural systems, the involved feedback loops are seldom, if ever, completely understood. Therefore, finding leverage points, monitoring them, and making predictions is more art than science. The range of accurate prediction is increasing into areas as complicated as the fluctuations of the stock exchange.
- Within an understanding of chaotic processes, individuals, companies, political organizations, and other agents can amplify their ability to influence

events by monitoring leverage points and feedback loops in order to anticipate likely bifurcations. At these space/time junctures, they have the ability to generate profound change with relatively little energy.

Hitler made use of such a bifurcation point in the turbulence of depression-era Germany. Marxist revolutionary theory also recognizes the power of concerted effort in turbulent social situations. Marxist revolutionaries have traditionally tried to disrupt the existing capitalist routine in order to create the turbulence that is required for drastic change.

Turbulence setting the stage for bifurcation creates the possibility for radical social change. At bifurcation points, competing programs and ideologies vie for pivotal dominance. In the turbulence of Europe after the Second World War, for example, communism, socialism, and capitalism engaged in combat for the hearts and governments of Western Europe, which resulted in various forms of social-welfare capitalism. In this case, communism did not dominate the resulting social system because it shared the arena with strong opposing ideologies. In related situations, such as depression-era United States, communism failed to prevail despite the expenditure of great energy because social turbulence had not mounted to the stage of bifurcation.

- Formal models of social systems, such as those proposed by Luhmann, are useful in predicting the abstract qualities that differentiating social systems will take, but cannot predict the time, shape, or extent of those differentiations.
- Material models of social systems, as cultural cognitive maps, for example, or as collective belief systems, are much more complicated than formal models. They require critical attention to matters of boundary, social strata, and so forth that are enumerated by Habermas. Attention to these matters is the equivalent of monitoring leverage points in cybernetic systems.
- We can make ourselves *victims* of cultural chaos and change by choice. We can also educate and empower ourselves to *master* chaos and direct change. We can couple attention to critical turbulence levels-as discovered by chance, cybernetic methods, or Habermas's criteria-with collective intellectual and political activity. We can use this dual awareness to guide us into a more human- and environment-friendly world.

This list of principles is not meant to be complete. A portrayal of the logical derivations among them is not attempted. They are the mere rudiments of a partially perceived systemic synthesis of our knowledge of our social world.

The three chapters that follow were generated by processes similar to those used in the previous two chapters, but did not employ any computer assistance. The chapters concern communication, cognition, and epistemology.

CHAPTER 20

COMMUNICATION

ABSTRACT

Communication is placed at the center of the social world by all the theorists explored in this research, but often with different meanings. Maturana and Varela consider communication to be third-order structural coupling (mutual social adaptation) and define it as mutually triggered, coordinated behaviors. Luhmann develops a theory of communication in which social relations, created by double contingency, generate the meaning of the world. Kampis explains non-symbolic signs. Goertzel presents a "cognitive law of motion," which functions as a universal law of communicative action. Kauffman applies the generalized grammar of polymer strings to explain the development of complex adaptive systems. Laszlo identifies the ultimate principle of closure and memory for self-organization in the interconnection provided by the Zero-Point Field (ZPF).

This chapter presents a coherent picture of systems-theoretical thought about communication. It weaves a tapestry of semiotics, interconnection, fuzzy sets, material implications, information-for, non-symbolic signs, language, and cognitive interaction. It encompasses interconnection on the quantum level and discourse on the sociocultural level.

SEMIOTICS

Charles Peirce said, "The entire universe is perfused with signs, if it is not composed exclusively of signs" (quoted in Sebeok, 1994, p. 14). This statement can be interpreted in several ways. In the standard interpretation, it expresses a kind of conceptual idealism. It says, "Our view of reality . . . entails an essential reference to mind . . . in its constitution" (Sebeok, p. 14). In this sense, Peirce's statement is equivalent to the "interpretativism" of Flood. It is similar to the Sapir-Whorf hypothesis that receives qualified approval from Goertzel. Sebeok agrees with this interpretation. He says that semiotics deals with "the role of mind in the creation of the world or physical constructs out of a vast and diverse crush of sense impression" (p. 15). This interpretation may unduly limit the scope of what Peirce was meaning to say.

The authors considered in this book advance a different kind of semiotics. For them, semiotic activity lies at the core of the universe's self-generativity. This interpretation presupposes the structure presented in the previous chapter-an interconnected, temporized, and reproducing world. Semiotic patterns inform and enable the continuous reproduction of the world. In this sense, signs inform the world from top to bottom and from beginning to end.

This interpretation is clearly expressed in Goertzel's "law of cognitive motion." It its every reproduction, a self-generating system recognizes patterns existing among its pre-reproductive patterns and those of its environment. It incorporates the newly recognized patterns into its "self." Then it begins a new reproductive cycle by mixing its new patterns with its environment, and repeats the process.

This kind of exposition of the pervasive reality of signs is not idealistic. It is cybernetic. It claims to explain how the universe really works.

INTERCONNECTION

Interconnection, as Laszlo uses it, expresses an extremely tenuous idea of communication. It is the minimal interconnection needed to make self-directed evolution possible. It connects processes in directional space-time by preventing the random cancellation of developing processes. By supplying this minimal memory to stochastic interactions, space-time interconnection lets those interactions move, over time, in a direction. This interconnective constraint enabled the evolution of ever more precise constraints such as the constraints that allow component-systems to evolve.

Laszlo's conception of this interconnection is a Zero-Point-Field holographic memory (ZPF). The minimal constraints of the ZPF supply a degree of circularity and consistency in the random crashes of quantum particles. Those crashes provide more than ample noise to prevent pure circularity and stagnation.

The ZPF is composed of Schrodinger waves with remarkable qualities (cf. Laszlo, pp. 43-44). In the hologram formed by their interfering wave patterns, time-varying information can be recovered in the proximity of its source. Because of the properties of these waves, quantum memory is universal *and* time-invariant. It is also selective and automatic. In other words, memories are collected without focusing mechanisms in areas that are topically and temporally proximate to the originating information.

Laszlo's ZPF-infused universe casts a rudimentary shadow resembling that of a component-system. As can be ascertained in the following table, it has elements that correspond to Goertzel's model of the mind.

Laszlo	Goertzel
Zero-Point Field (ZPF)	Structural Assoc. Memory (SAM)
Embodied Structural Couplings	Hierarchical Structure
Topical Access	Continuous Compositionality
Flows of Energy and Information	Flows of Energy and Information
Interconnection/Constraint	Circularity
Crashes	Noise
Continuity	Pattern Reproduction
Evolution	Cognitive Equation
Balance of Circularity and Openness	Balance of Circularity and Openness

A FUZZY SET

Communication, language, meaning, and information are fuzzy set ideas (L. Zadeh, 1982) whose prototype (1.0) is found in human conversation. Other forms of communication, for example, could be related to a hypothetical scale, in which chimpanzee interactions would rate (0.9), gray parrots (0.85), vervet monkeys (0.8), amoebas (0.5), pine trees (0.3), certain polymeric solutions (0.1), crystals (0.05), and quantum interactions (0.001) (cf., e.g., McNeill and Freibergber, 1993, pp. 34-39). The preceding section pushed the envelope of those ideas by applying Goertzel's universal equation and Luhmann's idea of meaning to inanimate systems (down to the 0.0... range on the fuzzy set scale) on the basis of Laszlo's universal holographic memory.

Kauffman's concept of random grammar contracts that hyper-extended envelope somewhat. Using this grammar, polymeric sets export and import strings of Boolean information, learn to cope with such exchanges, and structurally couple. This grammar offers promise of a universal grammar for modeling the mechanics of communication. It also conforms to predominant biological ideas about chemical messengers (cf., e.g., Losick and Kaiser, 1997; and Rossi, 1993).

Maturana and Varela stay much closer to the prototypical sense of these words. For them, communication is a special form of structural coupling in which members of the same species recurrently interact with each other; and language applies only to human communication. Meaning and the observer arise along with language in the phenomenological domain of description that is enabled by syntactic language. For them (also for Bickerton, Luhmann, and others), information is not a *thing* that can be communicated. It is a relationship of novelty that can be reacted to by an organism. Because it is not a thing, information cannot be re-presented. Nor can it be put into a tube so that someone (or something) can take it out at the other end.

MEANING

Kampis navigates mostly uncharted semantic waters. He works with non-symbolic signs and describes how material implications generate meaning by using "information-for."

Non-Symbolic Signs

Non-symbolic (referential) signs of communication are the very interactions (material causality) of the system. The interactions constitute the system.

In a non-symbolic system, information cannot be separated or abstracted from its carrier. Consider the many transformations that process sensory information. In these transformations the material content of the information disappears, but the information remains stable. In audition, for example, air compression and expansion transform into physical fluctuations of the tympanic membrane. What is common to both ends of this transformation? The compressions and expansions in the air have vanished. The membrane shares no physical components with the air. The common element is the pattern (meaning) of the contractions and expansions. The pattern transcends the momentary nature of the fluctuations of both air and membrane.

Similar transformations occur in the passage from tympanic membrane to hammer, anvil, and stirrup; and thence, to the cochlea where fluid motion is transferred to tiny hair cells on the basilar membrane attached to nerve endings that transduce mechanical energy into electrical-neural energy; thence, through the auditory nerve to the brain for further transformation. In all these transformations, the molecular signs go out of existence, but the pattern existing in those signs is reconstituted in a new medium of molecular signs.

Material Implications

Kampis call the emergence of life "a Second Big Bang" in which meaning makes its appearance. This *meaning results from material implications* in which "truly semantic relations occur with a semantics that cannot be reduced to syntax." What he means by this statement can be explained in the context of his thought as follows: meanings arise that are not the result of any formal implication. If his statement is understood in this way, it does not conflict with Goertzel's concept of syntax as it functions within the cognitive equation, because Goertzel's "syntax" is itself an emergent (not formal) component of a learning system.

According to Kampis, component-systems differ from computers that generate complexity by iterating operations with complex numbers. The computers operate within formal rules. They generate unpredictable, but deterministic patterns that can be expressed succinctly by identifying their iterated operations. In terms of mathematics, all the patterns generated by iterating the same formula are simple, because they are simply the next step in the application of the formula.

In addition, the result of applying the formula is an *event* that can be represented as a pixel in a screen of coordinates. The formula does not create the coordinates, the screen, or the memory that records the position of past and future pixels. We as experimenters, observers, and describers provide this entire framework. By providing this framework, we create the patterns and names (meanings). We generate the complexity on our phenomenal level; it does not exist in the mathematical algorithms.

It is different for component-systems. They actually create new meaning through their material implications. No formula determines the meanings that result from component interactions. They generate their ordered complexity on their own.

Where Do Meanings Exist?

Even though meanings are the result of selections made by us, there is a sense in which the meaning of component-systems exists "out there." In our cognition, we observe, project, and test likely unities in our ongoing selection of meanings. Component-systems prove to have unity (organized complexity, meaning) that is independent of our construction.

In our evolutionary history, these component-systems had reality without our having to name or shape them. Our descriptions of them do not reduce them to formulas. To adequately describe them, we would have to recount a history of the material interactions they have completed. In this way, their histories are analogues of our histories. Our patterns of learning and evolving are fractally similar to theirs.

Still, the complexity of an organism does not exist "out there" in the way that we describe it. It exists out there in the way that the organism lives. The organism does not understand the procedures of reproducing and increasing complexity that we ascribe to it. Our descriptions create the paradox of the incongruent perspective (cf. following chapter).

Information-About and Information-For

Our observations and descriptions create information *about* an organism. This kind of information is radically different from the information that the organism uses to in-form itself as it incorporates selections. The information that the organism embodies in staying alive is "information-for." Organisms know only what is needful *for* maintaining and improving their processes. Their knowledge is not speculative; it is practical. In the human realm, information-for corresponds to the multitude of unconscious activities that automatically maintain our material existence.

Organisms make their selections on the basis of information-for. They select "before the fall from grace" (Luhmann). That is, they select before they have created the "world" their unity of inside and outside (by language). In spite of the complicated, non-deterministic mathematics with which we model their processes, the organisms make selections of the utmost simplicity. They select what feels right. Their selections relate past successful selections to their present situation, but they are blissfully incapable of recounting their histories.

THE NATURE OF LANGUAGE

Most of the authors considered indicate that language is important in the creation of meaning, but they differ on the nature of what language is and what its influences are. Maturana and Varela restrict "language" to its prototypical sense: human communication with words. For them, meaning and the observer arise only with human language. Non-human species have no language; they are not observers of reality; they do not represent reality to themselves; they simply enact their lives.

Kampis works with the distinction between referential and non-referential information. He finds that proto-symbolic information (the beginning of symbolic language) arises with the appearance of a single living cell because a cell is a self-reproducing cycle of referential information. Such a cycle constitutes a simple cognitive map characterized by "the first steps towards symbols: permanence and relatability (Kampis, 1993, p. 145). Luhmann supports Kampis's position. He says that subject-object differentiation is present in every selection of an autopoietic system.

Bickerton defines language as a system of representation. He finds proto-representation in the behavior of plants and shows its evolutionary progression in animal species with special attention to the protolanguage of apes and small children.

If Kampis, Luhmann, and Bickerton are correct, then there is proto-representation, proto-observation, and proto-agency present in all living things. In addition, if Kauffman and Laszlo are correct, there are even more tenuous roots of representation, language, observation, and agency in polymeric communication and the zero-point field. If these evolutionary forebears of language exist, then our own evolutionary ancestors and our own biological subsystems possess both referential information and a modicum of non-referential information; they utilize both information-for and information-about. Our evolutionary ancestors' information-about is qualitatively different, however, by several levels of emergence, from our own.

Bickerton describes protolanguage as it exists in apes, very young children, and pidgin languages. He explains the emergence of truly human, syntactical language as a convergence of protolanguage with the treelike structure of mental representations. This convergence developed because of pressure for increased communication.

Habermas traces the emergence of language from gesture, to signal, to grammatical speech. Following the leads of Mead and Durkheim, he identifies the rules that enable these transitions and shows how these rules develop into social structures.

Luhmann agrees that rules developed around language and the early stages of society. But these systems of rules, for him, are merely positive ones. That is, they develop according to their own self-referential logic, in which past contingent enactions and judgments inform all incorporations of new information. They are not intrinsic to the nature of language. The norms that societies set up serve a functional purpose. They separate activities into conforming and deviant. By building a structure of such norms, a society provides itself with a guarantee of order.

For Luhmann, the essential element of language communication is the recognition that there is a difference between information and utterance. The essence of a pre-language signal is the identity of information and utterance, as in the mating rituals of animals. Signals evolve into language symbols when the participants in a communication realize that an utterance conveys something different from the utterance itself. Communication is, therefore, a process involving three selections: information, utterance, and understanding.

For Goertzel, language is a linguistic system that functions according to the cognitive law of motion. It consists of a structured associative memory (semantics) and a perceptual/motor hierarchy (syntax) which are mutually co-creative and which exhibit continuous compositionality.

Goertzel mediates the positions of Saussure and Peirce. In his semantics, Saussure considers two elements of meaning: the word as sign and the object as the signified. He finds that there is no direct connection between the sign and the signified. Peirce considers communication (semiosis) as a tripartite process involving an object, a sign representing that object, and a recipient of the sign who interprets its meaning. Goertzel agrees with Saussure that a word in a syntactic system is a relation between other words. In the broader context of communication, however, there is a third component: embodiment. With embodiment, language becomes a way of molding the world.

COGNITIVE INTERACTION

Non-Symbolic Interaction

Kauffman offers an instance of cognitive interaction. He describes polymer "agents" whose adaptations are near-sighted efforts to enable survival-short walks in evolutionary selection on varied fitness landscapes. Each agent adjusts to deformations of its landscape caused by the adaptive moves of other agents. These agents manage, as their best way of adapting, to maintain a mutual state of being poised between order and chaos. In this state of uneasy equilibrium, they try to predict each other's behavior. They create some order in processing the information that confronts them by assigning weights to differing clues and testing the effects of those assignments. In so doing, they progressively mold a closer attractor (image) of future behavior.

Kampis explores the logistics that enable human communication with language on the basis of referential and non-referential information. Referential (non-symbolic) information is the simple systemically universal and undifferentiated information necessary for autopoietic survival. Such non-symbolic information cannot be communicated through language.

Symbols

One might think that symbols are ideally suited for communication. This is not so. Non-referential information is communicable only within the system in which the relevant symbolic universe is available. Such a universe must have present all the replicative units that span the symbolic system, which presence is available only within the mind of a single person. In other words, units of non-referential information are communicable only in the brain of the person who possesses them.

To communicate between different minds, symbols alone are not suitable; they are not available or communicable in themselves. Only behaviors, not contents, are accessible by communication partners. The symbols of human language are important, however, because they supply names that activate similar mental models, superstructures, and replicative structures within the persons who use them.

In conversations, individually isolated cognitive maps are brought into a kind of match in some areas as people indicate their point of reference and its associated non-referential information. Through processes of intuition, trial, and error, the participants of conversations "begin to bring forth identical actions in identical situations. That is, the speakers behave as if their mental models were identical" (Kampis, 1993, p. 147).

Knots in the Evolutionary Tapestry

In the course of time, referential information (the essential non-symbolic knowledge of autopoietic and component-systems) developed the ever-present seeds of non-referential information (symbolic communication). In turn, symbolic communication led to the emergence of language as externally symbolized behavior. Language behavior facilitated trial and error experiments in which names came to be identified with similar mental models.

The ensuing process can be modeled as a component-system. Individual cognitive maps generated names that enriched the cultural language. In turn, cultural language, working under the constraint of reaching agreement on the names and rules of language, enriched individual cognitive maps. In this way individual and cultural cognitive maps cross-fertilized and cross-catalyzed each other for the enrichment of all. They generated a kind of a huge computer that processes global conversational information.

Variations on a Theme

Luhmann describes cognitive interaction in the context of double contingency. The problem of double contingency is resolved by creating the concepts of person, intelligence, and freedom. In this way, double contingency generates the unstable core structure around which social systems like language crystallize.

Cognitive interaction is a variety of interpenetration in which systems "converge in individual elements-that is they use the same ones-but they give each of them a different selectivity and connectivity, different pasts and futures" (Luhmann, p. 215). This convergence results in a momentary agreement that exists as an identical event for both parties, but means different things for them, and leads to different consequences.

Agreement emerges in cognitive interactions through the push and pull of perceptual understandings and verbal communications. The parties in interactions constrain themselves to certain norms of good behavior. They exchange verbal and non-verbal understandings. They readjust their own cognitive maps by incorporating bits and pieces of the maps portrayed in the interaction.

Goertzel's term that approximates "cognitive maps" is "belief systems." A belief system is a sort of mini-mind. It, like language, obeys the cognitive law of motion. It is constructed in language; therefore, it uses the same semantics. But it uses a different syntax, that is, different rules for putting the words together. On a global scale, the belief system that we call "external reality" is a mainstay of almost every culture. Cultural cognitive maps of a smaller scale are further subsets of language, each having its own syntax.

Campanella compares the cognitive and rational paradigms as they are used in political science. The cognitive paradigm centers on the real-life strategies that we use to learn and make decisions. It considers the human actor "not at rest but in action" (p. 242). It assumes that people form a simplified cognitive map of their world in order to cope with the confusing complexity of their lives and that they use this map to make choices that are not optimal, but are nevertheless satisfying.

RATIONAL INTERACTION

Habermas considers language to be a tool for reaching decisions. In simple consensual communication, the validity claims of truth, rightness, and truthfulness are assumed. In contested situations, discourse requires conventional rules for deciding which statements are true and what course of action is desirable. Discourse geared to understanding develops rules of logic and evidence. Discourse geared to action develops rules about who has the right to make decisions and command the allegiance of others, how rules are made, how they are to be obeyed, and how conflicts are settled. Communicative action, which works on the underlying norms of truth, truthfulness, and rightness, creates rules for attaining those norms in different situations, and, in the process, creates the normative/pragmatic structure of the social world.

The process of communicative action is rational from the perspective of its participants. They believe that they make rational choices that enable society to function and participants to achieve satisfactory situations. Rational action of this sort, for Habermas, does not require "consistent, value-maximizing choice" or strategies of "optimal choice and cost-benefit analysis in terms of a goals-means economy." For Habermas, these additional restrictions apply only to purposive rational action; they do not apply to communicative rational action.

Habermas deals with the way authority is legitimated and upholds the rights of ordinary people to defend themselves against arbitrary force. If people are in a hostile social situation, they must have available solid concepts of justice, such as "inalienable rights," compelling forms of argumentation, and forums in which to contend with their oppressors. Habermas presents the ideal speech situation as a "useful fiction" that rests upon quasi-transcendental ideals of truth, truthfulness, and rightness. He shows how those ideals are substrates of our social institutions.

Habermas passionately defends the causes of social justice and participatory democracy. For him, rational communicative action is essential, especially in the relativistic swamp of postmodernism. In a relativistic society where any cause appears as good as another, naked economic or political self-interest becomes the arbiter of truth and justice. Habermas sees modern society sinking back into the barbarity from the high ground that it had achieved by pursuing rationalistic ideals.

The juxtaposition of cognitive interaction, rational interaction, and communicative action creates a small tension that prompts a small synthesis. Just as most cognitive interactions involve a bit of rationality and logic, so rational communicative interactions involve a modicum of cognitive interaction. In practice, the ideal speech situation requires a great deal of mutual attention to the cognitive maps of its participants.

Perhaps the expression of the ideal speech situation should be amended to make explicit reference to the cognitive maps of its participants. In fact, the ideal speech situation is approximated, with this addition, in the design processes that are advocated by Banathy, Warfield, the open and critical systems theorists, and others.

THE EVOLUTION OF CULTURAL COGNITIVE MAPS

Maturana, Varela, and Goertzel portray the evolution of cultural meaning in similar terms.   Maturana and Varela emphasize how individual interactions with environments become standardized through communication. They say, "We are observers and exist in a semantic domain created by our operating in language where ontogenetic adaptation is conserved (1987, p. 211). Varela says that enactive cognition is "a history of structural coupling that brings forth a world" (1996, p. 206). Thus, the paths of cognitive enaction are not primarily individual enactions; they are a shared world of significance that a group has forged in a wilderness.

Goertzel considers the coevolution of individual and cultural cognitive maps in terms of internal and external dialogues. Internal maps (belief systems) function as mini-minds. They set up dialogues in our consciousness that enable the comparison and virtual activation of divergent perspectives and programs. They produce relationships among themselves that enrich the mind's ongoing self-reproduction through the cognitive equation.

Belief systems are shared when a similar dialogue occurs between the belief systems of at least two people. They become "collective reality" when they become the shared property of a group. Something exists in this reality if people recognize it as real *and* recognize that their collectivity also recognizes it as real.

Luhmann explains the dynamics of collective reality in terms of redundancy and double contingency. The social relations created by double contingency create the meaning of the world. "The generalized result of constant operation under the condition of double contingency is finally the social dimension of all meaning, namely, that one can ask for any meaning how it is experienced and processed by others"(p. 113).

In addition, conflicts force reflexivity upon societies and catalyze them to create media that expedite communication. Language, media of dissemination, and functional media such as money, bureaucratic power, positive law, and romantic rules of engagement are the result. These media are evolutionary achievements that interdependently ground the processing of information and increase the production of social communication.

CONCLUSION

The spectrum of communication, discussed in this chapter indicates that communication of some sort is present in the functioning of physical, organic, social, psychological, and sociocultural systems. The spread of communication over this spectrum indicates a similar spread of patterns of reproductive information, which constitute the temporal existence of those systems. In particular, this spread of communication indicates that social processes have evolutionary forbears.

This chapter on communication is closely linked to the following chapter on cognition. For that reason, an expansive array of communication and cognition conclusions will be withheld until after the next chapter.

The next chapter discusses the internal aspects of communication under the heading of cognition. The final chapter deals with the social aspect of cognition, how do we know that our cognitions are true (valid) under the heading of epistemology.

CHAPTER 21

COGNITION

ABSTRACT

This chapter synthesizes central ideas concerning cognition that are proposed by the authors under consideration. It draws together crucial distinctions that help us to understand the paradox of the incongruous perspective and the evolutionary processes that link elemental self-organization to our knowing about something in language. It unravels several theoretical knots that obstruct a commonsense acceptance of human intelligence as an evolutionary achievement. By unraveling theoretical knots this chapter highlights the universal self-organizing processes that generate punctuated cognitive evolution.

STATEMENT OF THE PARADOX

It is impossible for an organism to understand the procedures of reproducing and increasing complexity that we ascribe to it. For an organism, its complexity does not exist "out there." It exists out there only for us as observers and describers. Our descriptions create the "paradox of the incongruent perspective" (Luhmann, 1995). This paradox is evident in all explanations that try to explain the activity of component-systems in terms of categories that are imputed to be at work within them.

When the emergence of cellular life is explained as the conjunction of hypercyclic reproduction in a polymeric solution and compartmentalization within that solution (compartmented hypercycle; Eigen, 1992), no inference can be drawn that the involved polymers are aware of what they are doing; but there is some physical memory involved because the lack of such memory would make progressive complexity impossible (cf. Laszlo, 1995). Similarly, when Acrasiales amoebas congregate to form slime molds under the influence of cyclic AMP (cAMP) that then release spores that generate new amoebas, they are not aware of their need to reproduce, the effects of cAMP, the environmental factors that lead to the bifurcations in this process, or the cellular chemistry that keep them alive (Prigogine and Stengers, 1984, pp. 156-159).

In human social interactions, this paradox is observed in the differing attributions given to behavior by the actors and observers of that behavior. When a man is driving in traffic with his mate, for example, the man is intent on negotiating traffic while his mate often thinks that he is driving recklessly for some motive, such as, to prove his masculinity, to impress her, or to punish her (cf. Luhmann, 1995).

WHAT IS COGNITION?

"Cognition" is a fuzzy concept. In its evolutionary roots, cognition is the basal self-reference of component-systems. It is the pattern of reproduction. It is organisms doing their autopoiesis. Succinctly in the words of Maturana and Varela, "Knowing is doing." It is what organisms do to stay alive.

Cognition in its prototypical sense is the way that we humans know the world as observers through language. Our knowing has evolved from basal self-reference. Our form of cognition extends this basal self-reference to reflexivity, reflection, and rationality. Thus, the universal self-reference of component-systems has evolved into us human beings and our societies. Social cognition as society's self-reproduction in interactions, organizations, and society as a whole, is the product of evolving self-reference.

The explanations that we make of component-systems and their behaviors are formal simplifications of very complicated material processes. They explain what is happening in terms that make us feel that we understand. In this kind of "objective" description, we do not "simply trace how these systems experience themselves and their environment . . . We cover them over with procedures of reproducing and increasing complexity that are impossible for them to understand" (Luhmann, p. 56).

If we carefully distinguish the autopoietic perspective of the organism from our perspective as an observer, we can unravel this paradox. A similar attention to the phenomenal planes of action and the planes of observation is key to understanding the evolutionary steps between basal self-reference and human observation.

INFORMATION-FOR AND INFORMATION-ABOUT

Understanding the difference between the *information-about* an organism that we possess and the *information-for* that the organism uses is a useful first distinction in unraveling this paradox. Our observations and descriptions create *information-*

about an organism. This kind of information is radically different from the *information-for* that the organism uses to go on making the selections that it is.

Selection made on the basis of information-for is a way of describing the cognition of an organism that is very similar to Maturana and Varela's definition of knowledge as doing and to Varela's theory of enaction. These descriptions avoid the paradox of the incongruent perspective because they endeavor to explain behavior and evolution from an organism's point of view as it negotiates its continued existence. They also can be linked to less autopoietic descriptions of cognition because each enaction can be interpreted as an iteration of the cognitive equation (Goertzel, see below).

REPRESENTATION

Maturana and Varela

Even with an appreciation of this distinction, however, the question "What constitutes knowledge for an organism?" still requires explication. Maturana, Varela, Luhmann, Kampis, Goertzel, and Bickerton address this question.

Maturana and Varela say that knowledge is what an organism does to maintain its reproduction. For them knowledge is something that an organism enacts as it structurally couples with its environment (that it does not recognize as separate from itself) in such a way that organism and environment are both altered. In this situation, there is no pre-existing reality (internal or external) that can be represented in a symbol; the molecular construction that might model such a non-existent reality is unstable; and there is no receptive agent who might receive the message.

Maturana and Varela contend that information is not a *thing* that can be communicated. It is a relationship of novelty that can be reacted to by an organism. Because it is not a thing, information cannot be re-presented. Nor can it be put into a tube so that someone (or something) can take it out at the other end. They go further. They claim that organisms make no representations of themselves and their environments. They base this claim on an understanding of "representation" as "recovering or reconstructing extrinsic, independent environmental features" (Varela, Thompson, and Rosch, 1993, p. 136).

Luhmann

Other thinkers employ a broader understanding of "representation." From the participatory viewpoint, Luhmann finds a differentiation of self and environment present in every meaning-generating selection. From the objectivating viewpoint, Bickerton, Kampis, and Goertzel present cases for the existence of representations in all living things, at least in those that have neural systems (cf., e.g., Bickerton, 1990, pp. 16-17).

Luhmann says that an autopoietic system develops self-reference by defining itself against something. In its every selection, an autopoietic system differentiates its internal complexity from the greater complexity of the environment. It places emphasis, in other words, on its autopoiesis and considers environmental goings-on as inconsequential. Those environmental goings-on (differences) are important for the system only if they make a difference in its autopoiesis; that is, only if those differences are information for it. In this self-involvement, "the system acquires freedom and autonomy." It simplifies its own complexity while enlarging its ability to respond *ad hoc* to greater environmental complexities.

Autopoietic systems create distinctions beyond this autopoietic self-constraint. They gradually create distinctions in their environments, because some differences in the environment gradually become information for it. With such distinctions, they gradually create a more complex world of information in which they can function, survive, and thrive.

These distinctions for other animals and us are not clear-cut re-presentations of some pre-existing reality. They continually rearrange themselves even in our human conceptualizations. Over time, some of these tentative representations acquire redundancy through their repeated individual and communal utility. The redundancy of such distinctions and patterns of distinctions provide the lived reality of the system, that is, its cognitive map of its environment.

Luhmann develops the origin of distinctions and patterns in systems by producing a phenomenological reconstruction from within the autopoietic viewpoint. He does not speculate from the objectivating viewpoint about the biological basis of distinctions, patterns, and memories of organisms. Kampis, Goertzel, and Bickerton consider both the biology and the nature of the "representations" that embody such distinctions, patterns, and memories.

Kampis

The concept of referential information (Kampis) provides a way of unraveling this paradox further. Referential information has no process-independent scaffolding; it is contrasted with the kind of information that employs symbols. The non-symbolic signs of this information are neural patterns within organisms that interact with each other and the whole learning system. These multiple neural connections interact with each other in ways that maintain, alter, or replace existing event-patterns and thus produce new event-patterns. The non-symbolic signs of information are simply interacting components that constitute their ever-evolving meaning in their relationships. Such signs have no permanence or independence; they have the nature of molecular signs. They do accomplish semiosis, but they are not autopoietically aware of the object, sign, and interpretant that they embody. Only observers can overload them with such distinctions.

The semantics of this semiosis issues from material implications that cannot be reduced to description and syntax. The mind processes referential (non-symbolic) information in a replication process where the original information content disappears and new information is created. In the transition from sensation to perception, for example, the continuity between sensation and perception is maintained in patterns (or cognitive maps).

For Kampis, the patterned organization of cognition arises in a manner similar to the patterned organization of life. The replication of life produces semantic relations (of referential information) that cannot be reduced to syntax; these are relations of material co-causality. The meaning that arises from these relations is, in the first instance, the simple on-going of life. (To this point, Kampis is in agreement with Maturana and Varela). Even in this first instance, however, according to Kampis, a pattern of behavior is enacted that endures as a kind of memory.

Kampis sees cognitive maps as having both non-symbolic and symbolic components. They interact and organize and generate the mind.

> Cognitive maps can be conceptualized as intricate, hierarchically and heterarchically organized replicative superstructures built from permanent linkages of many elementary mental models, that is of many elementary cycles. The mind appears to be a continuously changing collection of such superstructures (Kampis, 1993, p. 145).

His description of mind is very similar to that of Goertzel.

THE COGNITIVE EQUATION

The cognitive equation models a hyper functional process in which a component-system at time t makes its components to be the patterns that existed in its predecessor system at time $t-1$. In this way, the system endures as a learning system that remembers its patterns of behavior.

In this ultra-simple formulation, the cognitive equation may seem ridiculously simple. On reflection, however, this "law of cognitive motion" is in the same category of daring as Newton's laws of physical motion. It should not be dismissed out of hand. It uses a sophisticated variation of the sort of multiple consecutive iterations that produce wonders such as the Mandelbrot set. Certainly simple operations, when they are iterated, can produce wondrous complexity.

Goertzel's brief, informal expression of the cognitive equation is:

(1) Let all processes that are "connected" to one another act on one another.

(2) Take all patterns that were recognized in other processes during Step (1), let these patterns be the new set of processes, and return to Step (1) (1994, p. 152).

This equation maps creative processes on all levels of reality, including the social. The creative progress of science, for example, begins with recognized patterns (theories), which are brought into question by novel information. In the uncertainty generated by non-conforming information (e.g., blackbody radiation in the 1890's), some scientists allowed "all the processes that are 'connected' to one another act on one another" (Goertzel), and they came to recognize a new pattern of information (Planck's constant, and eventually quantum physics). Then the process repeats itself; e.g., quantum physics confronts thermodynamics and generates the science of nonequilibrium thermodynamics.

The process of science, as conceived in terms of Goertzel's cognitive equation, which was just described, bears remarkable similarities to other explanations of knowledge and reality:

- The description of a dynamic component-system, provided by Csanyi and Kampis, which has physical components that are constantly assembling and disassembling *and* some components have functions that influence the probability of the genesis or survival of other components of the system (Csanyi and Kampis, 1991, p. 78). The cognitive equation specifies that the physical components are patterns existing among physical entities *and* the exact nature of the function they have on each other (pattern recognition).
- The Hindu trinity of Brahma, Vishnu, and Shiva that continually create, maintain and destroy the universe. Brahma creates. Vishnu maintains. Shiva destroys.
- In a variation, this process is described as the breath of Brahma. When Brahma breathes out he creates the world in all its articulated diversity. When he breathes in the world is gathered in undifferentiated unity. The in-breath is the process in which "all connected things act on each other (material interaction). The out-breath is the process in which particular patterns take shape.
- The description of Nirguna Brahman and Saguna Brahman by Shankara. Nirguna Brahman is Brahman without qualities; it is undifferentiated consciousness. Saguna Brahman is Brahman with qualities: he is creator and governor of the phenomenal world (cf. Reese, p. 526).
- *Wei wu wei* (doing without doing): The manner of acting that is portrayed in Zen archery and in Luke Skywalker's trust in the force. When acting in this manner, we let go of all our attempts to force reality into our concepts and programs. We go with the unconscious flow, letting all the patterns around us and in our minds act on each other. In this state, we do the super-pattern that emerges in our action. In this doing without doing, our higher body-mind enacts the super-pattern that it discerns in the plethora of conceptual, and performance patterns that fill our conscious mind.
- The oft-used model of the creative process, in which one fills one's head with information, sleeps on it, and finds unsuspected solutions, story lines,

syntheses, etc. In this case, the juxtaposed plethora of patterns are mixed together in sleep and allowed to cross-germinate. The cognition of the following morning consists of the pattern that was recognized among the patterns of the previous day (cf. Ghiselin, 1952).

Goertzel may have tapped into an archetype.

In Goertzel's conception, a single self-reproducing pattern cycle constitutes a simple cognitive map. This self-reproducing pattern cycle begins the process toward permanence and reliability of other more-complex patterns. These patterns are representations in a sense that is less strong than a re-presentation of a pre-given world. They are patterns that are associated with a stimulus. (These patterns are not reflexively grasped by organisms; they exist as material implications).

These patterns sever some aspects of reality from an amorphous background and thereby create meanings, cognitive maps, belief systems, and virtual worlds. These meanings (for animals) are maintained in nervous systems that as habitual patterns of functioning (those patterns which have proved successful).

ORIGINS OF LANGUAGE

Bickerton, who is a linguist, advances a theory that is complementary to that of Kampis and Goertzel. He traces the evolution of representations and categories in species. Categories (in the evolutionary linguistic system of Bickerton) are particular patterns of neural activity; they are concepts or protoconcepts.

Protoconcepts are, in their origin, non-referential information, based upon molecular (non-enduring) signs. They express semiotic information even though their hosts are not aware that they are communicating. Bickerton goes beyond this concurrence with Kampis. He traces the progression of protoconcepts, which form a species' Primary Representation System, to concepts, which form the Secondary Representation System of certain species of animals.

Some of these categories are innate as is revealed in a baby monkey's instinctual fear and avoidance of a snake. Activation of the category 'snake' links with other neural connections that automatically produce fear and avoidance. 'Snake' is a fuzzy category in its innate form: a naïve baby monkey can be excited by a waving piece of hose. 'Snake' can, however, be sharpened by experience; then it becomes a learned category. Such categories can be extended to other realities by verbal or preverbal metaphor.

The categories that animals and species create are intimately related to the kinds of differences that they can react to (knowing requires doing). If they cannot react to a difference, that difference is not information for them. They build their categories according to their evolutionary requirements: what they eat, what other things they need, what they fear, what their preferred strategies of survival are, and how those strategies interact with the strategies of other animals and species.

Categories serve needs for security, for quick understanding on the basis of signs
and sounds, and for means of abbreviation.

When an animal perceives something, it selects what seems to be the
appropriate category from the set of all categories in its reference system. The
categories and the category system are adaptable. All species seek to extend their
representations in ways that will be advantageous to them and to extend
generalizations over their experience. With humans, the search for regularity and
order is compulsive and a matter of self-awareness. The processes that we use to
impose order (binary division, parcelation, class membership, exhaustiveness,
causal primacy, and identification) are available to non-human species in varying
degrees.

Categories accumulate as patterns of activity in ganglia and ever-larger
nervous systems. (In human evolution, some of these categories enter spoken and
written language.) They create modules of neural activity that link with other
modules and integrate into ever-more-complex primary representational systems
(PRSs). Their accumulation builds an infrastructure of neuronal networks that
gradually integrates spatial matrices, physical behaviors, social activities, and
personality features of conspecifics within species-specific PRSs.

An animal's cognitive map (its PRS) is its system of reference as a whole.
It is the basis of language, which is understood as "a system of representation, a
means for sorting and manipulating the plethora of information that deluges us
during our waking life" (Bickerton, p. 5). The intermediate step between PRSs of
our evolutionary forebears and human language are the SRSs of those forebears as
they are manifested in signal communication and protolanguages.

Human cognition has its evolutionary basis in the learning processes and
cognitive maps of animals prior in the evolutionary tree and the history of the genus
homo. We have a basic-level preconceptual structure based upon inherited
"capacities for gestalt perception, mental imagery and motor movement. In
domains where we have no clearly discernible preconceptual structure, we import it
via metaphor" (Lakoff, 1980). Even the words we use to describe psychological
reality are metaphors built upon bodily experience. When we reason, we draw
analogies from the felt interrelations of our body and the relation of our body with
various external objects.

Language evolved when the bifurcating structure of our representational
system, based upon processing arrangements such as "x or not-x," was inserted into
our protolanguage to provide syntactical structure under the pressing urgency for
more flexible communication in our evolutionary past. This process was facilitated
by a concurrent evolution of our respiratory system that permitted us to exhale
through our mouths.

CONNECTING THE POLES

Progression from the simple knowing required for autopoietic survival to the knowing accomplished by syntactical language connects the poles of information-for and information-about. It exhibits the wondrous capabilities of self-organization in living things. It exalts this self-organizing process as the most truly creative part of us. This process generates our linguistic rationality. It continues to surpass our bounded rationality by opening up new avenues of rationality.

OBSERVATION, AGENCY, AND MEANING

Where theorists find the origin of the observer, agency, and meaning is dependent upon their notions of cognition and representation. Maturana and Varela state their position as follows:

> With language arises also the observer as a languaging entity; by operating in language with other observers, this entity generates the self and its circumstances as linguistic distinctions of its participation in a linguistic domain . . . Meaning [also] arises as a relationship of linguistic distinctions (Maturana and Varela, 1987, p. 211).

In the context of their thought, this quotation states that observation, agency, and meaning arise only with the emergence of human language. In other words, human observation, agency, and meaning lack evolutionary forebears.

Other authors take pains to identify such forebears. Bickerton could accept the above quotation, but only by using a concept of language that harks back to plant representations. Luhmann holds that every autopoietic selection (therefore the selections of every living thing) creates meaning, a self-environment differentiation, and a kind of agency. Kampis identifies self-reproducing cycles that transform non-symbolic (autopoietic) information into cognitive maps that evolve in complexity. While such maps may not be re-presentations of pre-existing realities, they do define species-specific virtual worlds for species. Goertzel develops the cognitive equation into a universal equation of self-generating systems in which beliefs and belief systems would have their analogues in plants and animals. The argumentative force of these other authors strongly disputes the position of Maturana and Varela that representation, observation, agency, and meaning are the preserves of human linguistic cognition.

This is not to say that human cognition and animal cognition exist in some kind of continuum. As Bickerton emphasizes, the differences between human language and animal communication, even that of apes, parrots, and dolphins, are qualitative; the discontinuities between them are immense. Human language has a syntax and semantics that has become self-reflective. It creates a system of reference that is qualitatively different from the cognitive maps of other animals. If terms are understood in their prototypical senses, human language does create human observers, human selves, and human meanings. These prototypical cases have evolutionary support, however, in non-prototypical languages, observers, selves, and meanings.

EXPECTATIONS AND NAMING

Meanings, in Luhmann's sense, provide the patterned stability upon which persons, roles, programs, values, norms, and cognitions are built. They make possible strategies of mutual anticipation, ambiguity, and generality that allow systems to survive and evolve. They are the patterns that drive the cognitive equation.

When an autopoietic system is faced with information, it makes a selection. This selection creates meaning. In concert with past meanings constituted by selection, this meaning constitutes the system's cognitive map. In its ongoing selections, the system builds a world of objects in the flux of its constantly reorganizing environment.

We assign linguistic categories to material implications as they enter the fringes of our awareness by naming them. Our naming of psychic and social realities is not as straightforward as our naming of physical objects because physical units are much more discrete. Many physical units are component- systems that have their own internal cohesion. Even when physical things are not component systems (consider a pile of rocks), they can be specified by their space-time coordinates. Psychic and social realities lack this material specificity and they can be recognized as autopoietic and/or component-systems only after considerable cogitation. As a rule, we do not fine-tune our naming and our interaction mode to suit the individuality of physical, psychic, or social realities. Instead, we use uniform interaction frames that approach those objects violently and nonspecifically.

Naming is fundamental to our human mode of language-mediated cognition. The process of naming is not, however, devoid of drawbacks and distortions:

- Nietzsche and Goertzel point out that, if we try to know anything by words alone, we "succeed in becoming conscious only of what is not individual, but average" (Nietzsche, 1974, p. 299)
- The same duo points out that names, as they are used in logic, are based upon a fundamental lie. "Logic is bound to the condition: assume there are identical cases [and identical things having the same name]. In fact, to make possible logical thinking and inferences, this condition must first be treated fictitiously as fulfilled" (Nietzsche, in *The Will to Power*, quoted by Goertzel, p. 58).
- We tend to reify names, giving them the status of "things." We do not remember that they are merely fuzzy sets, based upon metaphor and metonymy that more or less accurately identify individual things. Information at its root is a significant difference in our experience. We talk about it,

however as if it were a thing that could be stored, processed, sent, and received.

- We easily confuse the meanings of ideas that carry the same name and mistakenly make logical inferences based upon incorrect assumptions.

Even if names were not problematic in these ways, we still could not routinely rely on logical analysis to select and coordinate our beliefs. We simply do not have the inclination or the leisure required to logically process the tumult of differences that daily bombard us. Instead, we use a rough, approximate (usually unconscious) method that sorts out potential new beliefs in terms of how well they conform to our existing belief systems. If fragmentary beliefs do not fit into our existing belief systems, we discard them.

NEGATIONS AND CONTRADICTIONS

Negations are critical to our survival. They help us find the first thing to climb to get away from a bear. They provide simple binary choices for the weak but parallel processors of organic, mental, and social systems. They provide a reversibility (as a negation of a negation) that allows reversals of course during our bold forays into the unknown. They enable a system to embody the multiple contradictions of its immune system for dealing with contingencies.

Selection requires a structure that is a unity of contradiction. It requires a system to decide for one alternative and against others, and in the process, to hold all alternatives simultaneously as possible expectations. The meaning a system creates is defined by the meanings it rejects. In this unity of contradiction, a system also creates differences between itself and its environment.

When a system is faced with a contradiction, it reaches the limits of its observational powers. It cannot be sure of the success of its impending selection. It does not know whether its expectation will be fulfilled in that selection. Should it be conservative? Should it be bold? It cannot know by mere observation. In this situation, a system can only take action, hope for the best, and stay alert.

Contradictions knit together insecurities about expectations by indicating the tenuous nature of both tightly conservative and undisciplined selections. They enable a system to deal with successful *and* unsuccessful expectations. Before a system *does* something, it prepares for its jump into the unknown by its strategies of expectation. After it does something, it evaluates its behavior in the context of reversible negation. In this way, the temporal reproduction of the system is a continual reiteration of feedforward, action, and feedback. Contradictions in observation force action; action (or selection) creates new meanings that require new observations and strategies for facing new contradictions.

THE REALITY OF COGNITIONS

Our representations have no reality independent of our minds and languages. They do not re-present an existing reality that is present to us. In the course of evolution and ontogeny, we have fabricated these representations, imposed them on our experience, justified them, and come to rely upon them. With all these representations, however, and all our refinements of self-observation, self-description, reflection, and theories of reflection, we never achieve a privileged access to knowledge. We remain bound to self-observation.

Postulating Reality

While we are bound to self-observation, we are simultaneously bound to make selections in the complexity of our existence. In this stressful condition, we have to postulate unities based upon differences of some sort. In order to make a selection, we have to hold something to be true. It is the holding to the truth of something that enables our survival. The actual "truth" of what is held is worked out in subsequent history. Later on, we can judge our decision on the basis of its conformity to patterns (of nature, life, cognition, and communication) and its productivity.

A chief function of consciousness is to manufacture social reality (Nietzsche, 1974, pp.298-300; Goertzel, 1994, p. 108). Consciousness produces categories, narratives, models, and metaphors. It applies them to situations and cycles them through cognitive and perceptual processes. The postulated categories give shape to the world. The application of those categories and their recycling turns postulates into "reality."

As describers of an autopoietic system, we can say that it has a virtual world composed of representations that have helped it survive. Central to that world is the differentiation of system and environment. The interaction between what is within and what is beyond its power leads a system to note this difference. Constant interaction that leads the system back and forth over the self/other boundary gives system and environment the redundancy of reality. Other components of the virtual world achieve reality in similar ways.

Living Reality

As autopoietic realities ourselves, however, and situated in the middle of that autopoiesis (that is, not as observers), our representations *are* our reality. We can realize this truth in reflection upon self-reflection. We constantly act on the (unreflected) assumption of reality.

The sharp distinction between "our concept of reality" and "reality" is mistaken for two reasons. First, there are degrees of internal coherence present in the world. Component-systems have more internal coherence than non-component systems. Component-systems have an ordered complexity that is easy to enumerate, like a mother duck and seven ducklings in a row. Component-systems do exist in the world.

Non-component systems, on the other hand, exist as systems only in our information about them. *The Guinness Book of Records*, for example, as a list, can be identified only if one knows some or all of the entries in it. A similar situation holds for made-up psychological abstractions, such as "attention deficit hyperactivity disorder (ADHD)," which requires pages of definition in *DSM-IV* and inspires many volumes of scholarly dispute. Sociological notions, such as "progress," need similarly long and controversial descriptions.

Second, knowledge with its observations and differentiations is in neither the observer nor the object. It exists as shifting patterns of patterns that we produce, project on the world, and adjust in constant interaction. Even E=mc2 is a pattern produced by Einstein that we project on the world and constantly test. [It is almost universally confirmed, but seems not to apply to longitudinally propagating scalar (Tesla) waves (cf. Laszlo, 1995, p. 32).] The knowledge is in the interaction.

Our knowledge is emergent and exists in the plane of semantics. It "cannot be reduced to features already present in the object or in the subject" (Luhmann, p. 486). Semantics presupposes both "pregiven characteristics and projections into the environment that are relative to the system" (Luhmann, p. 486). The world of practical semantics is created, tested, and believed by autopoietic systems (living, psychic, social, and cultural) in a process that is co-determined by the projections it makes and the feedback it receives in response. It is the result of system-environment communication and the condition of double contingency.

Our knowledge of ourselves and our world progresses in an auto-catalytic fashion. It reorganizes material that has self-organized in matter and our line of vital and cultural evolution over four billion years. Some of this material is in reality external to us. Some of it is our internal semantics. This accumulated knowledge exists as our belief system and it follows the cognitive equation.

"Universe" and "World"

Goertzel's concept of "universe" is the hyperset containing both mind and physical reality. Luhmann's concept of "world" describes the unity of difference between system and environment. The two concepts bear a marked similarity.

According to Goertzel, the "universe" evolves according to the cognitive equation (which is a universal equation of self-organization). This universe is rooted in deep evolution as a collection of dual networks. It presupposes, therefore, universal interconnections and a structured universal memory of some sort, perhaps the holographic Zero-Point Field proposed by Laszlo. Goertzel's perspective is the "objective" one of the observer and model-maker. He believes that he is describing what really happened and what is going on.

Luhmann's perspective is more autopoietic. It involves the way we conceptualize. In the autopoietic viewpoint, we hypothesize that we never attain a privileged objective viewpoint; therefore, all our efforts in forming systems and constructing reality are self-referential. This self-referential closure opens the world to ever-new conceptualization.

The "world" that Luhmann describes is not some mythical world before the fall from grace, that is, before differences were introduced into it. His "world" is the world after the fall. It is the universal unit of conceptual closure subsequent to the difference introduced by language. It is everything we can think about and every way that we think about anything. It is an ultimate concept that contains all differences. It is the world as we do and can know it.

If we construct a subset of Goertzel's idea of the "universe" that we might call "concept universe," that subset would have the same domain as Luhmann's "world."

AN EVOLUTIONARY PERSPECTIVE

The intimate connections among our surviving, knowing, communicating, representing, and remembering are functions that constitute us as existing patterns of behavior, that is, as organisms, people, and social groups. The coherence and emergent complexity of these functions offers new baselines for understanding our social interactions and ourselves. As organisms, at the level of basic autopoiesis, surviving and knowing are essentially the same thing: they are the doing that is life. This elemental *doing* involves elemental communication and representation.

At its basis, communication is the semiotic component of the reproductive process, which underlies the organic world. It is the recognition of patterns, upon which successive generations of component-systems model themselves. The patterns themselves are existing functions and representations. This organic reproduction on the basis of patterns and some kind of physical memory can be postulated to exist in the non-organic world. In one form of such a physical memory, which is proposed by Laszlo, patterns are "remembered" at the quantum level. That is, they are physically retained in the holographic patterns of the quantum vacuum.

At the basic level of ongoing survival and evolution, the patterns of reproduction and existence are material implications. That is, they are influential functions that are not formalized in the description of some outside observer. Nevertheless, a modicum of internal observation is present in every autopoietic reproduction. An organism differentiates its less complex patterns from the overwhelming complexity of its environment. Also, in the hyper-abstraction of the cognitive equation, these processes of material cognitive reproduction can be formally stated. Thus, material implications, which escape description by any observer, are described in hyper-abstract formal terms. Parenthetically, this ability to state anything in a proposition conforms with Habermas's analysis of propositional speech acts.

With the aid of the insights and structured explanations provided by the authors represented in this research, an articulated sketch of the expanse of communication and cognition can be drawn. The stages of that sketch, which will not be attempted here, involve the following major elements:

- Holographic memory in the Zero Point Field of the quantum vacuum
- Structure as pattern reproduction
- The universe as a component-system that is built up as the four universal physical forces and the fifth force of quantum memory combine to excite particles and to combine them as component-systems
- Universal pattern reproduction in component-systems on the model of the cognitive equation
- The development of dual networks (hierarchical and heterarchical) that possess continuous compositionality in component-systems because of the action of the cognitive equation
- Material implications as the interactive functions of components
- Referential information, which has the nature of molecular signs, as the communicative link that binds material implications together
- Non-referential information as the reflexivity upon referential information that enables the process of symbolization
- The presence of self-reference in every selection of an autopoietic system
- The presence of the seeds of representation in plants
- The evolution of representations in the nervous systems of animals to the point where animals possess a Primary Representational System (PRS, cognitive map) of their internal and external environment
- The progress from PRS to Secondary Representational System (SRS, signals) in some species
- The progress within the SRSs of some species to "signal languages" and "protolanguages." In the genealogy of human language, ape use of words, "pidgin languages," and the language of babies are examples of protolanguage.
- The progress from protolanguage to fully human language in which the internal rules for ordering cognitive maps and survival thinking become explicit as syntax.
- The essence of human communicative action as the management of situations of double contingency

CONCLUSION

This list, like the last one, is not meant to be either exhaustive or articulated in its long and deep logic. It is given as a semi-ordered list of germinal ideas and structures that bear promise of synergistic organic growth. It indicates the profound relationships that exist between communication, cognition, and the structure of the universe.   It also indicates a possible framework for a systemic social analysis within a humanistic design.

Such a humanistic social analysis would unfold how communication and cognition, as understood above, are shaped by human intentionality and action. It would provide a convincing systemic rationale for human creativity and freedom.

*   *   *

If our concepts, theories, and beliefs do not re-present some existing reality that is present to us, then how can we assure ourselves of the validity of our ideas? That is the topic of the next chapter: epistemology.

CHAPTER 22

EPISTEMOLOGY

ABSTRACT

The considerations of the last chapter addressed the question: "How do we know something?" This chapter concerns epistemology: "How do we know that knowledge is reliable and valid?"

Natural epistemology is a necessary consequence in any theory that denies a priori assumptions. An assumption utilized in such an epistemology is recognized as being (a) the at-one-time hypothesis of someone (b) that has been tested and (c) has been found to be reliable. In a systemic, evolutionary account of the social world, the norms for judging a theory's validity have to self-consciously reflect the evolutionary processes that generate those norms.

This chapter develops several aspects of a self-referential epistemology. It considers the inherent circularity of knowledge and how its circle is broken. It explains the autopoiesis of knowledge and develops a self-referential epistemology. It relates self-referential epistemology to Quine's naturalized epistemology. It generates norms of validity on the basis of redundancy, test, and other criteria. It portrays a mimetic resemblance between self-referential logic and the scientific method.

CIRCULARITY

The Circularity of Knowledge

Conceptual knowledge is circular. We know everything in reference to something else. The science of semantics tends to make this a vicious circle. It contends that signs can be interpreted only in reference to each other. "Words are merely arbitrary symbolic conventions whose meanings are determined by other, equally arbitrary words" (Saussure, quoted by Artigiani, p.33). They do not directly signify anything in the external world. (cf. Lacan, p. 150).

The inescapable circularity of knowledge creates problems in knowledge theory. In the grand perspective of accumulated human wit, our knowledge produces all the units that it uses. In evolutionary history, our knowledge begins with the original formation of living cells. How do we advance given this circularity? How do we break out of the circle?

Asymmetry

We break the circle by creating asymmetries. We assert our will to power and simply declare something to be so. (When faced with a logical impasse, we just do something.) We tie our knowledge down to something that we project outside our self-enclosure. When we test our assertion repeatedly and find that it works for us, we add it to our reality bank.

Luhmann says that it is necessary to project differences, facts, and themes on the basis of basically arbitrary selections. These projected differences create the fictions that are needed for coping with the complexity of the world. Nietzsche and Goertzel find asymmetry even in the necessary fictions of logic. Nietzsche says that our will to power creates logic by declaring that unique things are equal. Goertzel says, "Logic is a lie, but a necessary one" (p. 58).

The Hermeneutic Circle

Hermeneutic science attends to how we interpret texts. How can we understand a text if all the criteria for understanding are interior to the text? How can we make sense in this circularity.

The hermeneutic circle is a virtuous circle. It is a process of assuming a general idea of the whole text and then continually adapting its meaning "because of the confrontation with the detailed parts of the text" (Polkinghorne, p. 227). Ricouer explains the circle in terms that Luhmann and Goertzel would appreciate. To break the circle, we guess first and then validate our guesses, changing them if necessary (Ricouer, p. 91).

Projected asymmetries break the circularity of knowledge. In the fabric of self-reference, they generate the realities and themes that enable interaction and evolution. The disciplined fictions of logic help us to a measure of referential transparency. Taken together, external and internal asymmetries create our worlds. They also place us on the edge of chaos where major breakthroughs in conceptualization and intimate nuances of understanding are possible.

Circular Epistemology

It is often objected that any natural epistemology involves a vicious circle in reasoning and creates self-sealing doctrines. Those objections as applied to self-organizing knowledge can be met as follows:

- All knowledge derives from autopoietic processes and involves a circular process.
- Epistemology cannot escape circularity because it is also an autopoietic creation.
- Autopoietic knowledge is not self-sealing because it must pass the justifiable criteria that are proposed by non-self-referential epistemologies.

REDUNDANCY

The circularity of knowledge precludes grounding its validity in final elements because all available elements have been produced autopoietically. Our theories and ourselves result from autopoietic processes. We cannot validate the authority of our theories from a position outside the autopoietic circle. Our knowledge is not grounded in direct observation of any object or process either. How then do we come to trust our ideas about our worlds and ourselves?

Luhmann explains that trust on the basis of redundancy. We trust ideas that have worked for us in the past, that are tried and true over and over, and that are shared by our social group. Those ideas that have worked in human history and the evolution of our race have the deepest redundancy for us, and they are most deserving of our trust.

Goertzel specifies how the elements of knowledge become redundant within the framework of his mathematical psychology. For him, a fact is an "interpretation which someone has used so often that they have come to depend upon it emotionally and cannot bear to conceive that it might not reflect 'true' reality" (Goertzel, p. 59). Facts are patterns that are produced by consciousness whose job is to produce reality. They are generated in a dynamic feedback process that cycles patterns through higher-level (unconscious) cognitive processes and middle-level perceptual (linguistic and phenomenological) processes. "A pattern only acquires the presence, the solidity that we call 'reality,' if it has repeatedly passed through this feedback loop" (Goertzel, p. 109).

THE AUTOPOIESIS OF KNOWLEDGE

Attaining, retaining, testing, and extending are autopoietic processes of knowledge. Knowledge continually reproduces itself by incorporating new patterns while retaining unity in its every reproduction. Luhmann says that autopoietic systems continually dissolve and reconstitute themselves while maintaining themselves as "accompanying" self-reference. In Goertzel's framework, a self-generating system reproduces at time **t** patterns that existed among its components at time **t-1**. In the redundancy of this autopoietic reproduction, a system gains a sense of self.

Autopoietic reproduction is not merely redundant. As Luhmann says, self-reference is not "a complete duplication of whatever functions as the self at any time" (Luhmann, p. 460). Self-organizing systems metabolize with outside energy and information. They expand their circle by coupling with information in their environments. A difference in its environment becomes information for the system when it selects an action that accommodates that difference.

EPISTEMOLOGY OF SELF-REFERENCE

Self-reference is a thematic asymmetry that we impose upon our reality. This imposition has major epistemological ramifications. If self-organization produces us and the world as we know it, then our knowledge continues that process of self-organization and our criteria for trusting that knowledge have developed in the same way.

Our epistemology, therefore, in the broad context of evolution, is also self-organized. It consists of pragmatic rules that have been developed and repeatedly tested for utility and logical coherence. These rules carry great weight because of the redundancy provided by their continual testing and their prominence in our cultural belief systems. In the final analysis, however, self-referentiality and circularity lie behind any rules we derive. Our rules originate as guesses and gain compliance because of their utility.

Any theory of knowledge, any epistemology, that aims for universality must apply to itself the standards that it applies to its field of study. If scientists of social theory find contingency and self-reference in the evolution of social processes, then they are bound to admit that their theories (which are particularized manifestations of evolution and social processes) are contingent and self-referential.

Even pragmatically universal principles, such as those proposed for communicative action by Habermas, are based upon the contingent evolution of our organisms and our language systems. Habermas, of course, recognizes (as does any post-modernist) that his standards or truth, rightness, and truthfulness are contingent creations of human history. Classical science fails to recognize this reality. It is caught in an insoluble paradox. It claims to be complete but it is "incapable of describing the scientists who produce it or of postulating existence without scientists to observe it (Artigiani, p. 51).

NATURALIZED EPISTEMOLOGY

Self-referential epistemology bears many similarities to Quine's "naturalized epistemology" (1969, pp. 69-90). The principal similarities are a denial of absolute norms, a recognition of circularity, and an emphasis on holism.

Quine holds that there is no role in epistemology for a philosophical theory that is independent of the sciences (cf. Hookway, 1994, p.465). A definition of this naturalized epistemology is provided by Solomon as follows:

> Epistemology is a science that identifies generalities (i.e., general laws) about the effectiveness of particular reasoning processes and methods that are applicable across all contexts-such as subject domain, stage of development of discipline, goals of inquiry, persons and groups conducting the inquiry (Solomon, 1995, p. 353).

Recent thinkers

> have broadened the scope of their naturalistic epistemological projects to include social (encompassing anthropological, historical, institutional, political, sociological) factors influencing reasoning. This broadening of scope is no longer viewed as controversial, but instead as a new and fruitful set of directions for research, dubbed "social epistemology" (Solomon, 1995, p. 354).

Self-referential epistemology emphasizes the systemic and evolutionary processes that create our social epistemology. In this, it follows a strain of development suggested by Quine, who suggested an "evolutionary epistemology" built upon blocks of inductive reasoning shaped in evolution (cf. Quine, 1969, p. 90).

Luhmann adds an important new consideration. "Quine . . . clearly emphasizes the connection between the 'naturalization' of epistemology and the acceptance of circularity, but he fails to see that reality is also structured circularly, independently of knowledge" (1996, p. 615). Because our reality, including our social and psychic realities, is the product of circular processes of self-reproduction, there is a conformance between our reasoning processes and the processes of nature, including our own evolution.

EPISTEMOLOGICAL RIGOR

Luhmann and Warfield offer complementary, requirements for attaining epistemo- logical rigor. Luhmann offers a self-referential theory of science and practical rules for attaining practical certainty. Warfield develops a complete science of generic design.

Luhmann

Luhmann believes that the "conventional theory of science in the natural sciences does not reckon with the epistemological problems of global theories" (1995, p. 615). A global theory must justify not only some particular facts but also its own theoretical framework. The importation of unexamined standards into such systems is unwarranted and debilitating. In particular, "concepts like emergence, self-reference, and entropy/negentropy . . . [have] a position of prime importance, which theories of science must honor, because they concern the genesis both of systems and of the possibility of observation" (Luhmann, 1995, pp. 480-481).

The circularity of epistemology that is present in global theories is forced upon us. "One cannot avoid them [circles]. One can sharpen them as a paradox and leave it at that. But one can also build them into the theory of science, for they contain precise instructions for self-control" (Luhmann, 1995, p. 482). The logic of this science goes as follows:

- The minimal requirement of theories is that they must "always be formulated so that their object is subject to comparison.
- "If they themselves appear among their objects, they must subject themselves to comparison.
- "As their own objects, they must continue to function under the pressure of comparison.
- "Whatever is attained in the system . . . [in the way of knowledge] must also prove its worth in the theory, however unpleasant (e.g., relativizing) the result of the self-comparison may turn out to be" (Luhmann, 1995, p. 482).

Self-referential circularity requires every science to acknowledge its roots. It requires physicists, for example, to acknowledge that they themselves and the way they conduct physics are physical processes. Also, "if one knows that all judgments are based on previously established categorizations, that is, rest on pre-judgments, then research into prejudice must recognize itself a research about itself" (p. 482).

Luhmann offers several ways that we use to assure ourselves that our knowledge is true. First of all, one supplies reasons for knowledge. Then since these reasons "merely transform the circle into an infinite regress," one places one's hopes in approximating reality evermore closely. "If one in turn justifies the reasons and keeps every step of this process open to critique and ready for revision, it becomes more improbable that such an edifice could have been constructed without reference to reality. The circularity is not eliminated. It is used, unfolded, de-tautologized" (1995, p. 479).

Luhmann also answers a similar question: "How can one guarantee that observation maintains contact with reality when it claims to be knowledge, even scientific knowledge?"

> First, one can move the site of knowledge claims from psychic systems to social systems. Social systems can be made independent of individual motives and reputations. The knowledge of social systems "can be subjected to its own conditionings, perhaps in the form of 'theories' and 'methods.' Knowledge can also be evaluated on its productivity and its ability to generate new knowledge. Second . . . One could force global theories "to test on themselves everything that they determine about their object" (1995, pp. 484-485).

Warfield

Warfield, basing himself on Peirce, integrates the embodied human knower in his a model of epistemological rigor. He poses four universal priors as the foundation of science: the human being, language, reasoning through relationships, and means of archival representation. A solid and productive science must attend to all of these priors. The natures of human beings, their abilities, and their limitations determine what science is possible and what its limitations are. Language, both specialized and everyday, needs to be understood in its relationships to other languages and to the things it represents. Science must have referential transparency; that is, it needs to make clear its relationships of meaning and logic. Finally, science must attend to the understandability and reliability of how it is recorded.

The "object language" of science is mathematical logic. It functions with utmost generality. It makes possible structures of deep and long inference and assists the human mind to overcome the limitations of short-term memory. It needs to be supplemented with a natural language that serves as its metalanguage.

In ordinary language, a word creates a single "fuzzy set" expression for the patterns that are related to its occurrence. "In other words, everything which relates to the word 'Ice' is part of the meaning of the word" (Goertzel, p. 69). Warfield proposes that definitions by relationship can supply semantic precision for sociological concepts. In these definitions, a concept's relationships to other factors are stated in terms of comparison, influence, time, space, and mathematics. He contends that the validity of any science depends upon the capacity of the scientific community to construct Definitions by Relationship for the full complex of relevant concepts involved in the science. Warfield actually constructs definitions by relationship for the key concepts of generic design. In so doing, he furnishes a model to other areas of social science.

The physical sciences have several methodological advantages over the social sciences. (1) They deal with discrete (numerable) entities or with ones that can be easily quantified. Therefore, (2) they can easily use numbers to order entities in a universally understood way. (3) Physical sciences do not have to attend to complicated definitions and histories of their concepts. (4) Physical science is standardized.

The human sciences do not have these advantages. In those sciences, many important qualities are not clearly distinguishable. The cool clarity associated with numbers cannot be transferred to those qualities merely by attaching quantifications to them. The social sciences lack standardization. They must attend closely to their core definitions and their Referential Transparency. In addition, they must attend to the abilities and limitations of the human beings doing social science and they must be clear about their use of object language and ordinary language.

Fortunately, science and mathematics do not require numbers. They merely require clarity. Validity in any science requires referential transparency. In addition to precise definitions, referential transparency requires strings of long and deep logic. Because much of social science lacks access to the "cool clarity" of numbers, it must specify its foundation, theory, methodology, and applications. It must explicate the circular relations between these stages by identifying (1) the postulates that intervene between foundation and theory, (2) the selection criteria that are used to create methodology from theory, (3) the roles and environment in which the methodology is applied, and (4) the evaluation feedback that applications supply toward improving the design.

CRITERIA OF VALIDITY

Epistemology establishes criteria for judging the validity, reliability, and probability of statements and systems of belief. Some criteria are almost universally accepted: coherence, compatibility with empirical results, openness to test, and ability to predict results. Some additional criteria are the following:

- Habermas says that the ideals of truth, truthfulness, and rightness are essential to both language and the social framework. In particular, these qualities are the essential, although counterfactual, bases for deciding matters of value and justice.
- Luhmann proposes quasi-absolutes that must be served by theories, although he does not make an issue of that fact. He says that problem of double contingency is solved when individuals grant freedom, intelligence, and personhood to each other. According to him, human culture rests upon the fragile foundation of that mutual respect. Under the presumption that the preservation of human culture is an essential good, these qualities of respect for individuality, freedom, and intelligence become essential goods that must be served by theories.
- Complexity theory sees a goal that is approached by all self-organizing systems-a position on the edge of chaos.
- Goertzel expresses this goal as maintaining a balance between circularity and dialogue.
- Evolutionary theorists aim to see physical and mental reality as part of a whole. They emphasize the imperatives of stewardship to the environment and future generations.
- Theorists of systems design emphasize the need of comprehensiveness in obtaining input from all the stakeholders of a situation.

From this list, it is evident that the criteria for judging theoretical validity go beyond the criteria of the scientific method.

SUMMARY

In self-referential epistemology, organisms use intuition (the recognition of pattern) in their every selection. They judge the "truth" of their selections on their productivity: whether they die, survive, or thrive. Organisms that live on the edge of chaos, balancing circularity and dialogue attain consistency in this kind of truth.

This epistemology illuminates blindspots that occur in mind-body (Descartes) and text-reality (Saussure) dualisms. It shows why we need our bodies to know. It explicitly states the kinds of language and reasoning that we need in social science. It employs criteria that have proven effective in the survival and progress of our race.

Epistemology, as implicit in systems theory, derives theoretical, practical, and ethical criteria of validity from the self-organization processes that create us. It displays a transparent evolutionary logic that connects us with our roots. It traces a self-organizing trail in which "Knowing as doing" progresses to become our communication, cognition, and epistemology. It identifies markers along the trail.

The validity claims of human cognition can be seen in an evolutionary perspective as extensions of pattern recognition, selection, survival, and productivity. Truth as internal coherence is an extension of autopoietic pattern recognition. Truth as authenticity (and truthfulness) is an extension of selections that conform to inner patterns. Truth as beauty is an extension of the coherence of newly emergent patterns with existing patterns. Truth as rightness is a selection that honors the solution to the problem of double contingency. Truth in the pragmatic sense is an extension of the tests of survivability and productivity. Other standards for true/ethical judgment and action, such as enlightened self-interest, concern for future generations, respect for the environment, and even "is it fun?" can easily be traced to evolutionary forebears.

NEW HORIZONS

Artigiani recounts a brief history of this century's scientific epistemology. Early in the century, scientists pursued Mach's ideal goal of science: to produce a "mimetic reproduction" of nature. Because of the anomalies of relativity and quantum physics, scientists gave up that ideal. Heisenberg, for instance, said "The atomic physicist has to resign himself to the fact that his science is but a link in the infinite chain of man's argument with nature, and that it cannot simply speak of nature itself (1955, p. 15). Bohr said, "We are suspended in language;" and "physics concerns what we can say about it in language (quoted in Artigiani, p. 41). Toward the end of the century, Prigogine's theory of dissipative structures reconceptualized physics in terms of evolutionary processes.

In this post-modern physics, nature observes itself in a manner analogous to that used by a scientist in a laboratory. "As science observes nature, nature is involved in continuous observation of itself" (p. 51). The correspondences are:

instrumental readings ~ perturbations;
laboratory conditions ~ environment;
alternative organization possibilities ~ bifurcations;
measurements ~ emergent structures.

In natural observation, evolving systems experience perturbations, react to them, and record their successful reactions as emergent structures.

At the end of the 20th century, the new science of emergent reality fulfills Mach's injunction (to produce a "mimetic reproduction") in a totally unexpected way. *The method of science is the method of nature.* "By proposing new models and metaphors, post-modern science mimics the processes by which nature self-organizes" (p. 55).

For Treumann, evolution is explained by "deterministic chaos, which is nothing else but . . . information dynamics" (p. 90). He finds a delightful surprise for science in the theory of information dynamics. He sees a common pattern at work in reductionistic science and determistic chaos. Reductionistic science finds itself dealing with deterministic chaos because it cannot avoid errors in measuring its subject matter. Deterministic chaos in nature arises because "nature is itself supplying these errors, because nature is itself measuring all the time" (Treumann, p. 94).

In other words, the measuring of science mimics the measuring of nature. As a result, science's description of the world is not only an idealization made by human minds. It is also a correct idealization because "[Nature] corrects itself and generates the chaos we are confronted with daily" (p. 94). The method of science is the method of evolution and, not surprisingly, it produces ever-more-precise representations of life.

INTERNAL EVALUATION OF THIS RESEARCH

The following considerations evaluate this research on the basis of the criteria that it has generated.

- **Self-Referentiality:** This research employs as a framing assumption the standard of employing only general and tentative standards of truth at the outset of its investigation. It develops articulated standards for applying systems theory to social processes by condensing the thought of systems theorists. In particular, it does not prematurely fix its epistemological criteria, but allows them to develop in the course of the research. Finally, it applies those criteria to itself.
- **Comprehensiveness:** This research considers a broad band of important systems theorists, but not, of course, all of them. In this openness with constraint, it fulfills the necessary requirements of all vibrant thought, which is

expressed, for example, as circularity and dialogicality (Goertzel) or as being on the edge of chaos (Kauffman).

- **The Requirements of Good Design:** In addition to addressing comprehensiveness, this research follows the prescriptions for good design that were developed in the enhancement pattern (chapter 19), especially its deep drivers.

  - It controls the velocity of the information (box 2) that it considers at any one time by breaking its task into discrete operations, and thus avoids information overload. In the preparatory stages, it generates individual summaries, precipitates key statements, progressively separates topics, and coalesces a limited number of standards. In the synthesis section of the research, it utilizes a tested design method, the CogniScope™ to simultaneously reduce the challenges of decision-making in very complicated contexts and speed the processing of that decision-making. In so doing, it introduces a new and powerful synthetic method to the realm of social theory. Good design is good design whether it exists in the practical or the theoretical realm. In the chapters that do not use computer assistance, I slow the virtual conversations down by breaking the synthesis into two stages: cluster-generation and synthesis-making.

  - It fosters innovative thought (box 3) by (a) considering the underlying structures exposed by systems theory in areas that do not relate specifically to social processes, (b) bringing many diverse perspectives into a relatively confined space (five chapters, many smaller subdivisions) to enable their synergy, (c) refraining from the premature closing that would have resulted from early attempts to cast the thoughts of various authors into a linear exposition.

  - This research did not *presume* beforehand that there was a consensus (box 4) about how systems theory might be applied to social processes. It did *assume* (guess or hypothesize) that there might be some kind of consensus, then it set out to test that assumption. Accordingly, it gave respectful attention to each author on his or her own expository turf. Only later, in the virtual conversations, did a basic consensus emerge.

- **Coherence:** This research is coherent in the way it matches its methods to its goals. It is consistent in its self-referentiality, and also in its openness to the contrary points of view expressed by the authors under consideration. It did not, of course, open itself to all the likely opposing points of view. Such openness, which is not possible in the confines of this book, awaits in the positive and negative reactions that this research may generate. The ultimate coherence of this research emerges only at its completion. At this point, one can survey what systems theory says about social processes in a broad and coherent way that enables a rational judgment on its applicability.

- **Compatibility with Empirical Results:** Highly theoretic research can rarely predict empirical results. It can be tested, however, on the basis of its compatibility with empirical science and its productivity in explaining diverse phenomena in a unified way. The systems concepts under consideration conform admirably with the new time physics of non-linear thermodynamics. They lay down universal laws of change. In particular, Goertzel's "law of cognitive motion," claims to be a universal law of time-directional change. At the present time this law is undergoing partial empirical test as "psynet AI" (self-generating, dual network, artificial intelligence; Goertzel, 1997, pp. 33-39). [Goertzel is presently working on this kind of AI in a commercial "Webmind" project] (Goertzel, 1999).
- **Respect for freedom, intelligence, and personhood:** This kind of basic human respect is exemplified in the enhancement pattern for design. In this pattern, every stakeholder has a voice and receives information at a respectful pace that does not induce "groupthink." Stakeholders are encouraged to generate novel insights and approaches. They are not coerced into agreement with a pre-existing consensus. They engage in debates that respectfully honor divergence.
- **Stewardship:** The enhancement pattern for design requires good stewardship of designer/stakeholders. Moreover, ideas related to cultural cognitive maps, collective belief systems, and lifeworld empower people to influence their worlds and design their futures.

CHAPTER 23

CONCLUSION

ABSTRACT

The preceding chapters present a coherent, dynamic model of a self-organizing world. Within this model is proposed a creative and ethical method of decision-making and design. In this concluding chapter, the relations between structure and process in the realms of knowledge and being are made explicit. The terrain this book has traversed is put into perspective. Finally, the new methodology that evolves in this book is crystallized. This methodology allows us to deal with enormous complexity, to relate and relate and relate ideas so as to drive out heretofore-unsuspected conclusions and syntheses. Therein lies the elegance and utility of this model.

STRUCTURE TO PROCESS

Our language starts with nouns; verbs come later. Our thinking processes follow the same course: things first; relationships later. This sequence has worked very well for us in our evolutionary past. It serves us well in our everyday lives. It allows us to focus on the relative simplicity of the practical matters that confront us and to generally ignore their surrounding context.

We cannot, of course, completely ignore the relationships between things, any more than we can discourse without using verbs. We usually try, however, to keep relationships simple in the manner of "A is like B," "A is included in B," "A is excluded from B," or "A causes B." In this way, we have generated the logical categories of identity, inclusion, exclusion, and linear causality.

Philosophy and science have followed this same sequence. They theorize about things and their relationships. Greek and Scholastic philosophy specialized in things and the static relationships between them using ordinary language. In the scientific revolution sparked by Descartes, Galileo, and Newton, the emphasis on things was supplanted by an emphasis on the mathematical relationships between them. The things in question were masses, levers, pulleys, points, and planets. The relationships were rates of descent, inertia, elliptical equations, gravity, etc. The language used was analytic geometry and calculus.

The scientific revolution of the 20th century had three stages: relativity, quantum mechanics, and non-equilibrium thermodynamics. The theories of relativity and the quantum developed early in the century. They unseated our presumption of pure objectivity and introduced indeterminacy into our calculations. In spite of their radicality, these theories tended to retain the idea that events in space and time were reversible in the manner of mathematical equations. The theory of non-equilibrium thermodynamics introduced a progressive arrow of time into physics. In this theory, matter self organizes itself in ways that are mapped in the mathematics of chaos and complexity.

In the history of our understanding of nature, we have moved along a continuum from structure to process. We have moved our emphasis from things, to static relations between them, to dynamic relations between them, to probabilistic relations among them, to the interrelated indeterminate relations that generate things. In summary, our understanding of nature has proceeded from nouns and things to relations and processes.

FROM PROCESS TO STRUCTURE

In physical evolution, the reverse is true: process generates structure. Any enduring structure, substance, or thing is at its root a self-reproducing process. On a practical level of reality, we talk of these self-reproducing systems as "things." It is important, however, especially in terms of theory and in turbulent situations, whether molecular or social, to remember that self-reproducing and self-organizing processes constitute these "things".

In arenas of rapid change, substances are ephemeral and self-organizing processes are substantial. For example, hardware and software appear and disappear monthly, but their underlying logics endure and metamorphose. At the quantum level, electrons appear and disappear in billionths of seconds, while the forces between them are constantly maintained. From beginning to end, relationships rule.

ITERATIVE SELF-REFLECTION

Systems theory is the science of relationships between wholes and parts, processes and structures. It is an integrative science. General Systems Theory aspired to integrate all the sciences, an aspiration that proved to be quixotic. As a result for the last three decades, systems theory has been applied to multiple disciplines; but in the process, it has not addressed integrating itself.

This book has consolidated multiple elements of systems theory into a cohesive integrated whole. It has applied systems theory to itself. It has integrated a fragmented integrative science—in accord with the ideals of GST.

In Chapters 1-17, it set the historical/evolutionary stage and summarized the work of numerous systems theorists. Chapter 18 applies Interactive Management (IM) methodology to the science of management. In so doing, it draws together a plethora of how-to ideas into an enhancement pattern that exhibits the dynamic relationships that exist among those ideas.

Three narratives in Chapter 19 illuminate the nature of our social world. They clarify the relationships between process and structure. They identify how self-organization occurs at strategic bifurcation points.

Chapters 20 to 22 manifest the self-referential nature of our knowledge and communication. They generate a mosaic that portrays the fundamental stages in our progress to literate and rigorous rationality. They show how structural coupling drives that process.

SELF-ORGANIZED CRITICALITY

A thousand and more statements were drawn together in the course of this investigation. Many of the ideas they contained had never before met each other. Their congregation created a necessary, but not sufficient, condition for creative bifurcation: critical mass.

The further necessary condition for creative bifurcation is intense interaction among those ideas. Together critical states of mass and interaction generate the Self-Organized Criticality (SOC; Bak) that generates multiple bifurcations. In this investigation, I used the following combination of approaches.

- Relating these ideas unconsciously in metaphoric maps. However, taken alone such unconscious processing often proves inadequate.
- Assisting these unconscious processes with conscious cognitive activity. In really complex situations such as the crush of statements generated in this investigation, such unassisted mental processing is easily overwhelmed. It must therefore be coupled with other methods.
- Using pen and paper along with methods of parcelation, comparison, saliency generation, and relationship accounting to assist cognitive efforts. These methods advance the synthesis process, but they are slow and stumble when faced with overwhelming complexity.
- Using well-designed computer programs in really difficult situations to expedite decision-making processes. These programs harness a rapid formal logic not available by other means and set the stage for creative bifurcations.

In the generation of this book, I carefully parceled hundreds of statements into five categories and whittled down their numbers on the criteria of novelty, saliency, and clarity. I employed the computer assistance afforded by the Interactive

Management (IM) methodology to create further affinity clusters and patterns of long and deep logic among those statements.

In the chapters on design and social structure, this combination of rationality and mechanical logic created interactions among the ideas massed in those statements. The combination of the mass of statements, intelligence (unconscious and conscious), and computer-assisted logic achieved self-organized criticality (SOC). The conclusions developed from this SOC came like ripe fruit falling from trees.

The truths concerning communication, cognition, and epistemology were developed without computer assistance, but they relied upon an IM-like logic. These truths did not drop into my lap. I had to climb trees in order to harvest them. SOC in this case required more exertion, but it was achieved.

If systems theory is applied to social processes in the manner exemplified in this book, it offers practical and ethical methods for advancing participatory democracy. Its use in theoretic arenas can facilitate cross-disciplinary conversations and progress toward disciplined consensus and disagreement. Systems theory, in its coherent view of the social world, offers a process-based understanding that avoids the logical and semantic difficulties that are created by structure-based conceptions of those realities.

AVENUES OF FUTURE RESEARCH

The synthesis that is described in this book invites future research in at least three areas: grounding, consolidation, and extension.

To adequately ground this synthesis, several research approaches are indicated:

- Detailed comparison of the self-referential epistemology sketched here with current theories of social epistemology, as they are indicated in the work of Hookway, Solomon, and others.
- Expansion of the concept of self-referential epistemology by including ideas of second order cybernetics as they are evidenced in the journal, *Cybernetics and Human Knowing*.
- Inclusion of explicit empirical applications of complex systems as they are reported in journals such as *Nonlinear Dynamics, Psychology, and Life Sciences*.
- Grounded criticism of the syntheses created in this book on the part of the authors surveyed, other systems theorists, and the community of sociologists and social philosophers. The content of such critique is only peripherally indicated in this book because of time and energy constraints. The detailing of such critiques would also be presumptuous because they are (for the most part) yet to be made. Forthcoming critiques will require detailed responses.
- Expansion and refinement of the self-organizing and self-referential concepts in this book to include the "new cybernetics" of Geyer and van der Zouwen,

the social entropy theory of Bailey (1990, 1994), the autopoietic work of Mingers (1995), the psychology of radiance (Combs, 1995), and similar work that is being produced.
- Consideration of recent influential publications such as Kondepudi and Prigogine (1998), Gabora (1999), Goertzel (1999), Gazzaniga (1998), Gazzaniga, Ivry, and Mangun (1998), Margulis (1998), Lakoff and Johnson (1999)

Research that aimed at consolidation would consider how other branches of systems thinking would add to or detract from this synthesis. Those other branches would include:

- cybernetic systems thought
- information theory
- information dynamics or
- deterministic chaos theory.
- semiotics

This synthesis also invites extension into closely related fields that can be illuminated by the systems perspective.

- These ideas can be extended back into management science.
- They can enlighten debate on public policy.
- They can pump life into psychological analysis and treatment.

*   *   *

Finally, systems theory has enormous present value in both practical and theoretical areas. It has staggering possibilities for our future.

REFERENCES

Abraham, R. (1991) Complex dynamical systems theory: Historical origins, contemporary applications. In Laszlo (1991), pp. 1-10.

Abraham, R. (1993). Erodynamics and cognitive maps. In Laszlo and Masulli, pp. 255-264.

Abu-Mostafa, Y.S. (1995). Machines that learn from hints. *Scientific American*, April, 1995, pp. 64-69.

Ackoff, R.L. (1981). *Creating the corporate future.* New York: Wiley.

Adorno, T.W. (1992). *Negative dialectics.* New York: Continuum.

Alexander, J.C. (1983). *The modern reconstruction of classical thought: Talcott Parsons.* Berkeley: University of California Press.

Alexander, J.C., and Colomy, P. (eds.). (1990). *Differentiation theory and social change.* New York: Columbia University Press.

Allen, P.M. and Lesser, M.J. (1993). Evolution, ignorance and selection. In Laszlo and Masulli, pp. 119-134.

American Psychiatric Association. (1998). *Diagnostic and statistical manual (4th ed.).* Washington, DC: Author.

American Psychological Association. (1984). *Publication manual (third edition).* Washington DC: American Psychological Association.

Argyris, C. (1982). *Reasoning learning and action.* San Francisco: Jossey-Bass.

Artigiani, R. (1991). Social Evolution: A nonequilibrium model. In Laszlo (1991), pp. 93-131.

Artigiani, R. (1993). From epistemology to cosmology: Post-modern science and the search for new cultural maps. In Laszlo and Masulli (1993) pp. 29-60.

Ashby, R. (1958). Requisite variety and its implications for the control of complex systems. *Cybernetica* 1.2, pp. 1-17.

Bailey, K.D. (1990). *Social Entropy Theory.* Albany: State University of New York Press.

Bailey, K.D. (1994). *Sociology and the new systems theory: Toward a theoretical synthesis.* Albany: State University of New York Press.

Bailey, K.D. (1995). The use of space in Living Systems Theory. *Systems Practice*, vol. 8. no. 1, pp. 85-106.

Bailly, F. (1994). The characterization of systems identity in the physical and the biological sciences. *World Futures*, vol. 42, pp. 11-19.

Bak, P. (1996). *How nature works.* New York: Springer-Verlag

Banathy, B.A. (1995). The 21st century Janus: The three faces of information. *Systems Research*, vol. 12, no. 4, pp. 319-320.

Banathy, B.H. (1989). The design of evolutionary guidance systems. *Systems Research*, vol. 6, no. 4, pp. 289-292.

Banathy, B.H. (1991). *Systems design of education.* Englewood Cliffs: Educational Technology Publications.

Banathy, B.H. (1993a). The cognitive mapping of societal systems: Implications for education. In Laszlo and Masulli, pp. 205-220.

Banathy, B.H. (1993b). Is the improvement of human condition our field? Making evolutionary science work for human betterment. *World Futures*, vol. 38, pp. 17-31.

Banathy, B.H. (1996). *Designing social systems in a changing world.* New York: Plenum Press.

Barglow, R. (1994). *The crisis of the self in the age of information: Computers, dolphins, and dreams.* New York: Routledge.

Barkow, J.H., Cosmides, L., and Tooby, J. (1992). *The adapted mind: Evolutionary psychology and the generation of culture.* New York: Oxford University Press.

Barth, J. (1984). *The Friday book.* New York: Putnam.

Bateson, G. (1972). *Steps to an ecology of mind.* New York: Bantam.

Bateson, G. (1987). Men are grass. In Thompson, W.I. *Gaia: A way of knowing.* Hudson, NY: Lindisfarne Press.

Bausch, K. (1997). The Habermas/Luhmann debate and subsequent Habermasian perspectives on systems theory. In *Systems Research and Behavioral Science*. *Vol. 14,*. pp. 315-330.

Bausch, K. (1998). A confluence of paradigm and technology. *Proceedings of the International Society for the Systems Science, 1998*, CD ROM.

Bausch, K. (1999). The applicability of systems theory to social processes. *Dissertation Abstracts International*.

Bausch, K. (2000a). Evolutionary Transcendence and Design. Accepted by *World Futures*.

Bausch, K. (2000b). The Practice and Ethics of Design. *Systems Research and Behavioral Science Vol. 17. pp.23-50.*

Beer, S. (1979). *The heart of the enterprise*. Chichester, UK: John Wiley.

Beer, S. (1985). *Diagnosing the system for organizations*. New York: John Wiley and Sons.

Bell, D. (1976). *The coming of the post-industrial society*. New York: Basic Books.

Bernstein, R. J. (ed.). (1985). *Habermas and modernity*. Cambridge: MIT Press.

Bickerton, D. (1990). *Language & species*. Chicago: The University of Chicago Press.

Bocchi, G. (1991). Biological evolution: The changing image. In Laszlo (1991), pp. 33-44.

Bocchi, G. (1991). Biological evolution: the changing image. In Laszlo E. (1991b) pp. 33- 54.

Bohm, D. (1980). *Wholeness and the implicate order*. London: Routledge & Kegan Paul.

Bohm, D., and Peat, F. D. (1987). *Science, order, and creativity*. New York: Bantam Books.

Bohm, D., and Hiley, B.J. (1993). *The undivided universe: An ontological interpretation of quantum theory*. New York: Routledge.

Boulding, K. (1956). General systems theory-the skeleton of science. In *General Systems*, 1, 11-17.

Bridges, W. (1991). *Managing transitions*. Reading: Addison-Wesley.

Briggs, J. and Peat, F. D. (1989). *The turbulent mirror: An illustrated guide to chaos theory and the science of wholeness*. New York: Harper & Row.

Brown, L.M., Tappan, M. B., Gilligan, C., Miller, B A., and Argyris, D. E. (1989). Reading for self and moral voice: A method for interpreting narratives of real-life moral conflicts and choice. In Packer, M.J. and Addison, R.B. (1989).

Buckley, W. (1967). *Sociology and modern systems theory*. Englewood Cliffs: Prentice-Hall.

Butz, M. (1997). *Chaos and complexity: Implications for Psychological Theory and Practice*. Taylor & Francis.

Campanella, M. L. (1993). The cognitive mapping approach to the globalization of world politics. In Laszlo and Masulli, pp. 237-254.

Cassirer, E. (1946). *Language and myth*. New York: Dover Publications Inc.

Checkland, P. (1981). *Systems thinking, systems practice*. New York: John Wiley & Sons.

Checkland, P. (1991). From optimizing to learning: A development of systems thinking for the 1990's. In Flood and Jackson (1991).

Checkland, P. and Scholes J. (1990). *Soft systems methodology*. New York: Wiley.

Chriss, J.J. (1995). Habermas, Goffman, and communicative action: Implications for professional practice. *American Sociological Review*, vol. 60, pp. 545-565.

Christakis, A.N. (1988). The Club of Rome revisited. In *General Systems*, vol. 31, pp. 35-38.

Christakis, A.N. (1993). The inevitability of demosophia. In *A challenge for systems thinking: The Aegean seminar*. Athens, Greece: University of the Aegean Press, pp. 187-197.

Christakis, A.N. (1996). A people science: The CogniScope™ system approach. In *Systems: Journal of Transdisciplinary Systems Sciences*. Vol. 1, pp. 16-19.

Christakis, A.N. land Shearer, W.L. (1997). Collaboration through communicative action: Resolving the systems dilemma through the CogniScope™ approach. (Unpublished, prepared for *Systems Research.)*

Churchman, C.W. (1961). *Prediction and optimal decision*. Englewood Cliffs, N.J.: Prentice-Hall, Inc.

Churchman, C.W. (1968). *The systems approach*. New York: Delacorte Press.

Churchman, C.W. (1971). *The design of inquiring systems*. New York: Basic Books.

Churchman, C.W. (1979). *The systems approach and its enemies*. New York: Basic Books.

Churchman, C.W. (1982). *Thought and wisdom*. Salinas, CA: Intersystems Publishers.

Cohen, A.P. (1994). *Self consciousness: An alternative anthropology of identity*. New York: Routledge.

Colapietro, V.M.. (1989). *Peirce's approach to the self: A semiotic perspective on human subjectivity*. Albany: State University of New York Press.

Cole, H.S.D. et al. (1973). *Models of doom: A critique of Limits to Growth.* New York: Universe Books.

Colomy, P. (1990). Revisions and progress in differentiation theory. In Alexander and Colomy, pp. 465-496.

Combs, A. (1996). *The radiance of Being: Complexity, chaos and the evolution of consciousness.* St. Paul, Minnesota: Paragon House

Corning, P.A. (1983). *The Synergism hypothesis: A theory of progressive evolution.* New York: McGraw-Hill.

Corning, P.A. (1995). Synergy and self-organization   in the evolution of complex systems. *Systems Research,* vol. 12, no. 2, pp. 89-121.

Courbage, M. (1983). Intrinsic irreversibility of Kolmogorov dynamical systems. *Physica.*

Crick, F. (1994). *The astonishing hypothesis.* New York: Simon & Schuster.

Csanyi, V. (1989). *Evolutionary systems and society.* Durham, N.C.: Duke University Press.

Csanyi, V. (1993). The biological basis of cognitive maps. In Laszlo and Masulli, pp. 23-28.

Csanyi, V. and Kampis, G. (1991). Modeling biological and social change: Dynamic replicative network theory. In Laszlo E. (1991b) pp. 77-92.

Csikszentmihalyi, M. (1993). *The evolving self: A psychology for the third millennium.* New York: Harper Perennial.

Dallmayr, F.R. and McCarthy, T.A. (eds.) (1977). *Understanding and Social Inquiry.* Notre Dame: University of Notre Dame Press.

Davies, P.C.W. (1975). Cosmological aspects of irreversibility. In Kubat and Zeman, 1975, pp. 11-27.

Deely, J. (1990). *Basics of Semiotics.* Bloomington: Indiana University Press.

Dennett, D.C. (1991). *Consciousness explained.* Boston: Little, Brown and Co.

Dennet, D.C. (1993). Book Review of *The Embodied Mind,* by Varela et al. *American Journal of Psychology,* vol. 106, no. 1, pp. 121-126.

Dennet, D.C. (1995). *Darwin's dangerous idea: Evolution and the meanings of life.* New York: Simon & Schuster.

Downing Bowler, T. (1981). *General systems thinking: Its scope an applicability.* Amsterdam: North Holland.

Drucker, P. (1989). *The new realities.* New York: Harper and Row.

Durkheim, E. (1984). *The division of labor in society.* New York: The Free Press.

Durkheim, E. (1995). *The elementary forms of religious life.* New York: The Free Press.

Edelman, G.M. (1992). *Bright air, brilliant fire.* New York: Basic Books.

Eigen, M. (1992). *Steps toward life.* New York: Oxford University Press.

Eisenstadt, S.N. (1990). Modes of structural differentiation, elite structure, and cultural visions. In Alexander and Colomy, pp. 19-51.

Eisler, R. (1991). Social shifts and phase changes. In Laszlo (1991), pp. 179-200.

Eisler, R. (1993). Technology, gender, and history: Toward a nonlinear model of social evolution. In Laszlo and Masulli, pp. 181- 204.

Eldredge, N. and Gould, S. (1972). Punctuated equilibria: An alternative go phyletic gradualism. In Schopf (Ed.). *Models in paleobiology,* San Francisco: Freeman, Cooper.

Espejo, R. (1990). Complexity and change: Reflections upon the cybernetic intervention in Chile, 1970-1973. *Sistemica,* 1.2.

Fay, B. (1987). *Critical social science: Liberation and its limits.* Ithaca: Cornell University Press.

Feenberg, A. (1994). The technocracy thesis revisited: On The Critique of Power. *Inquiry,* vol. 37, no. 1, pp. 85-102.

Flood, R.L. (1990). *Liberating systems theory.* New York: Plenum Press.

Flood, R.L. and Carson, E.R. (1988). *Dealing with complexity.* New York: Plenum Press.

Flood, R.L. & Jackson, M.C. (Eds.) (1991a). *Critical systems thinking: Directed readings.* New York: John Wiley & Sons.

Flood, R.L. & Jackson, M.C. (1991b). *Creative problem solving: Total systems intervention.* New York: John Wiley & Sons.

Flood, R.L. and Romm, N.R.A. (1996). *Critical systems thinking.* New York: Plenum Press.

Forrester, J.W. (1961). *Industrial dynamics.* Portland, Oregon: Productivity Press.

Forrester, J.W. (1969). *Urban dynamics.* Portland, Oregon: Productivity Press.

Forrester, J. (1971). *World Dynamics.* Cambridge: Wright-Allen Press.

Foucault, M. (1965). *Madness and civilization.* (R. Howard, Trans.). New York: Vintage Books. (Original work published 1961).

Foucault, M. (1972). *The archeology of knowledge.* (A.M. Sheridan Smith, Trans.). New York: Pantheon Books. (Original work published 1969).

Foucault, M. (1973). *The order of things.* (R. D. Laing, Trans.). New York: Vintage Books. (Original work published 1966).

Foucault, M. (1979). *Discipline and punish.* (A. Sheridan, Trans.). New York: Vintage Books. (Original work published 1975).

Foucault, M. (1984). *The Foucault reader.* (P. Rabinow, Ed. & Trans.). New York: Pantheon Books.

Foucault, M. (1988). *The care of the self.* (R. Hurley, Trans.). New York: Vintage Books. (Original work published 1984).

Fouts, R.S. and Fouts, D.H. (1993). Chimpanzees' use of sign language. In Cavalieri, P. and Singer, P. (1993). *The great ape project.* New York: St. Martin's Press.

Frankel, B. (1972). Habermas talking: an interview. *Theory and Society. 1.27*

Frankenberg, G. (1992). Disorder is possible: An essay on systems, laws, and disobedience. In Honneth, McCarthy, Offe, and Welmmer, pp. 17- 38.

Frantz, T.G. and Miller, C. (1993). An idealized design approach. *World Futures.* 36 (2-4): 83-106.

Freud, S. (1952). Beyond the pleasure principle. (C.J.M. Hubback, Trans.). In *Freud, No. 54, great books of the Western world,* pp. 639-663. Chicago: Encyclopedia Britannica, Inc. (Original work published 1920).

Fuenmayor, R. (1990). Systems thinking and critique: I. What is critique? *Systems Practice,* vol. 3, no. 6, pp. 525-544.

Fuenmayor, R. (1991). Between systems thinking and systems practice. In Flood and Jackson (1991).

Gabora, L. (1998). Autocatalytic closure in a cognitive system: A tentative scenario for the origin of culture. In *Psycoloquy. Http://www.cogsci.soton.ac.uk/cgi/psyc/newpsy?9.67.*

Gadamer, H-G. (1976). *Philosophical hermeneutics.* Berkeley: University of California Press.

Gallagher, W. (1996). *I. D.* also titled *Just the way you are.* New York: Random House.

Gardner, R.A., Gardner, B.T., and Van Cantfort, T.E. (eds.). (1989). *Teaching sign language to chimpanzees.* Albany: State University of New York Press.

Gazzaniga, M.S. (1992). *Nature's mind: The biological roots of thinking, emotions, sexuality, language, and intelligence.* New York: Basic Books.

Gazzaniga, M.S. (1998). *The mind's past.* Berkeley: University of California Press.

Gazzaniga, M.S., Ivry, R.B., Mangun, G.R. (1998). *Cognitive neuroscience: The biology of the mind.* New York: W.W. Norton.

Gell-Mann, M. (1994). *The quark and the jaguar.* New York: W.H. Freeman and Company.

Ghiselin, B. (ed.) (1952). *The creative process: A symposium.* New York: Bantam.

Giddens, A. (1985). Reason without revolution? Habermas's *Theorie des communicativen Handelns.* In Bernstein, pp. 95-121.

Giddens, A. (1987). *Social theory and modern sociology.* Stanford, CA: Stanford University Press.

Giddens, A. (1995). *Politics, sociology, and social theory.* Stanford: Stanford University Press.

Giorgi, A. (1970). *Psychology as a human science: A phenomenologically based approach.* New York: Harper & Row.

Giorgi, A. (1974). The relations among level, type, and structure and their importance for social science theorizing: A dialogue with Schutz. In Giorgi et al. *Duquesne Studies.* Vol. 3.

Gleick, J. (1987). *Chaos: Making a new science.* New York: Viking.

Goertzel, B. (1993). *The evolving mind.* Amsterdam: Gordon and Breach.

Goertzel, B. (1993a). *The structure of intelligence: A new mathematical model of mind.* New York: Springer-Verlag.

Goertzel, B. (1994). *Chaotic logic: Language, thought, and reality from the perspective of complex systems science.* New York: Plenum Press.

Goertzel, B. (1997). *From complexity to creativity.* New York: Plenum Press.

Goertzel, B. (1999). *Wild computing: Steps toward a philosophy of internet intelligence.* Http://goertzel.org/ben/wild.

Gould, S.J., and Eldridge, N. (1977). Punctuated equilibria: The tempo and mode of evolution reconsidered. *Paleobiology* 3:115-151.

Guastello, S.J. (1995). *Chaos, catastrophe, and human affairs: Applications of nonlinear dynamics to work, organizations, and social evolution.* Mahwah, NJ: Lawrence Erlbaum Associates.

Habermas, J. (1970). On systematically distorted communication. *Inquiry*, vol. 13, no. 3, pp. 205-218.

Habermas, J. (1971a). *Knowledge and human interests.* (J. Shapiro, Trans.). Boston: Beacon Press. (Original work published 1968).

Habermas, J. (1971b). *Toward a rational society.* (J. Shapiro, Trans.). Boston: Beacon Press. (Original work published 1961).

Habermas, J. (1974). *Theory and practice.* (J. Viertel, Trans.). London: Heinemann. (Original work published 1968).

Habermas, J. (1975). *Legitimation Crisis.* (T. McCarthy, Trans.).Boston: Beacon Press. (Original work published 1973).

Habermas, J. (1979). *Communication and the evolution of society.* (T. McCarthy, Trans.). Boston: Beacon Press. (Original work published 1976).

Habermas, J. (1984). *The theory of communicative action, volume one: Reason and the rationalization of society.* (T. McCarthy, Trans.). Boston: Beacon Press. (Original work published 1981).

Habermas, J. (1985). Psychic thermidor and the rebirth of rebellious subjectivity, pp. 67-77, and Questions and counterquestions, pp. 192- 216. In Bernstein.

Habermas, J. (1987). *The philosophical discourse of modernity.* (F.G. Lawrence, Trans.). Cambridge: The MIT Press. (Original work published 1985).

Habermas, J. (1989). *The theory of communicative action, volume two: Lifeworld and system.* (T. McCarthy, Trans.). Boston: Beacon Press. (Original work published 1981).

Habermas, J. (1991). A reply. In Honneth and Joas   (1991), pp. 214-264.

Habermas, J. (1992a). *Autonomy and solidarity.* (P. Dews, Ed.). New York: Verso.

Habermas, J. (1992b). *Postmetaphysical thinking: Philosophical essays.* (W.M. Hohengarten, Trans.). Cambridge: The MIT Press. (Original work published 1988).

Habermas, J. (1993a). *Moral consciousness and communicative action.* (C. Lenhardt & S.W. Nicholsen, Trans.). Cambridge: the MIT Press. (Original work published 1983).

Habermas, J. (1993b). *Justification and application: Remarks on discourse ethics.* (C.P. Cronin, Trans.). Cambridge: The MIT Press. (Original work published 1990).

Habermas, J. (1996). *Between facts and norms.* (W. Rehg, Trans.).Cambridge: The MIT Press. (Original work published 1992).

Habermas, J. and Luhmann, N. (1971). *Theorie der Gesellschaft oder Sozialtechnologie: Was leistet die Systemforschung?.* [Theory of society or social technology: What does systems research accomplish?]. Frankfurt: Suhrkamp.

Haken, H. (1981). Synergetics: Is self-organization governed by universal principles? In Jantsch (1981a), pp. 15-24.

Hameroff, S. & Penrose, R. (1996). Conscious events as orchestrated space-time selections. *Journal of Consciousness Studies. 3, no. 1*, pp.36-53.

Harman, W. & Horman, J. (1990). *Creative work.* Indianapolis: Knowledge Systems.

Harre, R. and Gillett, G. (1994). *The discursive mind.* Thousand Oaks: Sage Publications.

Hauser, M.D. (1996). *The evolution of communication.* Cambridge: MIT Press.

Hayles, K. & Luhmann, N. (1995). Theory of a different order: A conversation with Katherine Hayles and Niklas Luhmann. *Cultural Critique, Fall, 1995*, pp. 7-36.

Heidegger, M. (1987). *Nietzsche, vol. 3, The will to power as knowledge and as metaphysics.* San Francisco, Harper and Row.

Ho, M. & Saunders, P.T. (1979). Beyond neo-Darwinism: An epigenetic approach to evolution. *Journal of Theoretical Biology. 78*: 573-591.

Hobbes, T. (1963). *Leviathan.* In J. Somerville & R.E. Santoni (Eds.). *Social & Political Philosophy.* New York: Doubleday. (Originally published in 1651).

Hofstadter, R. (1945). *Social Darwinism in American thought.* Philadelphia: University of Pennsylvania Press.

Holmes, S. & Larmore, C. (1982). Translators' introduction. In Luhmann (1982).

Honneth, A. & Joas, H. (Eds.) (1991). *Communicative action: Essays on Jurgen Habermas's The Theory of Communicative Action.* Cambridge: The MIT Press.

Honneth, A., McCarthy, T., Offe, C., and Wellmer, A. (eds.) (1992). *Cultural-political interventions in the unfinished project of enlightenment._*Cambridge: The MIT Press.

Hooper, J. & Teresi, D. (1986). *Three pound universe.* New York: Jeremy P. Tarcher/Perigee.

Hookway, C. (1994). Naturalized epistemology and epistemic evaluation. *Inquiry.* 37, pp. 465-485.

Horgan, J. (1995). From complexity to perplexity. *Scientific American, June 1995*, pp. 104-109.

Horgan, J. (1995). The new social Darwinists. *Scientific American, October, 1995*, pp. 174- 181.

Horkheimer, M. & Adorno, T.W. (1993). *Dialectic of enlightenment.* New York: Continuum.

Hutchins, C.L. (1996). *Systemic thinking: Solving complex problems.* Aurora, CO: Professional Development Systems.

Jackendorff, R. (1987). *Consciousness and the computational mind.* Cambridge: MIT Press.

Jackendorff, R. (1993). *Semantics and cognition.* Cambridge: The MIT Press.

Jackson, M.C. (1991). Social systems theory and practice: The need for a critical approach. In Flood and Jackson (1991), pp. 117-138.

Jackson, M.C. (1992). *Systems methodology for the management sciences.* New York: Plenum.

Jantsch, E. (1975). *Design of evolution.* New York: Braziller.

Jantsch, E. (1980a). *The self-organizing universe.* Oxford: Pergamon Press.

Jantsch, E. (1980b). The unifying paradigm behind autopoiesis, dissipative structures, hyper-and ultra-cycles. In Zeleny (1980).

Jantsch, E. (Ed.). (1981a). *The evolutionary vision.* Boulder: Westview Press.

Jantsch, E. (1981b). Autopoiesis: A central aspect of dissipative self-organization. In Zeleny (1981).

Jay, M. (1985). Habermas and modernism. In Bernstein, pp. 125-139.

Jaynes, J. (1976). *The origin of consciousness in the breakdown of the bicameral mind.* Boston:Houghton Mifflin.

Jiayin, M. (1993). Transformations in the Chinese cognitive map. In Laszlo & Masulli, pp. 221-236.

Joas, H. (1991). The unhappy marriage of hermeneutics and functionalism. In Honneth & Joas (1991), pp. 97-118.

Johannessen, J-A., & Hauan, A. (1994). Communication--A systems theoretical point of view (third-order cybernetics). *Systems Practice, vol. 7, no. 1*, pp. 63-73.

Jones, C.J. (1980). *Design methods: Seeds of human futures.* New York: Wiley Interscience.

Jung, C.G. (ed.). (1964). *Man and his symbols.* New York: Dell Publishing Co.

Kampis, G. (1991). *Self-modifying systems in biology and cognitive science.* New York: Pergamon Press.

Kampis, G. (1993). On understanding how the mind is organized: Cognitive maps and the "physics" of mental information processing. In Laszlo and Masulli, pp. 135-150.

Kauffman, S.A. (1993). *The origins of order: Self-organization and selection in evolution.* New York: Oxford University Press.

Kosslyn, S.M. & Koenig, O. (1995). *Wet mind: The new cognitive neuroscience.* New York: The Free Press.

Kubat, L. & Zeman, J. (1975). *Entropy and information in science and philosophy.* New York: Elsevier Scientific Publishing Company.

Kuhn, T. (1970). *The structure of scientific revolutions,* (2nd ed.). Chicago: University of Chicago Press.

Lacan, J. (1977). *Ecrits.* New York: Norton.

Lakoff, G. (1987). *Women, fire, and dangerous things.* Chicago: The University of Chicago Press.

Lakoff, G. & Johnson, M. (1980). *Metaphors we live by.* Chicago: The University of Chicago Press.

Lane, D.C. and Jackson, M.C. (1995). Only connect! An annotated bibliography reflecting the breadth and diversity of systems thinking. *Systems Research*, vol. 12, no. 3, pp. 217-218.

Laszlo, A. (1990). Sociocultural systems: The cognitive dimension of cultural change. In *Revista International de Sistemas, vol. 2, no. 1.*

Laszlo, E. (1972). *Introduction to systems philosophy: Toward a new paradigm of contemporary thought.* New York: Gordon & Breach.

Laszlo, E. (1991a). *The age of bifurcation.* Philadelphia: Gordon and Breach.

Laszlo, E. (ed.). (1991b). *The new evolutionary paradigm.* Philadelphia: Gordon and Breach.

Laszlo, E. (1995). *The interconnected universe: Conceptual foundations of transdisciplinary unified theory.* New Jersey: World Scientific.

Laszlo, E. and Masulli I. (1993). *The evolution of cognitive maps.* Philadelphia: Gordon and Breach.

Leder, D. (1990). *The absent body*. Chicago: The University of Chicago Press.

Lettvin, J.Y. Maturana, H.R., McCulloch, W.S., & Pitts, W.H. (1959). What the frog's eye tells the frog's brain. *Proceedings of the IRE, vol. 47, no. 11*, pp. 1940-1959).

Levin, D.M. (1985). *The body's recollection of being*. Boston: Routledge & Kegan Paul.

Levinas, E. (1969). *Totality and infinity*. Pittsburgh: Duquesne University Press.

Levinas, E. (1989). *The Levinas reader*. Cambridge: Basil Blackwell.

Levi-Strauss, C. (1967). *Structural anthropology*. New York: Doubleday.

Lewin, K. (1947). Group decisions and social change. In T.M. Newcomb and E.L. Hartley. (eds.). *Readings in social psychology*. New York: Henry Holt.

Lewin, K. (1948). *Solving social conflicts*. New York: Harper & Brothers.

Lindenberg, K. & West, B.J. (1990). *The nonequilibrium statistical mechanics of open and closed systems*. New York: VCH Publishers.

Linstone, H. & Mitroff, I. (1994). *The challenge of the 21st century*. New York: State University Press.

Llewelyn, J. (1995). *Emmanuel Levinas: The genealogy of ethics*. New York: Routledge.

Lloyd, S. (1995). Quantum-mechanical computers. *Scientific American, October, 1995*, pp. 140- 145.

Losick, R. & Kaiser, D. (1997). Why and how bacteria communicate. In *Scientific American. February, 1997*, pp. 68-73.

Loye, D. (1991). Chaos and transformation: Implications of nonequilibrium theory for social science and society. In Laszlo (1991b), pp. 11-32.

Loye, D. (1993). Moral sensitivity and the evolution of higher mind. In Laszlo and Masulli, pp. 151-165.

Luhmann, N. (1982). *The differentiation of society*. (S. Holmes & C. Larmore, Trans.). New York: Columbia University Press. (Original works published 1971 & 1975).

Luhmann, N. (1990a). *Essays on self-reference*. New York: Columbia University Press.

Luhmann, N. (1990b). The paradox of system differentiation and the evolution of society. In Alexander and Colomy, pp. 409-440.

Luhmann, N. (1995a). *Social systems*. (J. Bednarz & D. Baecker, Trans.). Stanford: Stanford University Press. (Original work published 1984).

Luhmann, N. (1995b). The paradox of observing systems. *Cultural Critique, Fall, 1995*, pp. 37-55.

Lyotard, J-F. (1988). *The differend: Phrases in dispute*. (G. Van Den Abbeele. Trans.). Minneapolis, University of Minnesota Press. (Original work published 1983).

Lyotard, J-F. (1989). *The Lyotard reader*. (A. Benjamin, Ed.). Cambridge: Basil Blackwell.

Madison, G.B. (1988). *The hermeneutics of postmodernity*. Indianapolis: Indiana University Press.

Malaska, P. (1991). Economic and social evolution: The transformational dynamics approach. In Laszlo (1991), pp. 131-156.

Markely, O.W. & Harman, W. (1982). *Changing images of man*. London: Pergamon Press.

Maslow, A.H. (1954). *Motivation and personality*. New York: Harper & Row.

Maturana, H.R. (1980). Autopoiesis, reproduction, heredity, and evolution. In Zeleny (1980).

Maturana, H.R. (1981). Autopoiesis. In Zeleny (1981).

Maturana, H.R. & Varela, F.J. (1980). *Autopoiesis and cognition: The realization of the living*. Boston: D. Reidel Publishing Company.

Maturana, H.R. & Varela, F.J. (1987). *The tree of knowledge: The biological roots of human understanding*. Boston: New Science Library.

Masulli, I. (1993a). Toward a new historical consciousness. In Laszlo and Masulli, pp. 67-74.

Masulli, I. (1993b). Cognitive maps and social change. In Laszlo and Masulli, pp. 169-180.

Maynard, H. & Mehrtens, S. *The fourth wave*. San Francisco: Berrett-Koehler.

McCarthy, T. (1981). *The critical theory of Jurgen Habermas*. Cambridge: The MIT Press.

McCarthy, T. (1985). Reflections on rationalization in the *Theory of Communicative Action*. In Bernstein, pp. 176-192.

McCarthy, T. (1991). Complexity and democracy: or the seducements of systems theory. In Honneth and Joas (1991), pp. 119-139.

McFarland, D. (1985). *Animal behavior: Psychology, ethology, and evolution*. Menlo Park, CA: Benjamin Cumming.

McIntosh, D. (1994). Language, self, and lifeworld in Habermas's *Theory of Communicative Action*. *Theory and Society*, vol. 23, pp. 1-33.

McNeill, D. and Freiberger, P. (1993). *Fuzzy logic*. New York: Simon & Schuster.

McWhinney, W. (1992). *Paths of change*. Newbury Park: Sage.

Mead, G.H. (1962). *Mind, self, and society*. Chicago: The University of Chicago Press.

Meadows, D.H., Meadows, D., & Randers, J. (1972). *The limits to growth*. New York: Universe Books.

Meadows, D.H., Meadows, D., & Randers, J. (1974a). *Dynamics of growth in a finite world*. Cambridge: Wright-Allen Press.

Meadows, D.H., Meadows, D. & Randers, J. (1974b). *Toward global equilibrium: Collected papers for the Limits to Growth*. Cambridge: Wright-Allen Press.

Menzies, K. (1989). Talcott Parsons. In Kuper & Kuper pp. 573-574).

Merleau-Ponty, M. (1962). *The phenomenology of perception*. New York: The Humanities Press. (Original work published 1945).

Merleau-Ponty, M. (1964a). *The primacy of perception*. (Evanston: Northwestern University Press.

Merleau-Ponty, M. (1964b). *Signs*. (R.C. McCleary, Trans.). Evanston: Northwestern University Press. (Original work published 1960).

Merleau-Ponty, M. (1964c). *Sense and non-sense*. (H.L. Dreyfus & P.A. Dreyfus, Trans.). Evanston: Northwestern University Press. (Original work published 1948).

Merleau-Ponty, M. (1968). *The visible and the invisible*. (A. Lingis, Trans.). Evanston: Northwestern University Press. (Original work published 1964).

Mesarovic, M. & Pestel, E. (1974). *Mankind at the turning point: The second report to the Club of Rome*. New York: E.P. Dutton.

Michaels, M. (ed.). (1991, 1992, 1993, 1994). *Proceedings of the [first four] conference[s] of the Chaos Network*. Urbana, Illinois: people technologies.

Midgley, G. (1994). Ecology and the poverty of humanism: a critical systems perspective. *Systems Research, vol. 11, no.4*, pp. 67-76.

Miller, J.G. (1995). *Living systems*. New York: McGraw-Hill Book Company.

Miller, J.G. and Miller, J.L. (1995). Applications of living systems theory. *Systems Practice, vol.8. no. 1*, pp. 19-45.

Mingers, J.C. (1980). Towards an appropriate social theory for applied systems thinking. *Journal of Applied Systems Analysis, 7*.

Mingers, J. (1995). *Self-producing systems: Implications and applications of autopoiesis*. New York: Plenum Press

Mittroff, I. & Linstone, H. (1993). *The unbounded mind*. New York: Oxford University Press.

Monod, J. (1971). *Chance and necessity*. New York: Alfred A. Knopf.

Munch, R. (1990). Differentiation, rationalization, interpenetration: The emergence of modern society. In Alexander and Colomy, pp. 441-464.

Nadler, G. (1981). *The planning and design approach*. New York: Wiley.

Nadler, G. & Hibino, S. (1990). *Breakthrough thinking*. Rocklin CA: Prima Publishing.

Naess, A. (1995). Foreword. In Laszlo, 1995, pp. v-vii.

Nagel, T. (1986). *The view from nowhere*. New York: Oxford University Press.

Neurath, P. (1994). *From Malthus to the Club of Rome and back*. Armonk, NY: M. E. Sharpe.

Nietzsche, F. (1968). *The will to power*. (W. Kaufmann, Trans.). New York: Random House.

Nietzsche, F. (1974). *The gay science*. (W. Kaufmann, Trans.). New York: Random House. (Original work published 1887).

Nietzsche, F. (1977). *The Nietzsche reader*. (R.J. Hollingdale, Trans.). New York: Penguin Books.

O'Neill, J. (Ed.). (1976). *On critical theory*. New York: The Seabury Press.

Packer, M.J. and Addison, R.B. (eds.) (1989). *Entering the circle: Hermeneutic investigation in psychology*. Albany: State University of New York Press.

Pantzar, M. (1991). The evolutionary paradigm and   neoclassical economics. In Laszlo (1991), pp. 157-178.

Parsons, T. (1951). *The social system*. Glencoe, Illinois: The Free Press.

Parsons, T. (1966). *Societies: Evolutionary and comparative perspectives*. New York: Free Press.

Parsons, T. (1971). *The system of modern societies*. Englewood Cliffs, NJ: Prentice-Hall.

Parsons, T. (1977). *Social systems and the evolution of action theory*. New York: The Free Press. (Original work published 1968).

Parsons, T. and Shils, E. (1951). *Toward a general theory of action*. Cambridge:

Peccei, A. (1977). *The human quality.* Oxford: Pergamon Press.

Peirce, C.S. (1935). *Collected works of Charles S. Peirce.* Cambridge: Harvard University Press.

Penrose, R. (1994). *Shadows of the mind.* New York: Oxford University Press.

Pinchot, G. & Pinchot, E. (1993). *The end of bureaucracy and the rise of the intelligent organization.* San Francisco: Berrett-Koehler.

Polkinghorne, D. (1983). *Methodology for the human sciences.* Albany: State University of New York Press.

Pribram, K. (1995). Afterword. In Laszlo, 1995, pp. 143-146.

Prigogine, I. (1980). *From being to becoming.* New York: W.H. Freeman and Company.

Prigogine, I. & Stengers, I. (1984). *Order out of chaos.* New York: Bantam Books.

Quine, W.V. (1969). *Ontological relativity and other essays.* New York: Columbia University Press.

Rabinow, P. (Ed.). (1984). *The Foucault reader.* New York: Pantheon Books.

Richter, H.E. (1974). *Wachstum bis zur Katastrophe?* (Growth toward catastrophe?). Stuttgart: Deutsche Verlagsanstalt.

Ricoeur, P. (1970). *Freud and philosophy.* (D. Savage, Trans.). New Haven: Yale University Press. (Original works published 1961 & 1962).

Ricoeur, P. (1979). The model of a text: Meaningful action considered as a text. In P.Rabinow & W. Sullivan. (Eds.) *Interpretive social science: A reader.* Berkeley: University of California.

Rittel, H. & Webber, M. (1984). Planning problems are wicked problems. In N. Cross. (Ed.). *Developments in design methodology.* New York: Wiley.

Rock, P. (1985). Symbolic interactionism. In Kuper & Kuper, pp. 843-845.

Rorty, R. (1985). Habermas and Lyotard on postmodernity. In Bernstein, pp. 161-175.

Rosen, R. (1985). *Anticipatory systems.* New York: Pergamon.

Rossi, E.L. (1993). *The psychobiology of mind-body healing.* New York: W.W. Norton.

Sage, A. (1977). *Methodology for large-scale systems.* New York: McGraw-Hill.

Sanderson, S.K. (1990). *Social evolutionism.* Cambridge: Basil Blackwell.

Saunders, P.T. (1993). Evolution theory and cognitive maps. In Laszlo and Masulli, pp. 105-118.

De Saussure, F. (1959). *Course in general linguistics.* New York: Philosophical Library. (Original work published 1916).

Schnabel, P-E. (1985). Georg Simmel. In Kuper & Kuper, pp. 750-752.

Schull, J. (1991). Evolution and intelligence. In Laszlo, E. (1991b) pp. 55-76.

Schultz, D & Schultz, S.E. (1994). *Theories of personality.* Pacific Grove, CA: Brooks/Cole Publishing Company.

Schutz, A. & Luckmann, T. (1973). *The structure of the lifeworld.* Evanston: Northwestern University Press.

Scott, A. (1995). *Stairway to the mind.* New York: Springer-Verlag.

Searle, J.R. (1995). *The construction of social reality.* New York: The Free Press.

Searle, J.R. (1992). *The rediscovery of the mind.* Cambridge: The MIT Press.

Sebeok, T.A. (1994). *Signs: An introduction to semiotics.* Toronto: University of Toronto Press.

Senge, P.M. (1990). *The fifth discipline: The art and practice of learning organization.* New York: Double Currency.

Sertorio, L. (1991). *Thermodynamics of complex systems.* New Jersey: World Scientific.

Sheldrake, R. (1981). *A new science of life.* London: Blond & Briggs.

Siegel, H. (1995). Naturalized epistemology and "first philosophy." *Metaphilosophy. 26.1&2,* pp. 46-62.

Simon, H. (1969). *The science of the artificial.* Cambridge: The MIT Press.

Sixel, F.W., (1976). The problem of sense: Habermas v. Luhmann. In O'Neill, pp. 184-203.

Sless, D. (1978). A definition of design: Origination of useful systems. *Design methods and theories. 12(2),* pp. 123-130.

Smelser, N.J. (1959). *Social change in the industrial revolution.* Chicago: University of Chicago Press.

Smelser, N.J. (1968). *Essays in sociological explanation.* Englewood Cliffs, NJ: Prentice-Hall.

Smith, E.R. & Mackie, D.M. (1995). *Social Psychology.* New York: Worth Publishers.

Snell, B. (1982). *The discovery of mind in Greek philosophy and literature.* New York: Dover Publications.

Solomon, M. (1995). Naturalism and generality. *Philosophical Psychology. 8.4,* pp.353-363.

Somerville, J. & Santoni, R.E. (eds.) (1963). *Social and political philosophy: Readings from Plato to Gandhi*. New York: Anchor Books.

Stanley, S.M. (1975). A theory of evolution above the species level. *Proceedings of the National Academy of Science, USA. 72*, pp. 646-650.

Stapp, H. (1993). *Matter, mind, and quantum mechanics*. New York: Springer Verlag.

Stewart, I. (1989). *Does God play dice? The mathematics of chaos*. New York: Basil Blackwell.

Stone, I.F. (1988). *The trial of Socrates*. Boston: Little, Brown and Company

Thompson, J.B. (1984). *Studies in the theory of ideology*. Berkeley: University of California Press.

Tolman, E.C. (1932). *Purposive behavior in animals and man*. New York: Appleton.

Treumann, R.D. (1993). The cognitive map and the dynamics of information. In Laszlo & Masulli, pp. 77-98.

Ulrich, W. (1991). Critical heuristics of social systems design; & Systems thinking, systems practice, and practical philosophy: A program of research. In Flood and Jackson (1991), pp. 103-116 & 245-268.

Valle, R.S. and von Eckartsberg, R. (eds.) (1989). *Metaphors of consciousness*. New York: Plenum Press.

Varela, F.J. (1981). Describing the logic of the living. In Zeleny (1981).

Varela, F.J. (1993a). When is a map cognitive? In Laszlo and Masulli, pp. 99-104.

Varela, F.J. (1993b). Book review of *Consciousness Explained* by Daniel Dennet. *American Journal of Psychology, vol. 106, no. 1*, pp. 126-129.

Varela, F.J. (1994). A cognitive view of the immune system. *World Futures, vol. 42*, pp. 31-40.

Varela, F.J., Thompson, E., & Rosch E. (1991). *The embodied mind: Cognitive science and human experience*. Cambridge: The MIT Press.

Vickers, G. (1981). Some implications of systems thinking. In Open University systems group. (Ed.). *Systems Behavior*. London: Harper & Row.

Vickers, G. (1982). Social ethics. In O.W. Markley & W. Harmon, *Changing images of man*. London: Pergamon Press.

Von Eckardt, B. (1995). *What is Cognitive Science?* Cambridge: The MIT Press.

Von Foerster, H. (1979). Cybernetics of cybernetics. In K. Krippendorff, *Communication and control in society*. New York:

Warfield, J.N. (1976) *Societal systems*. New York: Wiley.

Warfield, J.N. (1994). *A Science of generic design*. Ames: Iowa State University Press.

Warfield, J.N. & Cardenas, A.R. (1994). *A handbook of interactive management*. Ames: Iowa State University Press.

Watson, J.D. (1980). *The double helix*. New York: W.W. Norton & Company.

Weisbord, M. (1992). *Productive workplaces: Organizing and managing for dignity, meaning, and community*. San Francisco: Berrett-Koehler.

Wellmer, A. (1985). Reason, utopia, and the *Dialectic of Enlightenment*. In Bernstein, pp.   35-66.

Wesson, R. (1994). *Beyond natural selection*. Cambridge: The MIT Press.

Wheatley, M.J. (1992). *Leadership and the new science*. San Francisco: Berrett-Koehler Publishers.

Wheeler, J.A. and Zurek, W.H. (Eds.). (1983), *Quantum theory and measurement*. Princeton: Princeton University

White, S.K. (1995). *The Cambridge companion to Habermas*. New York: Cambridge University Press.

Whitehead, A.N. (1966). *Modes of thought*. New York: Macmillan.

Whittaker, E.T. (190310). On the partial differential equations of mathematical physics. *Mathematische Annalen. 57*, pp. 333-355.

Wilber, K. (1995). *Sex, ecology, spirituality: The spirit of evolution*. Boston: Shambhala.

Wiley, R. (1994). *The semiotic self*. Chicago: The University of Chicago Press.

Winograd, T. and Flores, F. (1987). *Understanding computers and cognition*. New York: Addison-Wesley Publishing Company.

Wittgenstein, L. (1953). *Philosophical investigations*. Oxford: Basil Blackwell.

Zadeh, L. (1982). A note on prototype theory and fuzzy sets. *Cognition*, 12, pp.191-297.

Zeleny, M. (Ed.) (1980). *Autopoiesis, dissipative structures, and spontaneous social orders*. Boulder: Westview Press

Zeleny, M. (Ed.) (1981). *Autopoiesis: A theory of living organization.* New York: North Holland.
Zohar, D. (1990). *The quantum self.* New York: Quill/William Morrow.
Zurek, W.H. (ed.). (1990). *Complexity, entropy, and the physics of information.* Reading, Massachusetts: AddisonWesley Publishing Company.

Index

abduction, 159, 282
acceptable insecurity, 208, 337
Ackoff, 18, 65, 87, 103, 117, 129, 144, 312, 397
ADHD, 375
Adorno, 125, 194, 397, 402
"Agency", 22
AI, 45, 47, 390
Alexander, 11, 98, 101, 303, 397, 399, 403, 404
Allen, 278, 284, 285, 310, 312, 315, 397, 400, 404
Allport, 15, 202
American Psychological Association, 63, 64, 397
Anthropology, 274
Anticipatory structures, 210
Arendt, 114
Argyris, 164, 397, 398
Aristotelian space, 23, 24, 25
Aristotle, 23, 24, 110, 177, 198, 238
arrangements that serve a purpose, 240
arrow of time, 22, 23, 25, 31, 392
artificial intelligence, 2, 45, 242, 270, 390
artificial intersubjectivity, 270
Artigiani, 278, 279, 280, 380, 382, 387, 397
Ashby, 127, 168, 317, 397
asymmetrization, 225
atlas, 50
Attribute Profiles, 328
Attributes Field, 328
Attributes Profiles, 328
Augustine, 110
Austin, 69, 186
autocatalysis, 22
autogenesis, 34
autonomy, 11, 28, 35, 36, 40, 54, 58, 62, 74, 85, 190, 196, 198, 250, 268, 340, 366
autopoiesis, 2, 9, 15, 16, 19, 21, 28, 29, 33, 34, 35, 38, 39, 40, 41, 42, 47, 48, 67, 131, 175, 178, 179, 180, 181, 196, 199, 200, 214,

autopoiesis *(cont.)* 215, 216, 218, 222, 225, 227, 230, 274, 327, 328, 333, 334, 336, 340, 341, 344, 345, 346, 347, 364, 366, 374, 376, 379, 402, 404
autopoietic reproduction, 191, 218, 228, 347, 376, 382
average predictable detail, 284

Bacon, 1
Bailey, 395, 397
Bak, 393, 397
Banathy, 19, 95, 139, 140, 141, 142, 143, 144, 145, 146, 148, 149, 150, 152, 306, 310, 312, 313, 315, 319, 320, 321, 323, 360, 397
Barth, 278, 279, 397
basal self-reference, 200, 213, 222, 364
Bateson, 50, 180, 309, 397
Bausch, 299, 305, 328, 398
Be Your Enemy, 116
Beer, 18, 127, 313, 324, 398
before/after difference, 203
belief system, 210, 263, 264, 265, 267, 268, 269, 309, 315, 320, 333, 334, 335, 343, 359, 375
belief system, 262, 263, 264, 265, 266, 359
bell, 140
Belousov-Zhabotinsky, 25
Bènard, 25
Bentham, 111, 112
Bergson, 245, 246
Bhagavad-Gita,, 109
Bickerton, 41, 48, 50, 51, 52, 53, 54, 55, 58, 59, 241, 250, 309, 335, 339, 353, 356, 365, 366, 369, 370, 371, 398
bifurcations, 22, 25, 26, 242, 244, 276, 281, 310, 313, 324, 344, 347, 348, 349, 364, 388, 393